MOLECULAR CLOCKS AND LIGHT SIGNALLING

Novartis Foundation Symposium 253

MOLECULAR CLOCKS AND LIGHT SIGNALLING

2003

John Wiley & Sons, Ltd

This publication is designed to provide accurate and authoritative information in regard to
the subject matter covered. It is sold on the understanding that the Publisher is not engaged
in rendering professional services. If professional advice or other expert assistance is
required, the services of a competent professional should be sought.

Other Wiley Editorial Offices

John Wiley & Sons Inc., 111 River Street, Hoboken, NJ 07030, USA

Jossey-Bass, 989 Market Street, San Francisco, CA 94103-1741, USA

Wiley-VCH Verlag GmbH, Boschstr. 12, D-69469 Weinheim, Germany

John Wiley & Sons Australia Ltd, 33 Park Road, Milton, Queensland 4064, Australia

John Wiley & Sons (Asia) Pte Ltd, 2 Clementi Loop #02-01, Jin Xing Distripark, Singapore
129809

John Wiley & Sons Canada Ltd, 22 Worcester Road, Etobicoke, Ontario, Canada M9W 1L1

Wiley also publishes its books in a variety of electronic formats. Some content that appears
in print may not be available in electronic books.

Novartis Foundation Symposium 253
x+296 pages, 54 figures, 5 tables

Library of Congress Cataloging-in-Publication Data

Molecular clocks and light signalling / [editors, Derek J. Chadwick and Jamie A. Goode].
p. cm. – (Novartis Foundation symposium ; 253)
Includes bibliographical references and index.
ISBN 0-470-85283-6 (alk. paper)
1. Circadian rhythms–Congresses. 2. Photobiochemistry–Congresses. I. Chadwick,
Derek. II. Goode, Jamie. III. Series.
QP84.6.M653 2003
571.7′7–dc22   2003057596

British Library Cataloguing in Publication Data

A catalogue record for this book is available from the British Library

ISBN 0 470 85283 6

Typeset in $10\frac{1}{2}$ on $12\frac{1}{2}$ pt Garamond by Dobbie Typesetting Limited, Tavistock, Devon.
Printed and bound in Great Britain by T. J. International Ltd, Padstow, Cornwall.
This book is printed on acid-free paper responsibly manufactured from sustainable forestry,
in which at least two trees are planted for each one used for paper production.

Contents

Symposium on Molecular clocks and light signalling, held at the Novartis Foundation, London, 3–5 September 2002

Editors: Derek J. Chadwick (Organizer) and Jamie A. Goode

This symposium is based on a proposal made by Paolo Sassone-Corsi

Michael Menaker Chair's introduction 1

Russell G. Foster, Mark Hankins, Robert J. Lucas, Aaron Jenkins, Marta Muñoz, Stewart Thompson, Joanne M. Appleford and **James Bellingham**
Non-rod, non-cone photoreception in rodents and teleost fish 3
Discussion 23

Russell N. Van Gelder and **Aziz Sancar** Cryptochromes and inner retinal non-visual irradiance detection 31
Discussion 42

General discussion I 52

Xavier Bonnefont, Henk Albus, Johanna H. Meijer and **Gijsbertus T.J. van der Horst** Light signalling in *Cryptochrome*-deficient mice 56
Discussion 66

Satchidananda Panda, John B. Hogenesch and **Steve A. Kay** Circadian light input in plants, flies, and mammals 73
Discussion 82

Nicolas Preitner, Steven Brown, Juergen Ripperger, Nguyet Le-Minh, Francesca Damiola and **Ueli Schibler** Orphan nuclear receptors, molecular clockwork, and the entrainment of peripheral oscillators 89
Discussion 99

General discussion 102

Alec J. Davidson, Shin Yamazaki and **Michael Menaker** SCN: ringmaster of the circadian circus or conductor of the circadian orchestra? 110
Discussion 121

Nicolas Cermakian, Matthew P. Pando, Masao Doi, Luca Cardone, Irene Yujnovsky, David Morse and **Paolo Sassone-Corsi**   On the communication pathways between the central pacemaker and peripheral oscillators   126
Discussion   136

Paul E. Hardin, Balaji Krishnan, Jerry H. Houl, Hao Zheng, Fanny S. Ng, Stuart E. Dryer and **Nick R. J. Glossop**   Central and peripheral circadian oscillators in *Drosophila*   140
Discussion   150

Hitoshi Okamura   Integration of molecular rhythms in mammalian circadian system   161

John B. Hogenesch, Satchidananda Panda, Steve Kay and **Joseph S. Takahashi**   Circadian transcriptional output in the SCN and liver of the mouse   171
Discussion   180

Allan C. Froehlich, Antonio Pregueiro, Kwangwon Lee, Deanna Denault, Hildur Colot, Minou Nowrousian, Jennifer J. Loros and **Jay C. Dunlap**   The molecular workings of the *Neurospora* biological clock   184
Discussion   198

M. H. Hastings, A. B. Reddy, M. Garabette, V. M. King, S. Chahad-Ehlers, J. O'Brien and **E. S. Maywood**   Expression of clock gene products in the suprachiasmatic nucleus in relation to circadian behaviour   203
Discussion   218

Michael Rosbash, Ravi Allada, Mike McDonald, Ying Peng and **Jie Zhao**   Circadian rhythms in *Drosophila*   223
Discussion   232

Koyomi Miyazaki, Miho Mezaki and **Norio Ishida**   The role of phosphorylation and degradation of hPer proteins oscillation in normal human fibroblasts   238
Discussion   249

Achim Kramer, Fu-Chia Yang, Pamela Snodgrass, Xiaodong Li, Thomas E. Scammell, Fred C. Davis and **Charles J. Weitz**   Regulation of daily locomotor activity and sleep by hypothalamic EGF receptor signalling   250
Discussion   263

Emily Harms, Michael W. Young and **Lino Saez**   CK1 and GSK-3 in the *Drosophila* and mammalian circadian clock   267
Discussion   277

CONTENTSvii

Final general discussion   281

Michael Menaker   Closing remarks   285

Index of contributors   286

Subject index   289

Participants

Gregory M. Cahill   Department of Biology, University of Houston, Houston, TX 77204-5513, USA

Nicolas Cermakian   Douglas Hospital Research Center, 6875 La Salle Boulevard, Montreal (QC), H4H 1R3, Canada

Jay C. Dunlap   Department of Genetics, Dartmouth Medical School, 7400 Remsen, Hanover, NH 03755-3844, USA

Russell G. Foster   Department of Integrative & Molecular Neuroscience, Imperial College of Science, Faculty of Medicine, Charing Cross Hospital, Fulham Palace Road, London W6 8RF, UK

Susan Golden   Department of Biology, Texas A&M University, 3258 TAMU, College Station, TX 77843-3258, USA

Carla B. Green   Department of Biology, University of Virginia, Gilmer Hall, Charlottesville, VA 22903, USA

Paul E. Hardin   Department of Biology and Biochemistry, University of Houston, 369 Science and Research Building 2, Houston, TX 77204-5001, USA

Michael H. Hastings   Neurobiology Division, Laboratory of Molecular Biology, MRC Centre, Hills Road, Cambridge CB2 2QH, UK

Norio Ishida   Clock Cell Biology Group, Tsukuba Center, National Institute of Advanced Industrial Science and Technology (AIST), AIST Tsukuba Central 6, 1-1-1 Higashi, Tsukuba, 305-8566 Japan

Steve A. Kay   Department of Cell Biology, The Scripps Research Institute, 10550 North Torrey Pines Road, La Jolla, CA 92037, USA

Charalambous P. Kyriacou   Department of Genetics, Adrian Building, University of Leicester, University Road, Leicester LE1 7RH, UK

Cheng Chi Lee   715 E Department of Molecular & Human Genetics, Baylor College of Medicine, One Baylor Plaza, Houston, TX 77030, USA

Jennifer J. Loros   Department of Biochemistry, Dartmouth Medical School, Hanover, NH 03755-3844, USA

Michael Menaker (*Chair*)   Department of Biology, University of Virginia, Charlottesville, VA 22903, USA

Hitoshi Okamura   Division of Molecular Brain Science, Department of Brain Sciences, Kobe University Graduate School of Medicine, Chuo-ku, Kobe, 650-0017, Japan

Steven M. Reppert   Department of Neurobiology, University of Massachusetts Medical School, 55 Lake Ave North, Worcester, MA 01655, USA

Michael Rosbash   Howard Hughes Medical Institute, Department of Biology, Brandeis University, Waltham, MA 02254, USA

Paolo Sassone-Corsi   Institut de Génétique et de Biologie, Moleculaire et Cellulaire, CNRS-INSERM, Université Louis Pasteur, 1 Rue Laurent Fries, B P 163, Illkirch-Strasbourg, 67404, France

Ueli Schibler   Department of Molecular Biology, University of Geneva, 30, quai Ernest-Ansermet, CH-1211 Geneva 4, Switzerland

Amita Sehgal   Howard Hughes Medical Institute, University of Pennsylvania Medical Center, 233 Stemmler Hall, 35th & Hamilton Walk, Philadelphia, PA 19104, USA

Ralf Stanewsky   Institut für Zoologie, Lehrstuhl für Entwicklungsbiologie, Universitaet Regensburg, 93040 Regensburg, Germany

Joseph S. Takahashi   Howard Hughes Medical Institute, Department of Neurobiology & Physiology, Northwestern University, 2153 North Campus Drive, Evanston, IL 60208, USA

Gijsbertus T. J. van der Horst   Department of Cell Biology and Genetics, Erasmus University Rotterdam, PO Box 1738, 3000 DR Rotterdam, The Netherlands

Russell N. van Gelder   Department of Ophthalmology and Visual Sciences, Washington University Medical School, 660 South Euclid Avenue, Campus Box 8096, St Louis, MO 63110, USA

Charles J. Weitz   Associate Professor, Department of Neurobiology, Harvard Medical School, 220 Longwood Avenue, Boston, MA 02115, USA

Michael W. Young   Laboratory of Genetics, The Rockefeller University, 1230 York Avenue, New York, NY 10021, USA

Chair's introduction

Michael Menaker

Department of Biology, University of Virginia, Charlottesville, VA 22903, USA

Those of you who have been here before know that this is a special sort of meeting. There is no other meeting quite like this in terms of small size, time available to discuss major issues and the calibre of participants. It is my job to introduce the meeting by outlining the important unanswered questions. Starting with the light signalling end of things, we have known for almost 100 years now that organisms have ways of detecting light that are not obvious by simply looking at their morphology. That is, eyes are not the only photoreceptors; in some cases they are not even the main photoreceptors. This has been known since the work of von Frisch and Ernst Scharrer in the early part of the 20th century, when they showed that fish were capable of learning to respond to light without their eyes. This became, and has remained a theme in this field since it was discovered that the circadian system in general relies on extraretinal photoreceptors. In 1983, when Joe Takahashi was a graduate student and Pat De Coursey a visiting scientist in my laboratory, the three of us sat down and discussed how frustrating it was not to be able to find the extraretinal photoreceptors in the birds and reptiles that we had been working with for a long time. We knew they were in the brain but we had very little idea where. Out of that frustration, we decided to work on mammals in which it was clear that all photoreceptors were in the eye with the idea that it would be fairly easy to decide if rods, cones or both were involved. As you will hear later, this story is far from over, but we have made a lot of progress in the last 20 years. It has become increasingly interesting to ask where these light receptors are in mammals and how they operate. I should remind you that one of the people who worked in this field very early on was Donald Kennedy, who at that point was an invertebrate neurophysiologist working on clams. He published a beautiful paper in which he identified the photoreceptor in the siphon of the clam that enables the animal to withdraw its siphon when a shadow passes over it. He comments in the discussion of the Cold Spring Harbor Symposium devoted to biological clocks (XXV, 1960, p 268) that it is probably best not to assume that the obvious photoreceptive organs are doing all the photoreception. This was an extremely prescient comment. It now applies even to such things as the particular photoreceptive cells in the retina which one would assume are doing the photoreception, and which it now seems are not doing all of it. Light signalling remains a major theme in this field. Because the many different organisms on which we work have evolved in such diverse photic environments, it is not surprising that the light signalling mechanisms have already been found to be quite distinct and complex. There may be no more general principal than the one which is now clear: namely, that photoreceptors which are designed for image formation are not very useful in reporting light to the circadian system—photoreceptors with quite different properties are required. However the details may vary widely among different organisms.

It is interesting to note from the synopses of the various papers to be presented at this meeting, that there is not a great deal of focus on the molecular mechanisms that generate circadian oscillations. There is more emphasis on the way in which the oscillators in particular tissues within multicellular organisms are coordinated with each other and with the environment. Surely this is not because the molecular mechanism is completely understood, but I suspect that things may be getting a bit difficult in that area. People who have the fortitude to work through the present difficulties will undoubtedly be rewarded with further important discoveries.

What of the future for this field? It seems to me that there is a remarkable amount of physiology, behaviour and even ecology that cries out for explanation of temporal structure in mechanistic terms. In five years someone will probably propose a meeting in which the genomic information that is rapidly becoming available is used to explain some fundamental features of the physiology of mice. Hopefully, five years after that someone will propose another meeting applying this information to ecological questions, which clearly demand this kind of explanation. One of the last things that Colin Pittendrigh said to me, at the end of his life, was that he thought the term 'biological clock' had probably outlived its usefulness. I initially disagreed primarily because it has been such a useful term in generating interest. However, his point is well taken: it may be more useful to talk about 'temporal programs', not clocks *per se*. Thinking about temporal organization in terms of programs puts a somewhat different slant on things; one that suggests experiments that might not come up if one is thinking simply in terms of clocks. Perhaps after all this was rather a wise comment. But regardless of terminology, the field is thriving and will clearly continue to do so.

Non-rod, non-cone photoreception in rodents and teleost fish

Russell G. Foster, Mark Hankins, Robert J. Lucas, Aaron Jenkins, Marta Muñoz, Stewart Thompson, Joanne M. Appleford and James Bellingham

Department of Integrative & Molecular Neuroscience, Division of Neuroscience & Psychological Medicine, Imperial College Faculty of Medicine, Charing Cross Hospital, Fulham Palace Road, London W6 8RF, UK

Abstract. Until recently, all ocular photoreception was attributed to the rods and cones of the retina. However, studies on mice lacking rod and cone photoreceptors (*rd/rd cl*), has shown that these mice can still use their eyes to detect light to regulate their circadian rhythms, suppress pineal melatonin, modify locomotor activity and modulate pupil size. In addition, action spectra for some of these responses have characterized a novel opsin/vitamin A-based photopigment with a λ_{max} ~480 nm. Electrophysiological studies have shown that a subset of retinal ganglion cells are intrinsically photosensitive, and melanopsin has been proposed as the photopigment mediating these responses to light. In contrast to mammals, an inner retinal photopigment gene has been identified in teleost fish. Vertebrate ancient (VA) opsin forms a photopigment with a λ_{max} between 460–500 nm, and is expressed in a sub-set of retinal horizontal cells, and cells in the amacrine and ganglion cell layers. Electrophysiological analysis suggests that VA opsin horizontal cells are intrinsically photosensitive and encode irradiance information. In contrast to mammals, however, the function of these novel ocular photoreceptors remains unknown. We compare non-rod, non-cone ocular photoreceptors in mammals and fish, and examine the criteria used to place candidate photopigment molecules into a functional context.

2003 Molecular clocks and light signalling. Wiley, Chichester (Novartis Foundation Symposium 253) p 3–30

The evidence to-date shows that vertebrate photoreception is mediated by a closely related group of proteins termed opsins. These are G protein-coupled receptors characterized by their ability to bind a vitamin A-based chromophore (11-*cis*-retinal) via a Schiff base linkage using a lysine residue in the 7th transmembrane α helix (Fig. 1). The primary events of image detection by the rods and cones occurs with the absorption of a photon of light by 11-*cis*-retinal and its photoisomerization to the all-*trans* state (Burns & Baylor 2001, Menon et al 2001). Although photoreception is best understood in retinal rods and cones, photoreception is not confined to these structures. In non-mammalian

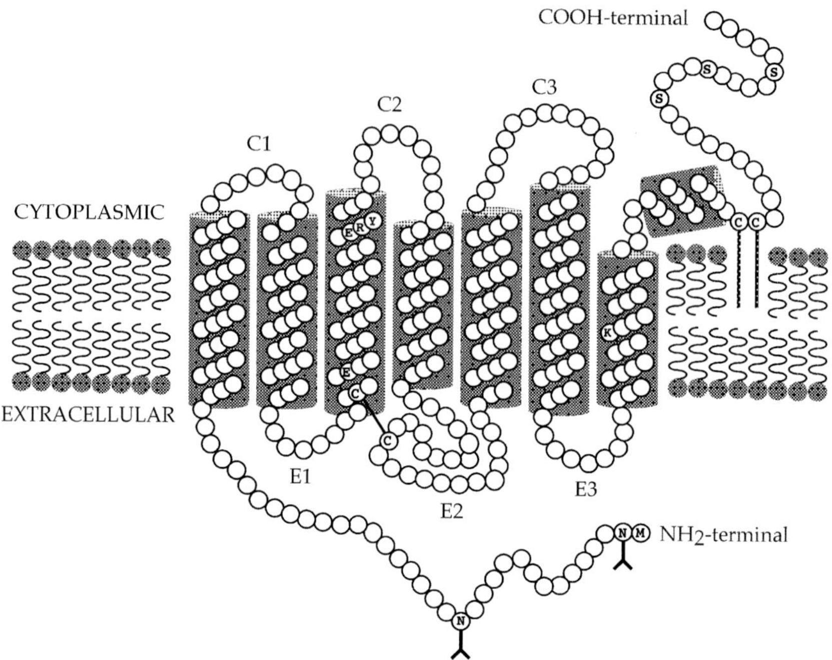

FIG. 1. A two dimensional model of the rod-opsin molecule, modified after Palczewski et al (2000). Features include the retinal attachment site (K296) in the seventh transmembrane domain; the Schiff base counterion (G113) in the third transmembrane domain; the disulfide bridge formed between C110 and C187; glycosylation sites (N2 and N15) at the amino terminal; palmitoylation sites (C322 and C323) at the C-terminal. Also indicated are the ERY triad crucial for transducin binding and activation in the third transmembrane domain, and serine residues (S) in the carboxyl terminal which are phosphorylated by rhodopsin kinase inducing quenching of the phototransduction cascade. Cytoplasmic (C1–C3) and extracellular (E1–E3) loops are indicated.

vertebrates, photoreception within the pineal and deep brain has been well documented (Shand & Foster 1999). More recently, evidence has emerged that cells within the inner retina appear to be directly sensitive to light in both teleost fish (Soni et al 1998, Kojima et al 2000) and rodents (Freedman et al 1999, Lucas & Foster 1999, Lucas et al 2001, Berson et al 2002). Furthermore, photoreception in the vertebrates is not confined to the structures of the central nervous system. Some classes of chromatophore such as the melanophores of the amphibian *Xenopus laevis* respond directly to light (Rollag 1996, Rollag et al 2000), as do the irridophores of some fish (Lythgoe et al 1984). Despite the diversity of these non-rod, non-cone responses to light, action spectra have implicated opsin/vitamin A-based photopigments in every case (e.g. Deguchi 1981, Foster et al 1985, Lucas et al

2001, Berson et al 2002). Molecular approaches have been partially successful in identifying these opsins, and recent studies have identified a range of new opsin families that have been variously linked to a variety of non-image-forming responses to light, including the regulation of circadian rhythms (Bellingham & Foster 2002). This review will consider non-rod, non-cone photoreception, and the extent to which candidate photopigment genes have been associated with this recently recognized form of ocular photoreception in rodents and teleost fish.

Vertebrate ancient (VA) opsin and inner retinal photoreception in teleosts

The first non-rod, non-cone opsin-based photopigment to be isolated from the retina of any vertebrate was vertebrate ancient (VA) opsin. VA opsin was first described as a result of the isolation of a cDNA coding for a 323 amino acid opsin-like protein from eye tissue of the Atlantic salmon (*Salmo salar*). Salmon VA opsin shares 37–41% amino acid identity with the classical retinal opsins and 43% identity with chicken P opsin (Soni & Foster 1997) (Table 1). *In vitro* expression of salmon VA opsin indicates that a functional photopigment (λ_{max} $\sim$460 nm) is formed on reconstitution with 11-*cis*-retinal. Significantly, *in situ* hybridization studies have shown that VA opsin is expressed in a subset of horizontal cells and cells in the amacrine cell layers (Soni et al 1998). These observations provided the first demonstration of photopigment expression within the vertebrate inner retina (Soni et al 1998). A second isoform of VA opsin was identified in two other species of teleost, the common carp, *Cyprinus carpio* (Moutsaki et al 2000) and the zebrafish, *Danio rerio* (Kojima et al 2000). These two second isoforms of VA opsin are characterized by a very long C-terminus (79 and 74 amino acids, respectively) in comparison to salmon VA opsin (13 amino acids) and have been termed VA-long (VAL) opsin. Furthermore, zebrafish VAL opsin possesses a λ_{max} of $\sim$500 nm when expressed *in vitro* (Kojima et al 2000). Additional *in situ* hybridization studies demonstrated that salmon VA opsin is also expressed in the pineal organ and habenular region of the brain (Philp et al 2000), and similar findings have been reported in zebrafish using immunocytochemistry to demonstrate the presence of VA opsin in the inner retina and brain (but not the pineal) of this species (Kojima et al 2000).

A key feature of both vertebrate and invertebrate opsin molecules is their ability to interact with a G protein, typically transducin, to initiate phototransduction (Ebrey & Koutalos 2001). The third cytoplasmic loop that connects α-helices V and VI contributes to the G protein binding and activation and is highly conserved amongst the rod and cone opsins (Fig. 1, Table 2). This conservation extends to the P opsin family. However, the third cytoplasmic loop of the VA

TABLE 1 Amino acid identity of verterbrate opsin transmembrane domains I–VII

Opsin Class	Rod	LWS	MWS	SWS	UVS/VS P	VA	PP	Ci	TMT	Enc	RGR	Per	Mel	Invert
Rod Opsin	—													
LW Opsin	42	—												
MWS Opsin	69	44	—											
SWS Opsin	52	43	55	—										
UVS/VS Opsin	46	41	50	51	—									
Pineal Opsin	46	50	49	52	48	—								
VA Opsin	36	43	39	42	40	43	—							
Parapinopsin	40	40	40	41	40	48	43	—						
Ciona Opsin1	42	44	39	41	40	44	43	44	—					
Teleost MT Opsin	33	35	35	36	35	39	37	39	35	—				
Encephalopsin	30	28	30	32	27	30	31	28	26	41	—			
RCR-Opsin	21	21	22	22	29	23	22	25	21	22	24	—		
Peropsin	29	26	26	26	22	27	27	29	27	23	31	25	—	
Melanospin	27	25	27	26	28	28	26	26	24	30	28	26	30	—
Invertebrate	27	25	26	24	24	26	23	28	24	29	26	22	29	42	—

Table showing the amino acid identity (%) encompassing transmembrane domains (α helices) I–VII, as defined by (Baldwin et al 1997), of representatives of the various vertebrate opsin classes. Sequence sources with associated GenBank accession numbers: Rod, chicken, D00702; LWS, chicken red, M62903; MWS, chicken green, M92038; SWS, chicken blue, M92037; UVS/VS, chicken violet, M92039; Pineal opsin, chicken, U15762; VA-opsin, salmon, AF001499; Parapinopsin, catfish, AF028014; *Ciona*, *Ci*-opsin1, AB058682; Teleost Multiple Tissue Opsin, *Fugu*, AF402774; Encephalopsin, human encephalopsin, AF140242; RGR-opsin, human, U14910; Peropsin, human, AF012270; Melanopsin, human, AF147788; Invertebrate, cuttlefish rhodopsin, AF000947.

TABLE 2 Alignment of the third cytoplasmic loop (C3) from representatives of photosensory opsin families and melanopsins

---

Photosensory Opsins

```
Human Rod         TVKEAAAQQQESATTQK
Mouse Rod         .................
Alligator Rod     .................
Xenopus Rod       ............L....
Chicken Rod       .................
Lamprey Rod       .......A.....S...
Zebrafish Rod     ............E...R
Zebrafish Exo     ..RA........E...R
Zebrafish LWS     AIHAV.Q..KD.ES...
Zebrafish MWS     ...A........ES...
Zebrafish SWS     .L.L..KA.AD..S...
Zebrafish UVS/VS  ALRAV....A..ES...
Chicken P-opsin   .LRA.....K.AD...R

Carp VAL-opsin    KLRKVSNTHGRLGNARK
Zebrafish VAL-opsin  .................
Salmon VA-opsin   .......--.D.....K.
Smelt VAL-opsin   .......--........
```

Melanopsins

```
Human     AIRETGRALQTFGACKGNGESLWQRQRLQ
Mouse     ........CEGC.ESPLRQ--RR.W....
Rat       ........CEGC.ESPLR---RR.W....
Zebrafish T..AA.KEIRELDCGETHKVYE----.M.
Xenopus   ...S...NV.KL.SYGRQ-SF.S.S--MK
Chicken   ...S...DV.KL.SCSRK-SF.S.S--MK
```

Alignment of the third cytoplasmic loop (C3), as defined by the recent rod-opsin model of (Palczewski et al 2000) from representatives of photosensory opsin families and melanopsins. This region forms a putative G-protein activation domain in the photosensory opsins. Identity is represented by a stop (.), whilst gaps are represented by a dash (–). Cone opsin classifications according to (Hunt et al 2001). GenBank accession numbers relating to the amino acid sequences used are: Human rod-opsin, U49742; chicken rod-opsin, D00702; alligator rod-opsin, U23802; *Xenopus* rod-opsin, L04692; lamprey rod-opsin, U67123-7; zebrafish rod-opsin, AF331797; zebrafish MWS opsin, AF109369; zebrafish SWS opsin, AF109372; zebrafish UVS opsin, AF109373; zebrafish LWS opsin, AF104904; chicken pinopsin, U15762; carp VAL-opsin, AF233520; zebrafish VAL-opsin, AB035276; salmon VA-opsin, AF001499; smelt VAL-opsin, AB074483; human melanopsin, AF147788; mouse melanopsin, AF147789; rat melanopsin, AY072689; zebrafish melanopsin, AY078161; chicken melanopsin, AY036061; *Xenopus* melanopsin, AF014797.

opsin family is very divergent from these opsin families, but is highly conserved within the class (Table 2). As a result, an alternative G protein has been proposed for these opsins (Soni & Foster 1997, Moutsaki et al 2000). The genomic structure of the VA opins is highly similar to the classical vertebrate visual opsins in that the position of the intron insertion sites 1, 3 and 4 are perfectly conserved, whilst intron 2 is shifted by 42 nucleotides in a $3'$ direction (Moutsaki et al 2000) (Fig. 2). Very recently, a third isoform of VA opsin (VAM) has been isolated from a smelt fish (*Plecoglossus altivelis*) which possesses a 24 amino acid C-terminal sequence but does not appear to arise from a expected splice site (Minamoto & Shimizu 2002). Thus three isoforms of VA opsin (VA, VAM and VAL) appear to exist, but the significance of these multiple isoforms remains to be determined. A genomic sequence from the marine lamprey (*Petromyzon marinus*) was isolated and termed lamprey P opsin (Yokoyama & Zhang 1997). A comparison of the lamprey P opsin gene and the VA opsin gene shows that they share the same intron insertion sites, which argues (coupled with its amino acid identity of 61–65% with VA opsins) that lamprey P opsin is actually a member of the VA opsin class (Moutsaki et al 2000). In view of the evolutionary position that lampreys occupy, lamprey P opsin is probably the evolutionary precursor of the teleost VA opsin family, and possibly the non-teleost P opsins (Moutsaki et al 2000).

In an attempt to study some of the functional properties of VA opsin photopigments in horizontal cells, we have utilized a cyprinid fish, the roach (*Rutilus rutilis*). This species was selected because, in contrast to zebrafish or salmon, the electrophysiological properties of roach horizontal cells have been characterized in considerable detail. We succeeded in isolating two forms of roach VA opsin (VA and VAL), and using *in situ* hybridization demonstrated expression in horizontal calls. The light responses recorded from these horizontal cells shows a novel depolarizing off-component that is seen after the usual off-response typical of horizontal cells. Significantly, the amplitude of this depolarization is dependent upon photon flux, and appears to code for environmental irradiance and duration. Rod and cone photoreceptor inputs to horizontal cells induce hyperpolarizing responses, whilst the novel component is depolarizing. As noted above, the third cytoplasmic loop of the VA opsin family is different from the classical opsin families (Table 2), and perhaps this difference relates to an interaction with a novel G protein. Furthermore, an action spectrum for this depolarizing off response predicts an opsin-based photopigment with a $\lambda_{\max}$ of ~ 477 nm. This photopigment response does not correspond to any of the known visual pigments of the roach, but is within the range reported for the VA opsins (460–500 nm). It will be necessary to compare the $\lambda_{\max}$ of functionally expressed roach VA opsin with these action spectrum results to establish that these novel responses are really driven by VA-opsin photopigment.

Inner retinal photoreception and opsin photopigments in rodents

Removal of the eyes in every mammal studied abolishes photoentrainment (Foster 1998). Because the rods and cones were the only known ocular photoreceptors, this led to the assumption that all photoreception can be attributed to these cells. Initial studies on *rd/rd* mice, which lack rod photoreceptors, and more recent studies on *rd/rd cl* mice, which lack all functional rods and cones, have provided overwhelming evidence that these classical photoreceptors are not required for photoentrainment (Foster 2002). By extension, the eye must contain at least one additional class of photoreceptor. In addition, studies on *rd/rd cl* mice have shown that the non-rod, non-cone photoreceptors do more than regulate the circadian system. They also contribute to both pupillary constriction and acute alterations in locomotor behaviour, and may be involved in a broad range of physiological and behavioural responses to light (Foster 2002).

The *rd/rd cl* mouse has provided a powerful model to characterize the ocular non-rod, non-cone photoreceptors of mammals using action spectrum techniques. The first completed action spectrum was for the pupillary light reflex PLR (Lucas et al 2001). The results demonstrated that the PLR in *rd/rd cl* mice is driven by a single opsin/vitamin A-based photopigment with a $\lambda_{max} \sim 479$ nm. The known photopigments of mice peak at ~ 360 nm (UV cone) (Jacobs et al 1991), ~ 498 nm (rod) (Bridges 1959), and ~ 508 nm (green cone) (Sun et al 1997a), and do not show any significant fit to the PLR action spectrum in *rd/rd cl* mice. Whether the same photopigment mediates all non-rod, non-cone ocular responses to light remains to be determined. Our recently completed action spectrum for phase-shifting circadian rhythms of locomotor behaviour in *rd/rd cl* mice has identified an opsin/vitamin A based photopigment with a λ_{max} at 481 nm (Fig. 3). The high degree of similarity between these two action spectra suggests that the same photopigment mediates both the PLR and circadian responses to light, and would exclude the involvement of a flavoprotein-based photopigment (Fig. 4).

Although rod and cone photoreceptors are not required for the regulation of the circadian system, this does not mean that the rods and cones play no role. Indeed, our recent studies have implicated these receptors directly. The action spectrum for phase shifting in wild-type mice is long wavelength-shifted compared to congenic *rd/rd cl* mice, and is well approximated by the absorption spectrum of murine rod ($\lambda_{max} = 498$ nm) (Bridges 1959) and m-cone ($\lambda_{max} = 508$ nm) photoreceptors. These results suggest that one or both of these photoreceptor classes contribute to entrainment under normal circumstances. This is an important observation because it suggests that it is going to be difficult to use gene-targeting techniques to test the candidacy of novel photoreceptor components unless the knock-out studies are undertaken in a rodless+coneless genetic background.

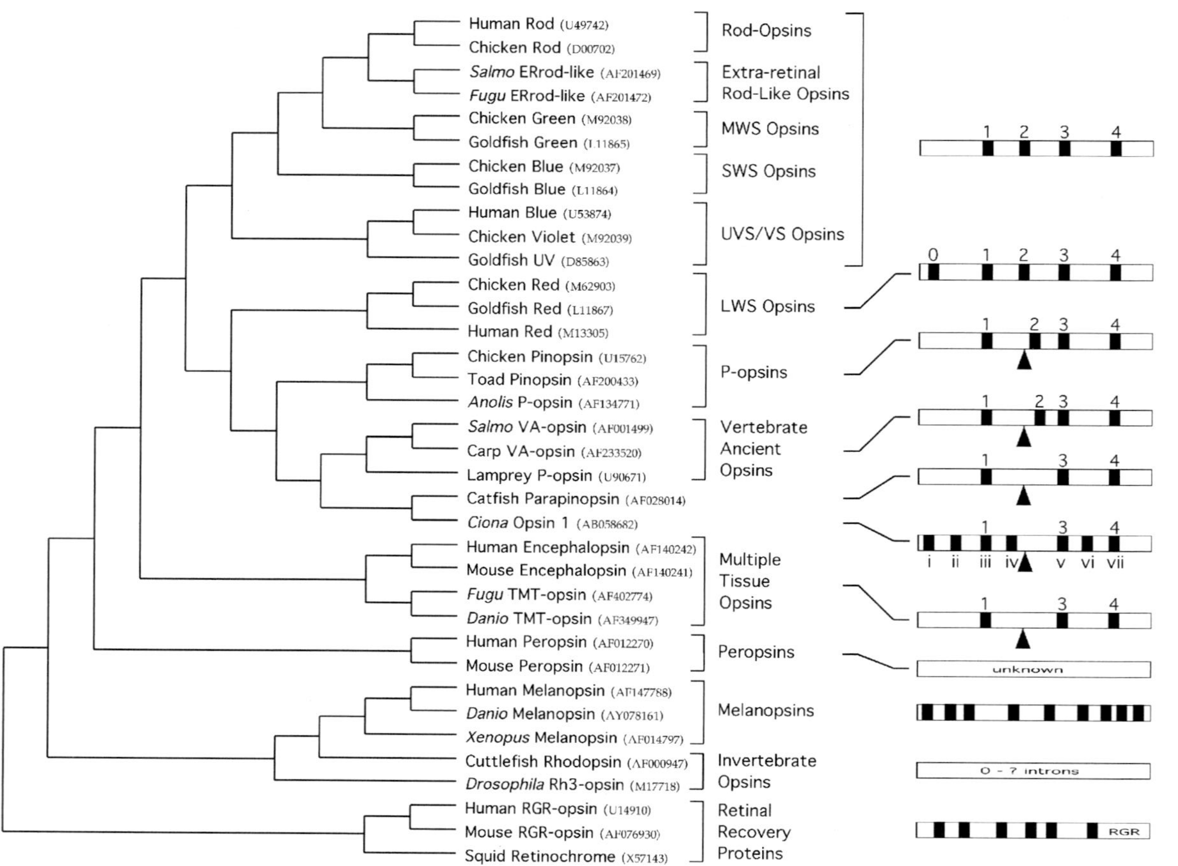
Human Rod (U49742)
Chicken Rod (D00702)
Rod-Opsins
Salmo ERrod-like (AF201469)
Fugu ERrod-like (AF201472)
Extra-retinal Rod-Like Opsins
Chicken Green (M92038)
Goldfish Green (L11865)
MWS Opsins
Chicken Blue (M92037)
Goldfish Blue (L11864)
SWS Opsins
Human Blue (U53874)
Chicken Violet (M92039)
Goldfish UV (D85863)
UVS/VS Opsins
Chicken Red (M62903)
Goldfish Red (L11867)
Human Red (M13305)
LWS Opsins
Chicken Pinopsin (U15762)
Toad Pinopsin (AF200433)
Anolis P-opsin (AF134771)
P-opsins
Salmo VA-opsin (AF001499)
Carp VA-opsin (AF233520)
Lamprey P-opsin (U90671)
Catfish Parapinopsin (AF028014)
Ciona Opsin 1 (AB058682)
Vertebrate Ancient Opsins
Human Encephalopsin (AF140242)
Mouse Encephalopsin (AF140241)
Fugu TMT-opsin (AF402774)
Danio TMT-opsin (AF349947)
Multiple Tissue Opsins
Human Peropsin (AF012270)
Mouse Peropsin (AF012271)
Peropsins
Human Melanopsin (AF147788)
Danio Melanopsin (AY078161)
Xenopus Melanopsin (AF014797)
Melanopsins
Cuttlefish Rhodopsin (AF000947)
Drosophila Rh3-opsin (M17718)
Invertebrate Opsins
Human RGR-opsin (U14910)
Mouse RGR-opsin (AF076930)
Squid Retinochrome (X57143)
Retinal Recovery Proteins
1 2 3 4
0 1 2 3 4
1 2 3 4
1 2 3 4
1 3 4
i ii iii iv v vi vii
1 3 4
unknown
0 - 7 introns
RGR

FIG. 2. Maximum parsimony phylogenetic tree showing the relationship of the opsins. To the right of the tree a diagrammatic representation of the genomic structures of the opsin classes is shown, with introns being represented by vertical black bars. The highly conserved structures of some of the opsin classes is very evident, with the positions of introns 1, 3 and 4 being invariant. The relative position of intron 2 in the non-visual opsins is indicated by an arrow when it is shifted in a 3′ direction (P and VA opsins) or absent (parapineal opsin and multiple tissue opsins). Neither melanopsin nor RGR share a common intron insertion site with those of the characterized photopigments, whilst the genomic structure of peropsin is unknown. Note that the tunicate *Ciona intestinalis* possesses an opsin (*Ciona* Opsin 1) that shares three perfectly conserved intron insertion sites (1, 3 and 4) with those of the vertebrate visual opsins. The melanopsin gene family differs markedly from all the opsin photopigments and may be attributed to either a unique line of photopigment evolution from an invertebrate-like ancestral gene or a non-photopigment role for this gene family. Opsin amino acid sequences were aligned using ClustalX 1.81 (Thompson et al 1997) and the maximum parsimony tree was calculated using Phylo_win (Galtier et al 1996) GenBank accession numbers for the sequences are indicated in parentheses.

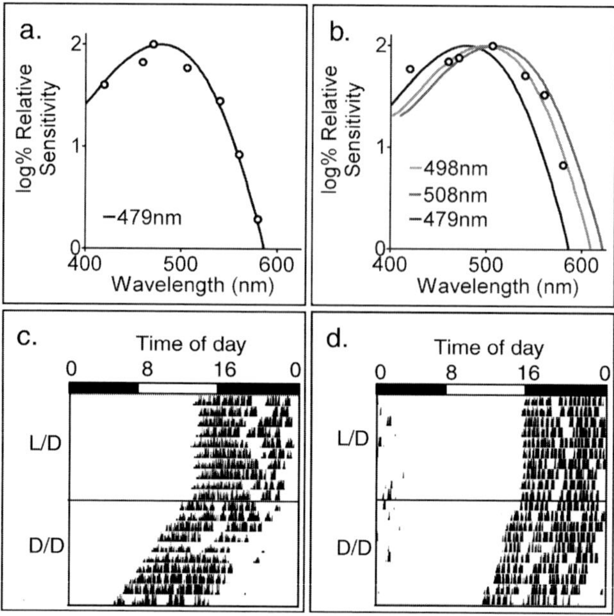

FIG. 3.   Action spectra for (a) *rd/rd cl* and (b) +/+ mice derived from irradiance response curves at seven wavelengths from 420–580 nm. The *rd/rd cl* spectrum corresponds to an opsin/vitamin A photopigment ($R^2 = 0.976$) with a λ_{max} at 481 nm. The +/+ action spectrum also describes an opsin photopigment ($R^2 = 0.896$), but with a λ_{max} of ~500 nm. (c, d) Representative actograms for (c) *rd/rd cl* and (d) +/+ mice exposed to an 8L/16D light cycle at low irradiance (below 25 μW/cm^2). Activity onset is significantly phase advanced relative to lights off in *rd/rd cl* mice (94.2 min ± 17.3, $n = 9$) compared to congenic +/+ mice (4.7 min ± 0.85, $n = 7$). The free-running period (*tau*) is indistinguishable in *rd/rd cl* (23.60 ± 0.08 h) and +/+ (23.66 ± 0.06 h).

Despite the difference in photoreceptive inputs into the suprachiasmatic nucleus (SCN) of +/+ and *rd/rd cl* mice, the sensitivity of the circadian system to acute 15 min. light pulses appears unaltered. These findings initially suggested that non-rod non-cone photoreceptors can fully compensate for the loss of rods and cones. However, examination of entrainment in *rd/rd cl* mice to full light cycles showed that this is not the case. When exposed to relatively dim light/dark cycles *rd/rd cl* mice entrain with a positive phase angle of ~94 min compared to +/+ mice (Figs 3c,d). Circadian formalisms suggest that an advanced phase angle of entrainment may be attributable to a reduction in either the amplitude of the phase-response curve or the intrinsic period (*tau*) of the circadian clock. Our data exclude these possibilities, suggesting that the loss of the rods and cones has an impact on the nature of the light information reaching the SCN.

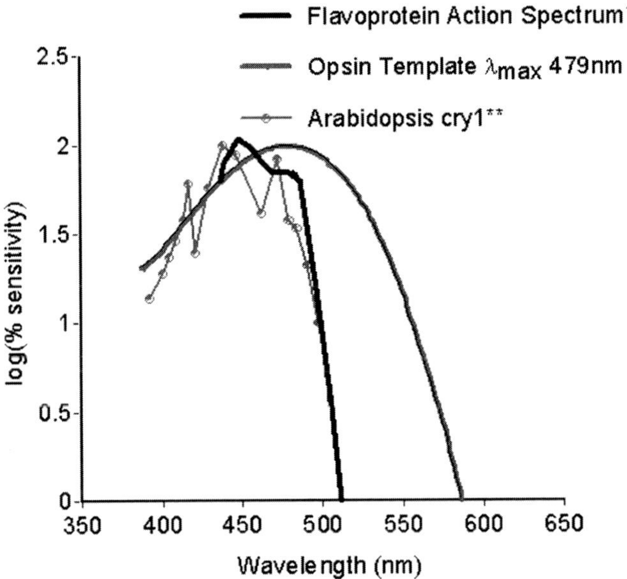

FIG. 4.   A comparison of the pupillary (Lucas et al 2001) and circadian action spectra which best fit an unidentified vitamin A-based pigment with a λ_{max} between 479–481 nm, with the absorbance spectrum of a flavoprotein photopigment, and the action spectrum for *Arabidopsis* CRY1. See text for details. *(Smyth et al 1988), **(Ahmad et al 2002).

David Berson's laboratory (Brown University) has recorded from retinal ganglion cells (RGCs) in the rat that were retrogradely labelled with fluorescent microspheres injected into the retino-recipient areas of the hypothalamus (Berson et al 2002). These cells were demonstrated as intrinsically light responsive, as their light-evoked depolarizations persisted in the presence of a cocktail of drugs known to block all retinal intercellular communication, and even continued when micro-dissected from the surrounding retinal tissue. Using this approach Brown's group went on to generate an action spectrum for the light-evoked depolarization. The data suggested a best fit to a retinal-based opsin photopigment template with a λ_{max} of 484 nm (Berson et al 2002), and in this regard the results are strikingly similar to pupillary and circadian responses in mice (Lucas et al 2001) and the modulation of cone ERG responses in humans (Hankins & Lucas 2002). Significantly, these RGCs express the novel opsin-like protein called melanopsin (Berson et al 2002, Hattar et al 2002, Provencio et al 2002). Melanopsin has therefore become a strong candidate for the photopigment opsin of the inner retina of mammals. The strength of melanopsin candidacy is considered below.

Melanopsin in the vertebrates

Poikilotherms such as fish and amphibia are able to change their external colouration by means of chromatophores (Oshima 2001). Some chromatophore classes appear to require the presence of retinoid in order to exhibit normal responses to light, e.g. the melanophores of the amphibian *Xenopus laevis* (Rollag 1996). Ignacio Provencio and Mark Rollag screened a *Xenopus* dermal melanophore cDNA library and isolated a clone with an open reading frame encoding a 534 amino acid protein that resembles an opsin. This opsin-like protein was termed melanopsin (Provencio et al 1998a). Rather than a glutamate at the putative counterion position, as seen in the rod, cone, P and VA opsins (Fig. 1), melanopsin has a tyrosine. In addition, the deduced amino acid sequence shares a relatively low level of identity with the known photopigment opsins ($\sim 27\%$) (Table 1). From a phylogenetic perspective, melanopsin is unique amongst the vertebrate opsins in that it co-segregates with the invertebrate opsins (Fig. 2). Over-expression of the melanopsin gene in cultured melanophores increased the sensitivity of these cells to light, suggesting that melanopsin must, at some level, be involved in regulating melanophore photosensitivity (Rollag et al 2000). Melanopsin is also expressed in the eye and brain of this species (Provencio et al 1998a). In the eye, melanopsin is expressed in the inner nuclear layer (in a region where horizontal cells are typically found), the retinal pigment epithelium (RPE) and the iris, whilst expression in the brain is localized to the magnocellular preoptic nucleus and the suprachiasmatic nucleus (Provencio et al 1998b).

Mammalian and avian orthologues of *Xenopus* melanopsin have recently been isolated (Provencio et al 1998b, 2000, Hattar et al 2002), and like *Xenopus* melanopsin, these melanopsins have a low identity to the visual opsins (Table 1), and have a tyrosine at the position of the putative counterion (Fig. 1). Expression of the mammalian melanopsins is reported to be restricted to the eye. In the primate retina, melanopsin is expressed in large numbers of cells in the ganglion cell layer and in cells of the inner nuclear layer which resemble the position of amacrine cells, in contrast to the murine retina where melanopsin is sparsely expressed in the ganglion cell layer and in even fewer cells in the amacrine cell layer (Provencio et al 2000). Melanopsin may also be expressed in the RPE/choroid of the human eye. These results were obtained by RT-PCR, and Provencio and colleagues have suggested that this may represent an artifact resulting from contamination during dissection (Provencio et al 2000). In the murine retina, the number and location of ganglion and amacrine cells that express melanopsin immediately suggested that melanopsin is expressed in those retinal cells that project to the SCN (Provencio et al 1998c). Two independent studies have subsequently confirmed this prediction. A combination of retrograde labelling and *in situ* hybridization showed that most of the RGCs that project to the SCN in the rat express melanopsin (Gooley et al

2001). Pituitary adenylate cyclase-activating polypeptide (PACAP) is expressed in the retina exclusively within the RGCs of the retinohypothalamic tract (RHT), and melanopsin was found to co-localize with PACAP in the retina (Hannibal et al 2002). As discussed above, it was further demonstrated that intrinsically photosensitive RGCs express melanopsin (Berson et al 2002, Hattar et al 2002). Collectively these correlations have been used to suggest that melanopsin is the novel photopigment mediating photoentrainment in rodents.

Recently we have succeeded in isolating a zebrafish orthologue of melanopsin which, like the mammalian melanopsins, appears to be expressed exclusively in the eye (Bellingham et al 2002). In common with the melanopsins from other vertebrate classes, zebrafish melanopsin shows some identity to the invertebrate opsins (Table 1). The surprising feature of all the melanopsins is their unexpectedly low level of identity to each other (Table 3). The known photosensory opsins (e.g. rod and cone) share a $\sim 40\%$ amino acid identity between opsin families, whilst members within an opsin family show a much higher level of identity of around 85%, even between the vertebrate classes (Bellingham & Foster 2002). Within a vertebrate class, identity is greater still. Human and mouse rod opsins share 94% identity at the amino acid level. Strikingly, this level of sequence conservation is not exhibited in the melanopsins. Comparison of the sequence spanning the seven transmembrane domains shows that mouse melanopsin is only 55% identical to *Xenopus* melanopsin, whilst zebrafish melanopsin is 68% identical to mouse and 54% identical to *Xenopus* melanopsin (Bellingham et al 2002) (Table 3). These levels of identity reduce significantly when the entire protein sequences are compared, mouse versus *Xenopus*, 35%; zebrafish vs. mouse, 48%; zebrafish vs. *Xenopus*, 36% (Bellingham et al 2002) (Table 3).

TABLE 3   Percentage identity between the known melanopsins

	Xenopus	*Zebrafish*	*Human*	*Mouse*	*Rat*	*Chicken*
Xenopus	−/−	**54**	**55**	**56**	**56**	**82**
Zebrafish	36	−/−	**65**	**68**	**66**	**54**
Human	34	47	−/−	**86**	**85**	**56**
Mouse	35	48	69	−/−	**96**	**57**
Rat	34	48	75	83	−/−	**56**
Chicken	60	35	34	36	35	−/−

Identities for complete sequences are in plain text, whilst those restricted to the sequence encompassed by α helices I–VII are in boldface. A similar analysis for the vertebrate rod-opsins would produce a percentage identity ranging between 75–95% across complete sequences, rather than the 34–75% seen for the melanopsins.

In the known photosensory photopigment opsins, high levels of amino acid identity are seen within an opsin family even though the photopigments have widely differing λ_{max}. For example, mouse and *Xenopus* UVS/VS opsins share 78% identity across the transmembrane domains but exhibit a 66 nm difference in their λ_{max} (359 nm and 425 nm respectively) (Starace & Knox 1997, Yokoyama et al 1998). It is not known whether any of the melanopsins can form a photopigment, and so their spectral maxima remain unknown. However, if we assume that *rd/rdcl* action spectra reported above (λ_{max} of $\sim$ 480 nm) (Lucas et al 2001), and the action spectra for melanophore aggregation in *Xenopus* (λ_{max} of $\sim$ 500 nm) (Batni et al 1996, Moriya et al 1996) are the product of a melanopsin photopigment, then their λ_{max} are rather similar. However, this similarity in their λ_{max} is achieved by a surprisingly low level of sequence conservation, with mouse and *Xenopus* melanopsin sharing only 56% identity across the transmembrane domains (Table 3).

The third cytoplasmic loop (Fig. 1) of the melanopsin family is poorly conserved between the rod and cone opsins and within the melanopsin family (Table 2). Although this region exhibits 96% identity between mouse and rat melanopsin, the same region between human and rodent melanopsin is only 48% identical. Between vertebrate classes, levels of identity for the melanopsin third cytoplasmic loop are lower still, ranging from 13–37% (Table 2). One exception is the 84% identity between *Xenopus* and chicken, which may be related to the phylogenetic co-segregation of these two melanopsins, and/or their conserved function (Bellingham & Foster 2002). This analysis suggests that the melanopsin family as a whole, and within a vertebrate class, may interact with a number of different G proteins, or, alternatively this diversity may reflect the fact that the melanopsins do not activate G proteins.

Taken together, these observations suggest that whatever the function of melanopsins they do not need to have a highly conserved protein structure, and in this respect differ markedly from the known photosensory opsins. Furthermore, the melanopsins do not share a conserved genomic structure with the photosensory opsins (Fig. 2), suggesting a different evolutionary lineage to the characterised photosensory opsins (Bellingham & Foster 2002).

Opsins can be photosensors or photoisomerases

The opsins probably perform a variety of different tasks, but their known roles are as photosensors or photoisomerases (Foster & Bellingham 2002). Photosensory opsins such as the rod and cone opsins, P opsin and VA opsin use light to activate a phototransduction cascade that ultimately results in a change in membrane potential of the photoreceptor cell. By contrast, photoisomerases are involved in photopigment regeneration. The best described photoisomerase is retinal G protein-coupled receptor (RGR) opsin. RGR is expressed in high concentrations in the RPE, has the Lys296 retinal attachment site (Fig. 1), a histidine at the Glu113 site (Fig. 1), and acts by harvesting the energy of a photon to photoisomerase all-*trans*-retinal into the 11-*cis*-retinal isoform. 11-*cis* retinal is then transported to the outer segments of the rods and cones where it is associated with a rod or cone opsin to regenerate a photopigment. Consistent with its non-photosensory role, RGR shares a relatively low level of amino acid identity (21–24%) (Table 1) and has a non-conserved genomic structure with the photopigment opsins (Hao & Fong 1996, 1999) (Fig. 2). Another suspected photoisomerase, peropsin (Sun et al 1997b), is similarly different from the photosensory opsins, sharing only $\sim 27\%$ amino acid identity with the photosensory opsins (Table 1), a non-conserved genomic structure, and along with melanopsin, a tyrosine at the putative counterion position. This comparison of the opsins would argue that the functionally related photosensory opsins share a close phylogenetic relationship based upon both high levels of amino acid identity ($\sim 40\%$) and a largely conserved genomic structure (Fig. 2).

As summarized in Table 4, RGR opsin, melanopsin and the photosensory opsins differ in their amino acid identity, genomic structure and in a number of critical residues. If melanopsin is a photosensory opsin, then it represents a quite separate line of photopigment evolution in the vertebrates. Alternatively melanopsin may function as a photoisomerase, acting to regenerate chromophore for an as yet unrecognised pigment. In this regard it is worth noting that rod-opsin and melanopsin have recently been shown to be co-expressed in *Xenopus* melanophores (Miyashita et al 2001). One interpretation of this finding is that rod opsin mediates the aggregation of the melanosomes, and this would agree with the aggregation action spectrum which has a λ_{max} 500 nm, whilst melanopsin may be governing melanophore dispersion (Miyashita et al 2001). Alternatively there may be a single rod opsin photopigment in the melanophores, and melanopsin supports rod opsin photopigment activity, perhaps acting as an RGR-like photoisomerase. Finally, of course, melanopsins may act as both photosensors and photoisomerases, and in this respect resemble the invertebrate photopigments, to which they share the highest level of amino acid identity (Table 1).

Placing candidate genes and photopigments into context

Studies on teleost fish and mammals have demonstrated the existence of non-rod, non-cone ocular photoreceptors. In the case of VA opsin in the roach, electrophysiological evidence suggests that one function of this photosensory photopigment is to modulate the activity of retinal horizontal cells. To what end remains unclear, but this fits with the general role of horizontal cells in the teleost

TABLE 4 Summary of the major differences between vertebrate photosensory opsins (rod, cone, **P** and **VA** opsin), photoisomerases (RGR opsins), and members of the melanopsin family

Feature	Photosensory opsins	Retinal G-protein coupled receptor (RGR)	Melanopsins
Known function	Image- and non-image forming light detection	Photoisomerase, harvesting light energy to convert all-*trans*-retinal to 11-*cis*-retinal	Unknown
Functional expression	Forms a photosensitive pigment capable of triggering a phototransduction cascade	Forms a photosensitive pigment that is incapable of triggering a phototransduction cascade	Unknown
Sites of expression	Ocular and extraocular photoreceptors	Retinal pigment epithelium and Müller cells	Ocular and extraocular photoreceptors
Lysine retinal attachment site (K296)	Yes	Yes	Yes
Glutamate Schiff base counterion (E113)	Yes	Histidine (H)	Tyrosine (Y)
Amino acid identity	Families share ≥40% identity between each other, and ≥70% within a family	∼22% with the known photosensory opsin families	∼27% with known photosensory opsin families, and ≥55% with each other
3rd cytoplasmic loop G protein (transducin) activation domain	Conserved in all except VA opsin. However, conserved in all VA opsins. Suggesting a common VA opsin G protein	Highly divergent from the known photosensory opsin families	Highly divergent, both between the known photosensory opsin families and within the melanopsin family
Genomic structure	Conserved (introns 1, 3 and 4) in the vertebrates and chordates	No introns conserved with the known opsin families	No introns conserved with the known opsin families
Phylogenetic position	Common ancestry in the vertebrates and chordates	Separate evolutionary lineage from the known photosensory opsins	Separate evolutionary lineage from the known photosensory opsins

retina as regulators of retinal activity in response to environmental irradiance. By contrast, studies in *rd/rd cl* mice have demonstrated that mammals use non-rod, non-cone photoreceptors for a broad range of irradiance detection tasks. Furthermore, action spectra for both pupillary constriction and circadian entrainment have defined a novel opsin/vitamin A-based photopigment. However, the identity of this opsin gene remains uncertain. Melanopsin is expressed within the intrinsically photosensitive retinal ganglion of mammals, and in the light-sensitive melanophores of *Xenopus*, but nothing is known of the functional properties of this opsin gene family. Moreover, the deduced structure of the melanopsins differ so markedly from the characterized photosensory and photoisomerase opsins, it is difficult to predict a role for this opsin in light detection (Table 4).

The relative ease with which genes can be isolated contrasts with the time it takes to determine their real function. In the case of photopigment genes the assignment of function has been based traditionally on a number of criteria. The primary criteria have been as follows. (1) The candidate protein should form a functional photopigment and be shown to alter its activity in response to photic rather than non-specific kinetic actions. For example, both infrared and ultraviolet energy can non-specifically activate a protein. In addition, it is often useful to show that an opsin can activate a transduction cascade to distinguish photosenory pigments from photoisomerases. (2) The candidate photopigment should have an absorbance spectrum that matches the action spectrum of the response in question. As discussed above if melanopsin forms a functional photopigment then its absorption spectrum should ideally match the action spectra for *rd/rd cl* responses to light. (3) The candidate molecule should be expressed in areas/cells defined as photoreceptors using physiological assays. The expression of melanopsin within the intrinsically photosensitive RGCs (Berson et al 2002), suggests that melanopsin is likely to play some role in the light-detecting capacity of these cells. Secondary criteria for photopigment identification would include: (4) genetic ablation of the candidate molecule. Two broad results are possible in knock-out studies. If the candidate gene provides the only photosensory input then the response will be abolished. If, however, there are multiple photoreceptor inputs, then gene ablation may result in an attenuated response or there may be no obvious phenotype. If attenuated, the action spectrum should be altered in a manner predicted by the absorbance spectrum of the photopigment. In the absence of the primary criteria 1–3 (above), gene ablation studies can only be used to correlate a gene with a light-dependent process, and will not distinguish between the loss of a photosensory pigment and/or loss of an element in the phototransduction process. (5) Chromophore identification or depletion. For example, 11-*cis*-retinaldehyde is only associated with opsin-based photopigments. 11-*cis*-retinaldehyde can be readily identified using HPLC, and its identification

(Foster et al 1993), apparent lack (Foster et al 1989) or depletion (Zatz 1994) has been helpful in defining the nature of photoreceptive pathways. Some care should be exercised when using this approach however, as chromophore depletion is not the same as chromophore loss. Even visual responses may be only moderately affected after severe chromophore depletion (Zimmerman & Goldsmith 1971). The identification of 11-*cis*-retinal can be significant, but failure to identify 11-*cis*, like all negative results, may not be meaningful. (6) Homology to known photopigment molecules. Much of the discussion in this paper has considered the possible role of opsins based on homology. While this can be informative, care has to be exercised, in the absence of criteria 1–3.

The vertebrate cryptochromes (CRYs) have been proposed as 'the' circadian photopigments (Sancar 2000). This claim appears somewhat premature as the vertebrate cryptochromes do not fulfil any of the primary criteria for photopigment identification. They have not yet been shown to form functional photopigments, action spectra have not been matched to absorption spectra, and they are not uniquely expressed in known photoreceptors. Of the secondary criteria, CRY knock-outs do not block light-induced clock gene expression (Okamura et al 1999), chromophore (FAD and FADH) depletion has not been undertaken and may not be practical, and deductions based upon homology have been misleading. For example, the homology of the vertebrate and plant cryptochromes is low. Indeed, animal cryptochromes are more similar to photolyases than the plant photopigment (Cashmore et al 1999). Finally, it has been reported that most CRY double-mutant mice develop a spontaneous ocular inflammatory phenotype (Van Gelder 2001), raising the possibility that any modification of light responses detected in CRY double-mutant mice may be the result of secondary effects of the inflammatory response. Currently nothing directly links the cryptochromes to a photopigment function in the vertebrates. Indirect evidence rests upon vitamin A-depletion experiments that have failed to abolish circadian responses to light in mice (Thompson et al 2001). Ignoring the weakness of the methodology, see (5) above, this reasoning cannot be taken as positive evidence for the involvement of a cryptochrome photopigment.

Melanopsin represents a promising candidate for the non-rod, non-cone photopigment in mammals, but we do not know whether melanopsin forms a functional photopigment, and should it do so, whether circadian action spectra will match its absorption spectra. These problems have been fully recognized by Ignacio Provencio and Mark Rollag who originally discovered this gene family, and their discussion of the role of the melanopsins has been suitably cautious (Provencio et al 2000, Provencio et al 2002). Regrettably the same caution has not been exercised by other researchers working on alternative candidates. Photobiologists have developed well-established criteria for the identification of photopigments, and for their assignment to a particular task (1–6). These criteria are demanding in both time and resources, but have proved their worth in many experiments in very diverse systems. Future studies aimed at identifying 'circadian photopigments' could only benefit from such approaches.

References

Ahmad M, Grancher N, Heil M et al 2002 Action spectrum for cryptochrome-dependent hypocotyl growth inhibition in *Arabidopsis*. Plant Physiol 129:774–785

Baldwin JM, Schertler GF, Unger VM 1997 An alpha-carbon template for the transmembrane helices in the rhodopsin family of G-protein-coupled receptors. J Mol Biol 272:144–164

Batni S, Scalzetti L, Moody SA, Knox BE 1996 Characterization of the *Xenopus* rhodopsin gene. J Biol Chem 271:3179–3186

Bellingham J, Foster RG 2002 Opsins and mammalian photoentrainment. Cell Tiss Res 309:57–71

Bellingham J, Whitmore D, Philp AR, Wells DJ, Foster RG 2002 Zebrafish melanopsin: isolation, tissue localisation and phylogenetic position. Mol Brain Res 107:128–136

Berson DM, Dunn FA, Takao M 2002 Phototransduction by retinal ganglion cells that set the circadian clock. Science 295:1070–1073

Bridges C 1959 The visual pigments of some common laboratory animals. Nature 184:727–728

Burns ME, Baylor DA 2001 Activation, deactivation, and adaptation in vertebrate photoreceptor cells. Annu Rev Neurosci 24:779–805

Cashmore AR, Jarillo JA, Wu Y-J, Liu D 1999 Cryptochromes: blue light receptors for plants and animals. Science 284:760–765

Deguchi T 1981 Rhodopsin-like photosensitivity of isolated chicken pineal gland. Nature 290:706–707

Ebrey T, Koutalos Y 2001 Vertebrate photoreceptors. Prog Retin Eye Res 20:49–94

Foster RG 1998 Shedding light on the biological clock. Neuron 20:829–832

Foster RG 2002 Keeping an eye on the time: the Cogan Lecture. Invest Ophth Vis Sci 43:1286–1298

Foster RG, Bellingham J 2002 Opsins and melanopsins. Curr Biol 12:R543

Foster RG, Follett BK, Lythgoe JN 1985 Rhodopsin-like sensitivity of extra-retinal photoreceptors mediating the photoperiodic response in quail. Nature 313:50–52

Foster RG, Schalken JJ, Timmers AM, De Grip WJ 1989 A comparison of some photoreceptor characteristics in the pineal and retina: I. The Japanese quail (*Coturnix coturnix*). J Comp Physiol A 165:553–563

Foster RG, Garcia-Fernandez JM, Provencio I, De Grip WJ 1993 Opsin localization and chromophore retinoids identified within the basal brain of the lizard *Anolis carolinensis*. J Comp Physiol A 172:33–45

Freedman MS, Lucas RJ, Soni B et al 1999 Regulation of mammalian circadian behavior by non-rod, non-cone, ocular photoreceptors. Science 284:502–504

Galtier N, Gouy M, Gautier C 1996 SEAVIEW and PHYLO_WIN: two graphic tools for sequence alignment and molecular phylogeny. Comput Appl Biosci 12:543–548

Gooley JJ, Lu J, Chou TC, Scammell TE, Saper CB 2001 Melanopsin in cells of origin of the retinohypothalamic tract. Nat Neurosci 4:1165

Hankins MW, Lucas RJ 2002 The primary visual pathway in humans is regulated according to long-term light exposure through the action of a nonclassical photopigment. Curr Biol 5:191–198

Hannibal J, Hindersson P, Knudsen SM, Georg B, Fahrenkrug J 2002 The photopigment melanopsin is exclusively present in pituitary adenylate cyclase-activating polypeptide-containing retinal ganglion cells of the retinohypothalamic tract. J Neurosci 22:RC191

Hao W, Fong HK 1996 Blue and ultraviolet light-absorbing opsin from the retinal pigment epithelium. Biochemistry 35:6251–6256

Hao W, Fong HK 1999 The endogenous chromophore of retinal G protein-coupled receptor opsin from the pigment epithelium. J Biol Chem 274:6085–6090

Hattar S, Liao HW, Takao M, Berson DM, Yau KW 2002 Melanopsin-containing retinal ganglion cells: architecture, projections, and intrinsic photosensitivity. Science 295:1065–1070

Hunt DM, Wilkie SE, Bowmaker JK, Poopalasundaram S 2001 Vision in the ultraviolet. Cell Mol Life Sci 58:1583–1598

Jacobs GH, Neitz J, Deegan JF 2nd 1991 Retinal receptors in rodents maximally sensitive to ultraviolet light. Nature 353:655–656

Kojima D, Mano H, Fukada Y 2000 Vertebrate ancient-long opsin: a green-sensitive photoreceptive molecule present in zebrafish deep brain and retinal horizontal cells. J Neurosci 20:2845–2851

Lucas RJ, Foster RG 1999 Neither functional rod photoreceptors nor rod or cone outer segments are required for the photic inhibition of pineal melatonin. Endocrinology 140:1520–1524

Lucas RJ, Douglas RH, Foster RG 2001 Characterization of an ocular photopigment capable of driving pupillary constriction in mice. Nat Neurosci 4:621–626

Lythgoe JN, Shand J, Foster RG 1984 Visual pigment in fish iridocytes. Nature 308:83–84

Menon ST, Han M, Sakmar TP 2001 Rhodopsin: structural basis of molecular physiology. Physiol Rev 81:1659–1688

Minamoto T, Shimizu I 2002 A novel isoform of vertebrate ancient opsin in a smelt fish, *Plecoglossus altivelis*. Biochem Biophys Res Commun 290:280–286

Miyashita Y, Moriya T, Yamada K et al 2001 The photoreceptor molecules in *Xenopus* tadpole tail fin, in which melanophores exist. Zoolog Sci 18:671–674

Moriya T, Miyashita Y, Arai J, Kusunoki S, Abe M, Asami K 1996 Light-sensitive response in melanophores of *Xenopus laevis*: I. Spectral characteristics of melanophore response in isolated tail fin of *Xenopus* tadpole. J Exp Zool 276:11–18

Moutsaki P, Bellingham J, Soni BG, David-Gray ZK, Foster RG 2000 Sequence, genomic structure, and tissue expression of carp (*Cyprinus carpio* L.) vertebrate ancient (VA) opsin. FEBS Lett 473:316–322

Okamura H, Miyake S, Sumi Y et al 1999 Photic induction of *mPer1* and *mPer2* in *Cry*-deficient mice lacking a biological clock. Science 286:2531–2534

Oshima N 2001 Direct reception of light by chromatophores of lower vertebrates. Pigment Cell Res 14:312–319

Palczewski K, Kumasaka T, Hori T et al 2000 Crystal structure of rhodopsin: a G protein-coupled receptor. Science 289:739–745

Philp AR, Garcia-Fernandez J-M, Soni BG, Lucas RJ, Bellingham J, Foster RG 2000 Vertebrate ancient (VA) opsin and extraretinal photoreception in the atlantic salmon (*Salmo salar*). J Exp Biol 203:1925–1936

Provencio I, Jiang G, De Grip WJ, Hayes WP, Rollag MD 1998a Melanopsin: an opsin in melanophores, brain and eye. Proc Natl Acad Sci USA 95:340–345

Provencio I, Jiang G, Hayes WP, Zatz M, Rollag MD 1998b Novel skin and brain opsin, melanopsin, is found in the chicken. Invest Ophth Vis Sci 39:S236 (abstract 1075)

Provencio I, Cooper HM, Foster RG 1998c Retinal projections in mice with inherited retinal degeneration: implications for circadian photoentrainment. J Comp Neurol 395:417–439

Provencio I, Rodriguez IR, Jiang G, Hayes WP, Moreira EF, Rollag MD 2000 A novel human opsin in the inner retina. J Neurosci 20:600–605

Provencio I, Rollag MD, Castrucci AM 2002 Photoreceptive net in the mammalian retina. This mesh of cells may explain how some blind mice can still tell day from night. Nature 415:493

Rollag MD 1996 Amphibian melanophores become photosensitive when treated with retinal. J Exp Zool 275:20–26

Rollag MD, Provencio I, Sugden D, Green CB 2000 Cultured amphibian melanophores: a model system to study melanopsin photobiology. Methods Enzymol 316:291–309

Sancar A 2000 Cryptochrome: the second photoactive pigment in the eye and its role in circadian photoreception. Annu Rev Biochem 69:31–67

Shand J, Foster RG 1999 The extraretinal photoreceptors of non-mammalian vertebrates. In: Archer SN, Djamgoz MBA, Loew ER, Partridge JC, Vallerga S (eds) Adaptive mechanisms in the ecology of vision. Kluwer Academic Publishers, Dordrecht, Netherlands, p 197–222

Smyth RD, Saranak J, Foster KW 1988 Algal visual systems and their photoreceptor pigments. Prog Phycol Res 6: 255–286

Soni BG, Foster RG 1997 A novel and ancient vertebrate opsin. FEBS Lett 406:279–283

Soni BG, Philp AR, Knox BE, Foster RG 1998 Novel retinal photoreceptors. Nature 394:27–28

Starace DM, Knox BE 1997 Activation of transducin by a *Xenopus* short wavelength visual pigment. J Biol Chem 272:1095–1100

Sun H, Macke JP, Nathans J 1997a Mechanisms of spectral tuning in the mouse green cone pigment. Proc Natl Acad Sci USA 94:8860–8865

Sun H, Gilbert DJ, Copeland NG, Jenkins NA, Nathans J 1997b Peropsin, a novel visual pigment-like protein located in the apical microvilli of the retinal pigment epithelium. Proc Natl Acad Sci USA 94:9893–9898

Thompson CL, Blaner WS, Van Gelder RN et al 2001 Preservation of light signaling to the suprachiasmatic nucleus in vitamin A-deficient mice. Proc Natl Acad Sci USA 98:11708–11713

Thompson JD, Gibson TJ, Plewniak F, Jeanmougin F, Higgins DG 1997 The CLUSTALX windows interface: flexible strategies for multiple sequence alignment aided by quality analysis tools. Nucleic Acids Res 25:4876–4882

Van Gelder RN 2001 Non-visual ocular photoreception. Ophthalmic Genet 22:195–205

Yokoyama S, Zhang H 1997 Cloning and characterization of the pineal gland-specific opsin gene of marine lamprey (*Petromyzon marinus*). Gene 202:89–93

Yokoyama S, Radlwimmer FB, Kawamura S 1998 Regeneration of ultraviolet pigments of vertebrates. FEBS Lett 423:155–158

Zatz M 1994 Photoendocrine transduction in cultured chick pineal cells: IV What do vitamin A depletion and retinaldehyde addition do to the effects of light on the melatonin rhythm? J Neurochem 62:2001–2011

Zimmerman WF, Goldsmith TH 1971 Photosensitivity of the circadian rhythm and of visual receptors in carotenoid-depleted *Drosophila*. Science 171:1167–1169

DISCUSSION

Loros: It seems to me that a melanopsin mutant would be useful. Are you going in this direction, and which system are you most interested in isolating a mutant from?

Foster: Yes, a mutant would be very interesting. However an alternative approach has been adopted by at least three laboratories that I am aware of who have developed a melanopsin knockout mouse, and we (Dr Robert Lucas and myself) are collaborating with one of these groups. Both circadian and pupillary response to light are attenuated to some degree in these animals. Although there is clearly an effect of ablating melanopsin it is not clear whether melanopsin is acting as a photosensory pigment or as a critical component in the phototransduction system. I am not sure if Mike Menaker or Steve Kay would like to comment further on this?

Schibler: What happens with the circadian rhythms?

Menaker: It is clear from what Russell Foster has told us that there are going to be other photoreceptors in the mammalian retina. In some sense we wouldn't expect a melanopsin knockout to abolish circadian responses.

Foster: This is critical. It will be important to combine the melanopsin knockout mice with a rodless+coneless mouse model. The *rd/rd cl* mouse would be the most useful as this phenotype is already well defined, but I understand other rodless+coneless models are also being developed.

Loros: So this work is all in mice.

Foster: Yes. I think if we are going to understand what melanopsin is doing a comparative approach will be valuable, and this is why Mark Hankins and I have started to work on zebrafish in parallel with mice. Although the transgenics and gene manipulation is not as sophisticated in fish, there are other things that can be done, particularly at the electrophysiological level. As I discussed in the presentation, Mark Hankins and I are keen to pursue this approach.

Rosbash: What are the various interpretations for the difference between the rodless coneless physiology and the wild-type? More generally, when I discussed these kinds of experiments with Aziz Sancar, he was always reminding me that the action spectrum on an animal is not the same as the action spectrum on a purified molecule. He was always critical of the facile interpretation of when you do an action spectrum with some behavioural output this defines the photopigment, in the same way that doing an absorption spectrum on a purified protein would.

Foster: I am surprised by Dr Sancar's comments. I'll deal with your general point first with some early and simple examples from humans. If you undertake an action spectrum for the human rod response using an electroretinogram (ERG) as an assay, this action spectrum predicts very precisely the absorption spectrum of the isolated pigment or from microspectrophotometry. Perhaps even more impressive is that CIE (Commission Internationale de l'Eclairage) of 1951 based the scotopic sensitivity of humans on action spectra derived from psychophysical experiments on normal human observers (See Crawford 1949, Wald 1945). The sensitivity curve used very closely approximates the rhodopsin template. In almost every case, if done properly—and this is the critical point, an action spectrum can reflect very beautifully the biochemistry of the isolated pigment. There are one or two things you have to be careful about. For example, if there are filters screening the pigment before the light is absorbed, then this can alter the apparent sensitivity of the pigment. However, if the filtering effect is known then corrections can be made, and if not, the distortion will be obvious because the irradiance response curves will not be univariant. You would not see a parallel series of irradiance response curves. It would be fair to say that an action spectrum is a true reflection of the absorption spectrum. Indeed, this has been the assumption made by the vision community and other photobiologists over the last 60 years, and part of the reason George Wald was awarded the Nobel Prize. It has been a remarkably powerful approach, which is why Dr Sancar's comments puzzle me.

Young: It is not so much a question of what the opsins are doing, but that of the nature of cryptochrome, which is a promiscuous protein. When do you expect to find cryptochrome by itself, as opposed to stuck with a dozen or so partners? What would you imagine that those partnerships might do to assays of this sort? Presumably these *in vitro* assays that have been done in *Arabidopsis* and algae have been quite distinct from the *in vivo* situation.

Kay: Cryptochromes are particularly problematic in matching action spectra to absorption spectra.

Foster: Yes I agree, there has been much discussion about the absorption spectrum of the cryptochromes because of the two potential chromophores, but what is fascinating is that the normalized flavoprotein and CRY1 action spectrum I showed in my presentation (see Fig. 4) shows a striking similarity with that action spectrum for *Arabidopsis* CRY1 as published by Margaret Ahmad and colleagues (Ahmad et al 2002).

Kay: That was a good experiment that Margaret Ahmad did. She over-expressed cryptochrome and then used levels of light that are much lower and which don't initiate a response in the wild-type but do in the over-expressor. This is how she was able to determine a CRY1-specific action spectrum.

Foster: Tony Cashmore has consistently argued that we simply can't know what the absorption spectra of the cryptochromes will be. Others have argued differently. The empirical evidence of a comparison between a flavin absorption spectrum and that of the new CRY1 over-expression by Ahmad and colleagues suggests that there is a close correlation.

Menaker: Russell Foster, if you think about the rodless coneless mouse as a good model since it is likely to have perhaps only one photopigment, and if you assume that this photopigment is not melanopsin, what is there available that fits the action spectrum?

Foster: Nothing. We simply don't know. What I am proposing is that we should keep a reasonably open mind. Melanopsin is clearly the best candidate we have, but for the reasons I outlined in my talk it doesn't seem quite right. Melanopsin is certainly expressed in the right place. But this means it could be either a photosensory pigment or acting as a local isomerase. Its structure predicts neither. For example, let us turn the argument around. Let us pretend that the original aim was to look for a photoisomerase that is likely to be present in the intrinsically photosensitive ganglion cells. After finding melanopsin, one could make just as strong a case, if not stronger, that melanopsin was the photoisomerase, because its structure is so very different from the photosensory opsins. Of course you could argue that the melanopsins are so different because they are expressed in non-traditional photoreceptors. But this argument doesn't hold. The VA opsin photopigments are not expressed in rod or cone cells, but are expressed within the inner retina. Yet these functional photosensors share the key features of the rod and cone opsins.

Green: It seems that the comparison between these opsins that you have presented is a little bit skewed in terms of vertebrate opsins. If you compare melanopsin to invertebrate opsins it actually looks very much like a visual opsin in several respects.

Foster: I agree, the melanopsins have been suggested as possible photopigments partly on the basis of their similarity to invertebrate opsins (Table 1). Furthermore, another invertebrate-like feature of the melanopsins is their possession of a tyrosine rather than a glutamate at the putative counterion position. Many invertebrate opsins do have a tyrosine in this position, but not all—some use phenylalanine. Furthermore, the melanopsins are not unique amongst the vertebrate opsins in possessing a tyrosine 'counterion'—for example, the putative retinal isomerase peropsin has a tyrosine in the counterion position. But unlike the photoisomerases, the third cytoplasmic loop of the melanopsins is very large. In RGR and peropsin this is very short, so there are big differences between the melanopsins and the photoisomerases too. Again, I am not saying that melanopsin can't be a photopigment, only that its structure is not obviously like a photopigment.

Kay: Don't you see that same type of divergence among other G protein-coupled receptors that are not opsins? I don't think this is a special feature.

Foster: Absolutely. The point I was trying to make is that the known photosensory opsins are highly conserved, whilst the whole opsin family shows considerable divergence.

Rosbash: Has that lysine substitution been done?

Foster: Yes, and you can't form a photopigment. Mutations have also been done on the glutamate and in most cases the glutamate is needed for a stable photopigment. But this is not an essential feature. The invertebrates do it perfectly well with a tyrosine, and *Xenopus* short-wavelength pigments can use an aspartate. I am not saying that melanopsin is not a photopigment, but if it is, it becomes even more interesting because it would represent a completely divergent line of photopigment evolution in the vertebrates. In addition, as the 3rd cytoplasmic loop is so different from the conventional opsins, and so different within the different members of the melanopsin family, it is impossible to predict the nature of the phototransduction cascade.

Green: It is worth noting that the invertebrate-like opsins can reisomerize their chromophore themselves and do not need a separate photoisomerase.

Foster: This is a good point, and that is why I said that it is possible that melanopsin does both. However, part of the reason why melanopsin was originally considered as a photopigment when it was isolated from *Xenopus* melanophores by Iggy Provencio and Mark Rollag, was that no other opsins were thought to be expressed in melanophores. We now know from the work of Asami's group (Miyashita et al 2001), that melanopsin is co-expressed with rod opsin. And of course, melanophore aggregation is nicely described by a rod-like opsin action spectrum.

Stanewsky: A channel opsin has recently been isolated from algae which can function as a circadian receptor without even talking to a G protein. Have people looked in mammals to see if this exists there?

Foster: Everyone is trawling the databases for opsins and related molecules. Although some of the opsins are so different that they would not necessarily be identified by conventional searches.

Stanewsky: These channel opsins are very different.

Foster: That's true, and potentially very interesting. On the topic of missing potential candidates. Greg Cahill told me that he has failed to find zebrafish melanopsin in the zebrafish databases, despite the fact we have just isolated and published the full sequence (Bellingham et al 2002). I suspect there could be some very interesting opsin-like molecules lurking undiscovered.

Cahill: I have a question relating to the human variants. If you have both a photopigment and a photoisomerase that are necessary for the response, you shouldn't be able to get a univariant action spectrum. Is this true? I thought you were suggesting the possibility that the reason the melanopsin knockouts had a subtle effect was that melanopsin was necessary as a photoisomerase.

Foster: I was suggesting that an attenuated response in a melanopsin knock-out could not distinguish between the ablation of melanopsin acting as a photosensory pigment or as an element of the phototransduction system. For example, destroying the chromophore regeneration system could have the same effect as destroying the photopigment itself.

Cahill: Then you would not agree with the possibility that photoisomerase would contribute to your intensity of response curves.

Foster: I see what you are getting at. But let me think about it.

Dunlap: You showed data on melanopsin expression from zebrafish, indicating that there is a lot in the eye, a little in the brain and virtually nothing anywhere else.

Foster: On the basis of the protection assay I showed I would say that we have no evidence for melanopsin outside the eye. But it might be in the brain or pineal at low levels. We have preliminary evidence that there may be different forms of melanopsin in the teleosts, and that they may be expressed in different sites.

Dunlap: The point is, there is very little in the somatic cells at all. Yet Whitmore et al (2000) showed normal circadian photoreception in the heart. This suggests that melanopsin cannot be the circadian photoreceptor in the fish and there must be something else. What is it?

Foster: I can suggest something! As you say isolated organs and cell lines from zebrafish exhibit circadian oscillations in clock gene expression that can be entrained to a 24 h light/dark cycle, and the mechanism underlying this cellular photosensitivity is unknown. Vivi Moutsaki in my lab has identified a novel opsin, *tmt-opsin*, that has a genomic structure characteristic of vertebrate photopigments, an amino acid identity equivalent to the known photopigment opsins, and the essential residues required for photopigment function. Significantly *tmt-opsin* is expressed in a wide variety of neural and non-neural tissues, including a zebrafish embryonic cell line that exhibits a light entrainable clock. Collectively our data suggest that *tmt-opsin* is a strong candidate for the photic regulation of zebrafish peripheral clocks. But I know this view is not shared by all. Paolo, would you like to comment?

Sassone-Corsi: In the experiment you showed it is clear that in all those receptors there is extensive alternative splicing. In an RNAse protection assay, picking the probe is crucial.

Dunlap: So you think that melanopsin is being expressed in the somatic cells?

*Sassone-*Corsi: What I am saying is that you might miss expression in the brain, for instance, if it is only in a few cells. You should also do *in situ* hybridizations.

Dunlap: Has anyone looked for melanopsin in the somatic cells by an alternative assay?

Foster: We have performed RT-PCR for melanopsin in brain and other tissues but failed to find it. As I mentioned, I think TMT-opsin is a better candidate.

Schibler: Has anyone used multi-unit electrode approaches to study the intrinsically photosensitive ganglion cells?

Foster: This would be tricky with a normal mammalian retina because of the presence of the rods and cones, although possible on a rodless+coneless mouse retina. In David Berson's studies he identified the ganglion cells projecting to the SCN using dye injections into the SCN that retrolabelled the ganglion cells of the retina. He then recorded from these identified cells. But of course there may be more than one population of intrinsically photosensitive inner retinal neuron.

Schibler: Have you tried taking these cells out?

Foster: Rob Lucas, Jim Bellingham and I are working on a model that will not just ablate the melanopsin gene, but ablate the melanopsin expressing ganglion cells. If we succeed, then we could establish how the rod and cone inputs get to the circadian system. Addressing the question of whether these inputs go through the novel ganglion cells or via some other route to the SCN? The input need not necessarily go through the retinohypothalamic tract (RHT). An ablation of the RHT cells would be very powerful.

Van Gelder: How do you reconcile the action spectrum data with the vitamin A-depletion data that we have presented? As the perspective for this, in *Neurospora*, entrainment of the clock is resistant to vitamin A metabolic defects and in *Drosophila* it is resistant to vitamin A depletion. We found a similar result in the mammal with near-complete vitamin A depletion of the eye, and no attenuation in the response, as measured by immediate-early gene induction in the SCN.

Foster: I am not suggesting that all circadian photopigments have to be opsin-based. That would be silly, the point I have tried to make is that action spectra in both retinal and extraretinal photoreceptors have always described an opsin/vitamin A based photopigment. As to vitamin A depletion, we have considerable experience of using HPLC to look for retinoids, and published the first papers on this topic using diode array detectors. What struck me with your study (see p 31–51) was that the HPLC assays that you used to look for vitamin A depletion were relatively insensitive. At the sensitivities that you were using, you would not have seen chromophore in the pineal of a bird or a fish. Your assays could certainly show that there is a reduction in chromophore content, but you can't say that there is no chromophore left. Indeed, I seem to remember that you do show low levels of chromophore. Furthermore, Iggy Provencio has argued that if melanopsin is behaving like an invertebrate photopigment then it will not release its vitamin A during the photopigment re-bleaching cycle. As a result it would be very difficult to deplete the vitamin A of the novel receptors. The trouble is that your interpretation of your data rests on a complete depletion of the chromophore, and that is not the case.

Van Gelder: We know that they are at least 500-fold reduced.

Foster: In view of the concentration of chromophore in the visual system, a 500-fold reduction of chromophore in the eye may have very little effect on the inner retinal photoreceptors. It is striking that *Drosophila* carotenoid-depletion experiments only reduce the visual ERG but did not abolish it.

Young: As I recall, weren't the depletions done over several months?

Van Gelder: 10 months in our studies.

Young: So what is the half-life of melanopsin protein, if you are talking about scavenging function?

Foster: I am not thinking of melanopsin as a scavenger, but as a potential photoisomerase, or perhaps even a retinoid transport protein.

Young: You were talking about it holding on to a photoreceptor.

Foster: Yes, in the same way that invertebrate opsins are thought to hold on to their chromophore during the photopigment regeneration cycle.

Young: What is the half-life of these proteins? If this is taking place over several months I have a hard time imagining melanopsin as a sponge that holds on to chromophore for this long.

Foster: I don't know the mechanism whereby chromophore is retained. The bottom line is that we know chromophore is only depleted in flies and mice, and that most of the retinoid will be part of the visual system anyway. There must be some other way of retaining vitamin A. In fact in mammals, there are a whole range of potential vitamin A binding proteins, like IRBP in the pigmented epithelium, that could serve to mop-up and act as a chromophore sink.

Menaker: There is a real irony here. When we first started looking at rodless mice the argument was that there would be a few rods left. Now what's going on is that people are arguing that there are a few molecules of vitamin A left. It is a quantitative question. In Russell Van Gelder's vitamin A-deprived mice there may or may not be enough vitamin A left to support this function. We really don't know.

Foster: As it turns out those initial criticisms were correct, and this raises an important general issue. You cannot base a conclusion on 'negative data'. You can't prove a negative. When we made the *rd/rd cl* mouse, we found no evidence for functional rods and cones using exquisitely sensitive techniques. But we could not prove that they were none there. The reason we were so excited by the pupillary action spectrum results we published last year, and the new phase shifting results in *rd/rd cl* mice that I showed today, is because even if there were a few rods and cones left, the responses we are getting can't be due to these receptors because the action spectra describe a completely novel opsin-based pigment. These results illustrate the power of using action spectrum approaches.

References

Ahmad M, Grancher N, Heil M et al 2002 Action spectrum for cryptochrome-dependent hypocotyl growth inhibition in *Arabidopsis*. Plant Physiol 129:774–785

Bellingham J, Whitmore D, Philp AR, Wells DJ, Foster RG 2002 Zebrafish melanopsin: isolation, tissue localisation and phylogenetic position. Brain Res Mol Brain Res 107:128–136

Crawford BH 1949 The scotopic visibility function. Proc Phys Soc Lond B62:321–334

Miyashita Y, Moriya T, Yamada K et al 2001 The photoreceptor molecules in *Xenopus* tadpole tail fin, in which melanophores exist. Zool Sci 18:671–674

Wald G 1945 The spectral sensitivity of the human eye: a spectral adaptometer. J Opt Soc Am 35:187

Whitmore D, Foulkes N, Sassone-Corsi P 2000 Light acts directly on organs and cells in culture to set a vertebrate circadian clock. Nature 404:87–91

Cryptochromes and inner retinal non-visual irradiance detection

Russell N. Van Gelder*†, Aziz Sancar‡

*Departments of Ophthalmology and Visual Sciences, †Molecular Biology and Pharmacology, Washington University Medical School, CB# 8096, 660 S. Euclid Avenue, St Louis, MO 63110, and ‡Department of Biochemistry and Biophysics, University of North Carolina School of Medicine, 405 Mary Ellen Jones, CB# 7260, Chapel Hill, NC, 27599, USA

Abstract. Nearly all circadian clocks have free-running periods that differ significantly from 24 hours. To maintain synchrony with the 24 h day, the mammalian circadian clock is reset by light. Unlike other animals, mammalian circadian entrainment occurs exclusively via the eyes and optic nerves. Remarkably, the classical photoreceptors — the rods and cones — are not necessary for photic entrainment. Instead, a subset of inner retinal ganglion cells are directly photoresponsive and transmit photic information specifically to brain centres involved in irradiance detection, including the master circadian pacemaker in the suprachiasmatic nucleus of the hypothalamus. The photopigment(s) responsible for inner retinal phototransduction are unknown. Several lines of evidence constrain candidate photopigments. First, near-total vitamin A depletion does not diminish retinohypothalamic signalling. Second, loss of cryptochrome function in retinal-degenerate mice substantially decreases photic signalling to the suprachiasmatic nucleus, and markedly decreases pupillary light responses. Third, vitamin A depletion of cryptochrome mutant mice leads to loss of photic signalling to the suprachiasmatic nucleus. These findings suggest a model where either classical photopigments or inner retinal photopigments are sufficient for non-visual irradiance detection.

2003 Molecular clocks and light signalling. Wiley, Chichester (Novartis Foundation Symposium 253) p 31–51

Nearly all circadian clocks in nature run with free-running periods significantly different than 24 hours; in order to be useful timekeepers, these clocks must be continually reset to the 24 h day (Pittendrigh 1993). The circadian clocks of all organisms show similar responses to light, being largely photo-insensitive during the day (or when the organism's clock is in the subjective day), delaying the phase of the clock in response to light in the early part of the night, and advancing the phase in response to light in the late part of the night.

Circadian phase shifting in *Drosophila* and *Neurospora* has long been recognized to be resistant to vitamin A depletion (Sargent et al 1966, Zimmerman & Goldsmith 1971) but sensitive to abnormalities in flavin metabolism (Paietta & Sargent 1981), which has led to the search for a flavin-based circadian blue light

"

photopigment. Soon after the discovery of cryptochromes as flavin-based candidate photopigments in *Arabidopsis* (Ahmad & Cashmore 1993), Sancar and colleagues discovered mammalian homologues and suggested that these genes may function in circadian phototransduction (Hsu et al 1996, Sancar 2000). In a forward genetic screen in *Drosophila melanogaster*, Hall, Rosbash and colleagues discovered a gene that rendered the mutant fly's clock insensitive to short pulses of light (Emery et al 1998, Stanewsky et al 1998, Emery et al 2000). This gene was found to be a homologue of the *Arabidopsis* gene cryptochrome. *Drosophila* cryptochrome is necessary for normal phase entrainment of flies and appears to function in a cell-autonomous manner in the fly's central pacemaker neurons (Emery et al 2000). *In vitro, Drosophila* cryptochrome appears to function as a photopigment, and is able to show light-dependent binding to its partner, Timeless, in a yeast two-hybrid system (Ceriani et al 1999, Rosato et al 2001).

In mammals the central circadian pacemaker is located in the suprachiasmatic nuclei (SCN) of the hypothalamus. Photic information reaches the SCN through the eyes; neither enucleated mice (Freedman et al 1999) nor *math5* (Wee et al 2002) mutant mice (which lack a well developed optic nerve) are able to entrain to external light–dark (LD) signals. Remarkably, the classical photoreceptors — the rods and cones — are not necessary for photic entrainment of the clock (Freedman et al 1999). Severe outer retinal degenerate mice (*rd/rd cl*) show normal phase shifting and entraining responses to light. Several other light-dependent phenomena are preserved in these mice, including photic suppression of pineal melatonin (Lucas et al 1999) and pupillary responsiveness (Lucas et al 2001). Recently, directly photoresponsive retinal ganglion cells have been discovered (Berson et al 2002). At least a subset of these cells contain an opsin-family member, called melanopsin (Provencio et al 2000), and melanopsin-containing cells appear to project exclusively to brain areas involved in non-visual irradiance detection tasks (Hattar et al 2002).

Mammals have two cryptochrome homologues (Hsu et al 1996); these genes are also expressed in the inner retina (Miyamoto & Sancar 1998). We have sought to evaluate the roles of these genes in non-photic light signalling to the circadian clock and other non-visual irradiance detection tasks. We find that the classical rods and cones and the cryptochromes have functionally complementary roles, and that mice lacking both the classical photoreceptors and the cryptochromes have markedly reduced photosensitivity for circadian phase shifting and for pupillary light responsiveness.

Materials and methods

Generation of retinol-binding protein knockout (Quadro et al 1999), cryptochrome knockout (Thresher et al 1998, Vitaterna et al 1999), and *rd/rd;mCry1$^{-/-}$;mCry2$^{-/-}$* (Selby et al 2000) mice was by standard embryonic stem cell insertional targeting, as described. Behavioural and physiological tests performed on these mice included measurement of free-running and entrained circadian rhythms (using Actimetrics hardware and software, Evanston, IL), photic induction of immediate early and *mPeriod* genes in the SCN by *in situ* hybridization (Vitaterna et al 1999, Selby et al 2000), and measurement of pupillary responses. The latter was performed in unanaesthetized mice using infrared video recording of pupillary responses, essentially as described (Lucas et al 2001). Melanopsin staining and pseudorabies virus tract tracing were performed in *rd/rd;mCry1$^{-/-}$;mCry2$^{-/-}$* and *mCry1$^{-/-}$;mCry2$^{-/-}$* mice essentially as described (Pickard et al 2002, Provencio 2002).

Results and discussion

Do mammals require a retinal-based photopigment for photic signalling to the SCN?

To determine if a vitamin A-based pigment is required for photic signalling from the eye to the circadian clock, we utilized mice homozygous for a targeted null allele of retinol-binding protein (RBP) (Quadro et al 1999). Mice lacking RBP cannot mobilize vitamin A stores from the liver to peripheral tissues; *RBP$^{-/-}$* mice raised on vitamin A-free diets thus lose ocular retinal, thereby disabling opsin-based photoreception. Electroretinograms in these animals become unrecordable by 4 months of age (Quadro et al 1999). We raised *RBP$^{-/-}$* and wild-type mice for 10 months on a vitamin A-free diet and measured photic induction of the *mPer* genes as a marker for ocular–SCN signalling. Despite undetectable retinal in the eyes of these mice (signifying at least 500-fold depletion of ocular retinal stores compared with *RBP$^{-/-}$* mice raised with vitamin A, or wild-type animals raised on vitamin A-free chow), no significant decrement in light-induced *mPer* gene expression in the suprachiasmatic nuclei was seen in these mice, even at limiting irradiance levels (Thompson et al 2001). This suggests that either retinal-based pigments are not necessary for inner retinal phototransduction, or that an opsin-based inner retinal photopigment is extremely resistant to vitamin A depletion.

Mice lacking cryptochromes lose free-running circadian rhythms but maintain photic input to the SCN

Two research groups have independently produced targeted-inactivation 'knockout' cryptochrome alleles for both cryptochrome genes in mice (Thresher et al 1998, van der Horst et al 1999, Vitaterna et al 1999). Loss of *mCry1* results in mice with short free-running periods, while loss of *mCry2* results in lengthened free running periods. When mice lack both cryptochromes, no free-running circadian rhythms can be appreciated. However, these mice show normal masking activity (i.e. clock-independent modulation of behaviour by light) in LD cycles (Mrosovsky 1999, 2001).

Photic induction of *mPer* genes has been examined in mice lacking cryptochromes, and—although quantitatively reduced even in single cryptochrome gene mutants (Thresher et al 1998)—remains qualitatively intact in *mCry1*$^{-/-}$; *mCry2*$^{-/-}$ animals, suggesting that the cryptochromes are not necessary for photic signalling to the SCN (Okamura et al 1999, Vitaterna et al 1999).

Retinal-degenerate mice lacking cryptochromes show
markedly decreased photic sensitivity for behaviour and
photic induction of SCN gene expression

Several previous studies have suggested that, under normal conditions, the action spectrum for photic phase shifting in rodents is consistent with a rod- or cone-based pigment (Takahashi et al 1984, Nelson & Takahashi 1991, Provencio & Foster 1995). To determine whether the classical photoreceptors were acting in parallel with a cryptochrome-requiring pathway, we bred *rd/rd* mice with cryptochrome null-allele mice to create *rd/rd;mCry1*$^{-/-}$;*mCry2*$^{-/-}$ mice (Selby et al 2000). There was no effect of the cryptochrome mutation on the course of retinal degeneration in these mice. Unlike *mCry1*$^{-/-}$;*mCry2*$^{-/-}$ mice, which show normal behavioural responses to light, most *rd/rd;mCry1*$^{-/-}$;*mCry2*$^{-/-}$ mice were arrhythmic under LD cycle conditions. Additionally, compared to either *mCry1*$^{-/-}$;*mCry2*$^{-/-}$ or *rd/rd* mice, *rd/rd;mCry1*$^{-/-}$;*mCry2*$^{-/-}$ mice showed markedly reduced photic induction of the immediate-early gene c-*fos* in response to light pulses, suggesting that these mice had reduced signalling to the SCN.

Several hypotheses may be invoked to explain the decreased photoresponsiveness seen in the *rd/rd;mCry1*$^{-/-}$;*mCry2*$^{-/-}$ mice. First, the cryptochrome mutation might induce a developmental anomaly leading to loss of the retinohypothalamic tract. We performed tract tracing experiments with a green fluorescent protein-labelled Bartha pseudorabies virus to look at the retinohypothalamic tract in *mCry1*$^{-/-}$;*mCry2*$^{-/-}$ as well as wild-type animals (Pickard et al 2002, Van Gelder et al 2002). Both showed normal numbers of labelled cells indicating that the retinohypothalamic tract in these animals is intact. Second, loss of cryptochrome might cause loss of expression of another photopigment in the eye. Melanopsin has been suggested as a candidate inner retinal photopigment (Gooley et al 2001, Hattar et al 2002, Provencio 2002). We therefore tested *rd/rd;mCry1*$^{-/-}$;*mCry2*$^{-/-}$ mice for melanopsin immunostaining, and found normal distribution of melanopsin immunoreactivity in these mice.

Interestingly, a subset of about 30% of the *rd/rd;mCry1$^{-/-}$;mCry2$^{-/-}$* mice still showed some behavioural photoresponsiveness in an LD cycle. While under limiting irradiance conditions, photic induction of c-*fos* in the SCN is reduced by 10–100-fold in *rd/rd;mCry1$^{-/-}$;mCry2$^{-/-}$* mice, about 20% of normal peak levels of immediate early gene induction were seen in these mice following exposure to very bright light. These results demonstrate that there is a photopigment spared in the *rd/rd;mCry1$^{-/-}$;mCry2$^{-/-}$* animals.

rd/rd;mCry1$^{-/-}$;mCry2$^{-/-}$ mice show markedly reduced pupillary responsiveness

The decreased behavioural photoresponsiveness in cryptochrome-mutant mice is somewhat difficult to interpret since these mice also lack free-running circadian rhythms. Therefore, one is measuring masking, not circadian responses. Although masking is also preserved in retinal-degenerate animals (Mrosovsky et al 2000), the neural and molecular mechanisms of masking are not as well understood as those of circadian rhythms.

Since pupillary responses are also preserved in retinal-degenerate mice (Keeler 1927, Lucas et al 2001), we sought to determine whether these responses were compromised in *rd/rd;mCry1$^{-/-}$;mCry2$^{-/-}$* mice. Compared with wild-type mice, *rd/rd cl* mice show substantially reduced photosensitivity, with half-maximal constriction in *rd/rd cl* occurring at nearly 100-fold higher irradiance than in wild-type animals. The peak of sensitivity is about 479 nm (Lucas et al 2001). *rd/rd* mice similarly show an approximately 1.5 log decrease in pupillary light responsiveness. Compared with *rd/rd* mice, *rd/rd;mCry1$^{-/-}$;mCry2$^{-/-}$* mice show another log decrement in sensitivity, with half-maximal pupillary responses occurring at $\sim 6 \times 10^{13}$ photons/cm^2/s for 470 nm light. However, at very bright light ($> 1 \times 10^{14}$ photons/cm^2/s) there is still pupillary responsiveness in these mice. This is similar to the low level induction of immediate-early genes seen in the *rd/rd;mCry1$^{-/-}$;mCry2$^{-/-}$* mice under bright lighting conditions.

The nature of the preserved photoreceptor and photopigment(s) in the *rd/rd;mCry1$^{-/-}$;mCry2$^{-/-}$* mice is not yet known. It is possible that incomplete retinal degeneration has left a small subset of cones that are providing the photic information to the clock and pupil (Garcia-Fernandez et al 1995). Alternatively, another inner retinal photopigment may be functioning. To examine whether the residual photosensitivity in the *rd/rd;mCry1$^{-/-}$;mCry2$^{-/-}$* mice might be dependent on vitamin A, we examined the immediate-early gene induction of *RBP$^{-/-}$;mCry1$^{-/-}$;mCry2$^{-/-}$* mice. Following total vitamin A depletion, these mice show no induction of immediate-early genes in the SCN (C. L. Thompson, C. P. Selby, A. Sancar, unpublished results). This demonstrates that all ocular photopigments are either vitamin A dependent or cryptochrome dependent.

Murine cryptochrome 1 shows light-dependent binding to PIAS
in a heterologous expression system

Possibly the best evidence to date that *Drosophila* cryptochrome functions as a photopigment is the finding that this protein binds the clock protein Timeless in a light-dependent fashion when expressed in the yeast two-hybrid system (Ceriani et al 1999, Rosato et al 2001). Although initial attempts to find light-dependent binding of mammalian cryptochrome were unsuccessful (Griffin et al 1999), these experiments tested only a small number of potential binding partners. We have performed a yeast two-hybrid screen for binding partners for murine cryptochrome 1. After screening more than 300 000 clones, three clones were identified as showing a light-dependent interaction with mCRY1. Two independent clones encoding protein inhibitor of activated STAT (PIAS) were recovered, as well as one clone each of WD-repeat containing protein 9 (WDR9) and 6-pyruvoyl-tetrahydropterin synthase (6PTS). The PIAS and WDR9 interactions were seen primarily in dark-grown cultures, while the 6PTS interaction was seen primarily in the light. These results suggest that mammalian cryptochrome can undergo light-dependent conformational changes that influence its ability to bind potential protein partners.

Making sense of non-visual photoreception

We do not yet have a full understanding of the photopigments and signal transduction mechanisms underlying photic entrainment of circadian rhythms and other non-visual irradiance detection tasks. However, the following findings constrain possible models for circadian phototransduction (summarized in Table 1):

- The outer retinal photoreceptors (rods and cones) are not necessary for photic signalling to the SCN and pupillary constriction centres
- A subset of melanopsin-containing retinal ganglion cells are directly photosensitive
- Near-total vitamin A depletion of the eye does not significantly decrease photic signalling to the SCN, as measured by photic immediate-early gene induction
- Retinal-degenerate mice lacking cryptochromes show markedly decreased behavioural photoresponses and pupillary responses, while non-degenerate mice lacking cryptochromes show intact photic signalling
- Mice lacking cryptochromes and ocular retinal show no photic induction of immediate early genes in the SCN

Several models may be hypothesized that fit within these experimental constraints. First, it appears that pigments in both the outer and inner retina are

TABLE 1 Summary of non-visual irradiance detection in wild-type and mutant mouse strains

Measure	Wild-type	rd/rd	mCry1$^{-/-}$; mCry2$^{-/-}$	rd/rd; mCry1$^{-/-}$; mCry2$^{-/-}$	RBP$^{-/-}$	RBP$^{-/-}$; mCry1$^{-/-}$; mCry2$^{-/-}$
Circadian entrainment/ masking	+++	+++	+++	+	++	0
Photic gene induction in SCN	+++	+++	++	+	+++	0
Pupillary responsiveness	+++	++	+++	+	?	?

Pluses indicate relative qualitative effects noted in experiments. 0, no response; ?, experiment not performed.

sufficient to provide signalling to the SCN and pupillary response centres, but neither necessary. The most economical model for explaining this finding would be the existence of a through-signalling pathway between the outer retina and the light-sensitive retinal ganglion cells (Fig. 1, top). The photosensitive retinal ganglion cells do appear to have broad dendritic arbors that may allow such through-signalling (Berson et al 2002, Provencio 2002). Although the cells themselves are thus wired in series, effectively this arrangement works in parallel, where photoreception by either outer retinal pigments or inner retinal pigments results in photic information being transmitted to the brain.

While the rhodopsin and cone opsins likely provide the photopigments for the outer retinal branch of this pathway, the nature of the inner retinal photopigment is not known. The finding that photic signalling is intact in vitamin A-depleted eyes, but lost in vitamin A-depleted eyes lacking cryptochromes suggests that cryptochromes are both sufficient and necessary for photic signalling to the SCN in the absence of vitamin A-based pigments. Are mammalian cryptochromes themselves photopigments? This is the most parsimonious explanation of the data. Several lines of evidence suggest that mammalian cryptochromes are photopigments. First, mammalian cryptochromes contain both flavin adenine dinucleotide and pterin, which serve as chromophores in related photolyase proteins (Hsu et al 1996). Second, mammalian cryptochromes have highly conserved sequence homology to *Drosophila* and *Arabidopsis* cryptochromes. Cryptochromes in both these genera have been shown to be subject to light-dependent protein–protein interactions or post-translational modifications (Ceriani et al 1999, Rosato et al 2001, Shalitin et al 2002). While mammalian cryptochromes did not behave identically to *Drosophila* cryptochromes in terms

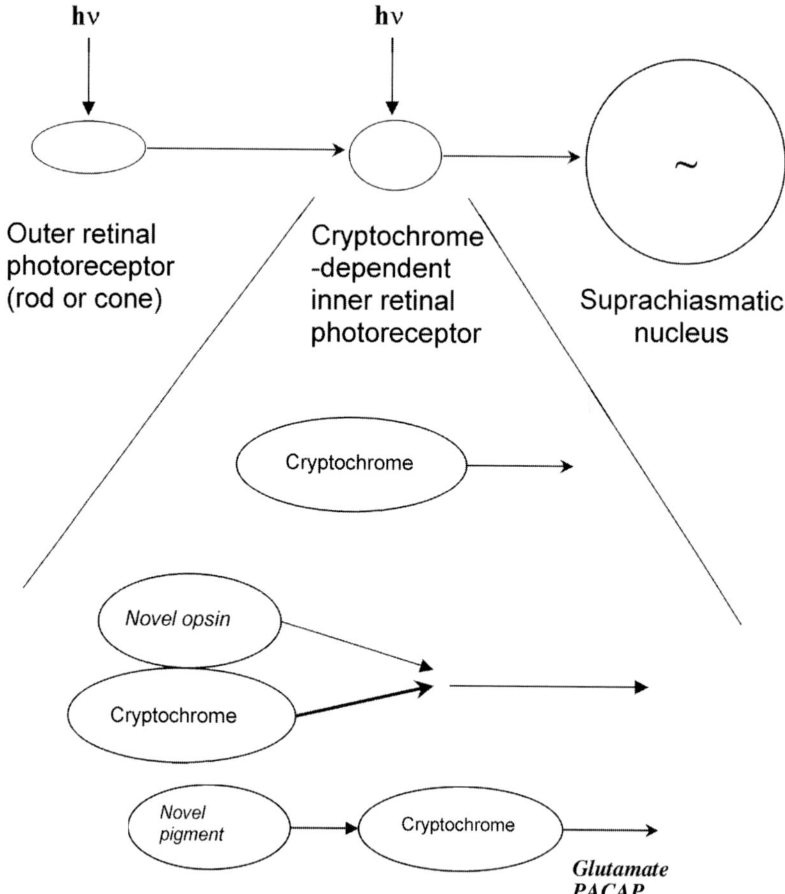

FIG. 1. Alternative models for non-visual irradiance detection. Top: through-signalling model for signalling to the suprachiasmatic nucleus (SCN). Outer retinal photopigments can transmit signals through the inner retinal photoreceptor cells projecting to the suprachiasmatic nucleus; in their absence, the inner retinal photoreceptors are sufficient for retinohypothalamic signalling. Bottom: alternative models for inner retinal photoreception. Cryptochromes may be acting as photopigments alone (top), or in parallel with a vitamin-A depletion resistant opsin (that would be active only under bright light conditions). Alternatively, cryptochromes may be a necessary component for signalling from another, as-yet-unidentified photopigment.

of light-dependency in two-hybrid interactions (Griffin et al 1999), a negative result in a heterologous system does not rule out the possibility that these proteins have photopigment properties. Our recent preliminary results with mammalian cryptochrome two-hybrid interactions suggests that murine cryptochromes are capable of light-dependent protein–protein interactions.

Third, cryptochromes are highly expressed in retinal ganglion cells (Miyamoto & Sancar 1998). (It is important to note, though, that cryptochromes likely require inner-retinal specific partners for their photoreceptive function, since cryptochrome-containing SCN tissue slices are not directly light-entrainable; Herzog & Huckfeldt 2003.) However, these lines of experimentation offer only indirect evidence for a photopigment function for cryptochrome. Definitive demonstration of cryptochromes' photopigment properties awaits biochemical purification and *in vitro* reconstitution of mammalian cryptochromes.

It is theoretically possible that cryptochromes do not function as photopigments, but are required either for the production of another photopigment, or for the signal transduction pathway of another photopigment. Such a pigment would need to be fully resistant to severe vitamin A depletion (since photic immediate early gene induction in the SCN is fully preserved in $RBP^{-/-}$ mice but lost in $RBP^{-/-};mCry1^{-/-};mCry2^{-/-}$ mice) and so is unlikely to be opsin-based. No candidates for such a pigment have been proposed to date, and no obvious homologues of other plant flavin-based photopigments (i.e. NPH and NPL; Sakai et al 2001) have been identified in the human or mouse genome projects.

The finding that $rd/rd;mCry1^{-/-};mCry2^{-/-}$ mice retain some photoresponses to very bright light (i.e. pupillary responsiveness and photic induction of c-*fos* in the SCN) demonstrates that photopigments are still present in these mice. Since photic immediate-early SCN gene induction is lost in vitamin A-depleted $RBP^{-/-};mCry1^{-/-};mCry2^{-/-}$ mice, these pigment(s) are likely opsin-based. It is possible that these photopigments are found in the few non-degenerated cones of these mice (Garcia-Fernandez et al 1995), but more likely this photopigment is located in the inner retina. Melanopsin immunoreactivity has a remarkable distribution in the inner retina, occurring exclusively in cells projecting to non-visual irradiance response centres of the brain such as the SCN and olivary pretectal nuclei, and melanopsin-containing cells are directly photoresponsive (Berson et al 2002, Hattar et al 2002). It is certainly possible that an opsin-based inner retinal pigment is responsible for the preserved photosensitivity seen in $rd/rd;mCry1^{-/-};mCry2^{-/-}$ mice. What might be the relationship between vitamin A-dependent and cryptochrome-dependent inner retinal photoreceptive pathways? It is theoretically possible that these pathways function in parallel or in series (Fig. 1, bottom). The finding of fully preserved immediate-early gene induction in vitamin A-depleted $RBP^{-/-}$ mice strongly suggests a parallel pathway; the only other explanation for this result would be if the inner retinal opsin-based pigment were highly resistant to vitamin-A depletion. However, this 'depletion-resistance' hypothesis is at odds with the finding that the residual photoresponses seen in $rd/rd;mCry1^{-/-};mCry2^{-/-}$ mice are lost in $RBP^{-/-};mCry1^{-/-};mCry2^{-/-}$ animals.

Conclusions

The entirely unexpected richness of circadian photoreception in mammals has led to a multifaceted assault on its mechanism, from photobiology through systems biology. The identification of the photopigment(s) of the inner retina identifies one of the end-points of this effort. Most experimentation addressing this question to date has been directed toward establishing constraints on the nature of these pigments. Genetic approaches have been very fruitful in both *Drosophila* and mammals. The results of experiments using genetic and dietary disruption of vitamin A metabolism, retinal degeneration models, and targeted disruption of cryptochrome genes, coupled with physiological and behavioural assays, has clearly demonstrated a critical role for the cryptochrome genes in non-visual phototransduction, separate from its central role in circadian rhythm generation.

While physiologic action spectra can provide data suggesting the presence of novel photopigments (Lucas et al 2001), their utility in the absence of detailed biochemical knowledge of the photopigment in question is limited (i.e. see Ahmad et al 2002). This can lead to inappropriate exclusion of photopigment candidates on the basis of extrapolated assumptions (for instance, comparison of algal flavin-based chemotropic action spectra and murine pupillary responses; Foster 2002). Ultimate demonstration of cryptochromes as photopigments will require detailed biochemistry and photochemistry akin to the analyses applied to photolyase (Jorns et al 1987, Sancar et al 1987a, Sancar et al 1987b) and phototropins and related proteins (Christie et al 1998, Crosson & Moffat 2001, Sakai et al 2001).

References

Ahmad M, Cashmore AR 1993 HY4 gene of *A. thaliana* encodes a protein with characteristics of a blue-light photoreceptor. Nature 366:162–166

Ahmad M, Grancher N, Heil M et al 2002 Action spectrum for cryptochrome-dependent hypocotyl growth inhibition in *Arabidopsis*. Plant Physiol 129:774–785

Berson DM, Dunn FA, Takao M 2002 Phototransduction by retinal ganglion cells that set the circadian clock. Science 295:1070–1073

Ceriani MF, Darlington TK, Staknis D et al 1999 Light-dependent sequestration of TIMELESS by CRYPTOCHROME. Science 285:553–556

Christie JM, Reymond P, Powell GK et al 1998 *Arabidopsis* NPH1: a flavoprotein with the properties of a photoreceptor for phototropism. Science 282:1698–1701

Crosson S, Moffat K 2001 Structure of a flavin-binding plant photoreceptor domain: insights into light-mediated signal transduction. Proc Natl Acad Sci USA 98:2995–3000

Emery P, So WV, Kaneko M, Hall JC, Rosbash M 1998 CRY, a *Drosophila* clock and light-regulated cryptochrome, is a major contributor to circadian rhythm resetting and photosensitivity. Cell 95: 669–679

Emery P, Stanewsky R, Helfrich-Forster C, Emery-Le M, Hall JC, Rosbash M 2000 *Drosophila* CRY is a deep brain circadian photoreceptor. Neuron 26: 493–504

Foster RG 2002 Keeping an eye on the time: the Cogan Lecture. Invest Ophthalmol Vis Sci 43:1286–1298
Freedman MS, Lucas RJ, Soni B et al 1999 Regulation of mammalian circadian behavior by non-rod, non-cone, ocular photoreceptors. Science 284:502–504
Garcia-Fernandez JM, Jimenez AJ, Foster RG 1995 The persistence of cone photoreceptors within the dorsal retina of aged retinally degenerate mice (rd/rd): implications for circadian organization. Neurosci Lett 187: 33–36
Gooley JJ, Lu J, Chou TC, Scammell TE, Saper CB 2001 Melanopsin in cells of origin of the retinohypothalamic tract. Nat Neurosci 4:1165
Griffin EA Jr, Staknis D, Weitz CJ 1999 Light-independent role of CRY1 and CRY2 in the mammalian circadian clock. Science 286:768–771
Hattar S, Liao HW, Takao M, Berson DM, Yau KW 2002 Melanopsin-containing retinal ganglion cells: architecture, projections, and intrinsic photosensitivity. Science 295: 1065–1070
Herzog ED, Huckfeldt RM 2003 Circadian entrainment to temperature, but not light, in the isolated suprachiasmatic nucleus. J Neurophysiol, in press
Hsu DS, Zhao X, Zhao S et al 1996 Putative human blue-light photoreceptors hCRY1 and hCRY2 are flavoproteins. Biochemistry 35:13871–13877
Jorns MS, Baldwin ET, Sancar GB, Sancar A 1987 Action mechanism of *Escherichia coli* DNA photolyase. II. Role of the chromophores in catalysis. J Biol Chem 262:486–491
Keeler CE 1927 Iris movements in blind mice. Am J Physiol 81:107–112
Lucas RJ, Freedman MS, Munoz M, Garcia-Fernandez JM, Foster RG 1999 Regulation of the mammalian pineal by non-rod, non-cone, ocular photoreceptors. Science 284: 505–507
Lucas RJ, Douglas RH, Foster RG 2001 Characterization of an ocular photopigment capable of driving pupillary constriction in mice. Nat Neurosci 4:621–626
Miyamoto Y, Sancar A 1998 Vitamin B2-based blue-light photoreceptors in the retinohypothalamic tract as the photoactive pigments for setting the circadian clock in mammals. Proc Natl Acad Sci USA 95:6097–6102
Mrosovsky N 1999 Masking: history, definitions, and measurement. Chronobiol Int 16:415–429
Mrosovsky N 2001 Further characterization of the phenotype of mCry1/mCry2-deficient mice. Chronobiol Int 18: 613–625
Mrosovsky N, Salmon PA, Foster RG, McCall MA 2000 Responses to light after retinal degeneration. Vision Res 40:575–578
Nelson DE, Takahashi JS 1991 Sensitivity and integration in a visual pathway for circadian entrainment in the hamster (*Mesocricetus auratus*). J Physiol 439:115–145
Okamura H, Miyake S, Sumi Y et al 1999 Photic induction of mPer1 and mPer2 in cry-deficient mice lacking a biological clock. Science 286:2531–2534
Paietta J, Sargent ML 1981 Photoreception in *Neurospora crassa*: correlation of reduced light sensitivity with flavin deficiency. Proc Natl Acad Sci USA 78:5573–5577
Pickard GE, Smeraski CA, Tomlinson CC et al 2002 Intravitreal injection of the attenuated pseudorabies virus PRV Bartha results in infection of the hamster suprachiasmatic nucleus only by retrograde transsynaptic transport via autonomic circuits. J Neurosci 22:2701–2710
Pittendrigh CS 1993 Temporal organization: reflections of a Darwinian clock-watcher. Annu Rev Physiol 55:16–54
Provencio I, Foster RG 1995 Circadian rhythms in mice can be regulated by photoreceptors with cone-like characteristics. Brain Res 694:183–190
Provencio I, Rodriguez IR, Jiang G, Hayes WP, Moreira EF, Rollag MD 2000 A novel human opsin in the inner retina. J Neurosci 20:600–605
Provencio I, Rollag MD, Castrucci AM 2002 Photoreceptive net in the mammalian retina. This mesh of cells may explain how some blind mice can still tell day from night. Nature 415:493

Quadro L, Blaner WS, Salchow DJ et al 1999 Impaired retinal function and vitamin A availability in mice lacking retinol-binding protein. EMBO J 18:4633–4644

Rosato E, Codd V, Mazotta G et al 2001 Light-dependent interaction between *Drosophila* CRY and the clock protein PER mediated by the carboxy terminus of CRY. Curr Biol 11:909–917

Sakai T, Kagawa T, Kasahara M et al 2001 *Arabidopsis* nph1 and npl1: blue light receptors that mediate both phototropism and chloroplast relocation. Proc Natl Acad Sci USA 98:6969–6974

Sancar A 2000 Cryptochrome: the second photoactive pigment in the eye and its role in circadian photoreception. Annu Rev Biochem 69:31–67

Sancar GB, Jorns MS, Payne G, Fluke DJ, Rupert CS, Sancar A 1987a Action mechanism of Escherichia coli DNA photolyase. III. Photolysis of the enzyme-substrate complex and the absolute action spectrum. J Biol Chem 262:492–498

Sancar GB, Smith FW, Reid R, Payne G, Levy M, Sancar A 1987b Action mechanism of Escherichia coli DNA photolyase. I. Formation of the enzyme-substrate complex. J Biol Chem 262:478–485

Sargent ML, Briggs WR, Woodward DO 1966 Circadian nature of a rhythm expressed by an invertaseless strain of *Neurospora crassa*. Plant Physiol 41:1343–1349

Selby CP, Thompson C, Schmitz TM, Van Gelder RN, Sancar A 2000 Functional redundancy of cryptochromes and classical photoreceptors for nonvisual ocular photoreception in mice. Proc Natl Acad Sci USA 97:14697–14702

Shalitin D, Yang H, Mockler TC et al 2002 Regulation of *Arabidopsis* cryptochrome 2 by blue-light-dependent phosphorylation. Nature 417:763–767

Stanewsky R, Kaneko M, Emery P et al 1998 The cryb mutation identifies cryptochrome as a circadian photoreceptor in *Drosophila*. Cell 95:681–692

Takahashi JS, DeCoursey PJ, Bauman L, Menaker M 1984 Spectral sensitivity of a novel photoreceptive system mediating entrainment of mammalian circadian rhythms. Nature 308:186–188

Thompson CL, Blaner WS, Van Gelder RN et al 2001 Preservation of light-signaling to the suprachiasmatic nucleus in vitamin-A deficient mice. Proc Natl Acad Sci USA 98:11708–11713

Thresher RJ, Vitaterna MH, Miyamoto Y et al 1998 Role of mouse cryptochrome blue-light photoreceptor in circadian photoresponses. Science 282:1490–1494

van der Horst GT, Muijtjens M, Kobayashi K et al 1999 Mammalian Cry1 and Cry2 are essential for maintenance of circadian rhythms. Nature 398:627–630

Van Gelder RN, Gibler TM, Tu D et al 2002 Pleiotropic effects of cryptochromes 1 and 2 on free-running and light-entrained murine circadian rhythms. J Neurogenet 16:181–203

Vitaterna MH, Selby CP, Todo T et al 1999 Differential regulation of mammalian period genes and circadian rhythmicity by cryptochromes 1 and 2. Proc Natl Acad Sci USA 96:12114–12119

Wee R, Castrucci AM, Provencio I, Gan L, Van Gelder RN 2002 Loss of photic entrainment and altered free-running circadian rhythms in math5$^{-/-}$ mice. J Neurosci 22:10427–10433

Zimmerman WF, Goldsmith TH 1971 Photosensitivity of the circadian rhythm and of visual receptors in carotenoid-depleted *Drosophila*. Science 171:1167–1169

DISCUSSION

Sehgal: Has anyone tried to do any physiology on the ganglion cells from $mCry1^{-/-};mCry2^{-/-}$ mice?

Van Gelder: We are doing this now in collaboration with Dr David Berson, but don't yet have results.

Lee: In your triple $rd/rd;mCry1^{-/-};mCry2^{-/-}$ mutants which don't show any entrained patterns, what happens when you increase the light intensity?

Van Gelder: We have increased it up to about 500 lux, where we find a few animals show entrainment where they didn't under dim light, but the majority of animals remain unresponsive. This is not the same level of light intensity that they are subjected to in the photic gene induction studies. We haven't taken the intensity up to this level yet in the behavioural studies.

Schibler: A suggestion. What you would like to do is knock out cryptochromes in the ganglion cells of the retina. This is difficult. However, what you may be able to do is to rescue the knockout mice with a transgene expressed in the SCN, for example.

Van Gelder: The problem is that there is, to my knowledge, no pan-SCN marker that will label all SCN cells. Any marker you pick is going to be expressed only in a subset of cells such as VIP-expressing or vasopressin-expressing cells, or a similar subset. We are approaching this problem the opposite way, which is to try to rescue the ocular phenotype with recombinant adenoviral associated virus (AAV) expressing cryptochrome. Preliminary data from a number of groups show that one can express green fluorescent protein (GFP) stably in the inner retina after an intravitreal injection. We are now making the cryptochrome AAV2 virus to try to do it this way.

Kay: Have you also thought about doing RNA knockdown?

Van Gelder: This would also potentially be a good approach, if we could get sufficient expression in the inner retina.

Kay: In your two-hybrid interactions, do you know which CRY1 domain is participating?

Van Gelder: No, we used the full length CRY1 in the fusion protein.

Kay: Is there any evidence of protein-level light-dependent lability in mammalian cryptochrome?

Van Gelder: Not that I am aware of, but I don't know how thoroughly this has been investigated.

Rosbash: The comparable *Drosophila* experiments argue more closely for what you are observing. That is, that the interactions would take place in the dark and are destroyed in the light.

Van Gelder: That is the photolyase model. In photolyase, the damaged DNA is bound in the dark, and then dissociates following the light-mediated enzymatic repair of the thymidine dimer. Photolyases are believed to be the molecular ancestors of cryptochromes.

Sassone-Corsi: Is the PIAS3 interaction occurring in the right place?

Van Gelder: PIAS3 is ubiquitously expressed and is famous for being pulled out of multiple two-hybrid screens. We have not yet immunostained, but I would be fairly confident that PIAS3 is going to be expressed all over the retina.

Young: What about the other CRYs in your *in vivo* precipitation?

Van Gelder: We haven't looked for CRY2. On Western blot our CRY1 antibody works reasonably well and our CRY2 antibody doesn't.

Rosbash: What is the source of protein for the co-immunoprecipitations?

Van Gelder: Mouse brain extract.

Foster: So is the prediction that the brain could be photosensitive?

Van Gelder: I don't think so, because the cryptochromes are expressed in a superset of the known photoresponsive cells. Cryptochromes are probably expressed in many of the cells that David Berson showed were not photosensitive in the inner retina. This argues that there must be a cell-type-specific partner that interacts with cryptochrome to confer light sensitivity to the intrinsically photosensitive retinal ganglion cells. We already know of two markers that seem specific in these cells: PACAP (pituitary adenylate cyclase activating peptide) and melanopsin. My colleague Erik Herzog has looked carefully for evidence that SCN slices can be entrained by external light/dark cycles, but has not found entrainment, even including near UV light.

Menaker: We have done this with a lot of peripheral organs too in the rat model, and we don't get any entrainment.

Rosbash: In a wild-type animal, under normal conditions, what is providing 80–90% of the oomph?

Van Gelder: On the basis of our data I can't tell you whether rods, cones, or inner retinal photoreceptors are dominant. Russell Foster's data showing entrainment angle abnormalities in the rodless-coneless animals suggest that the outer retinal cells do play a role in circadian entrainment. Teasing this apart will be tricky. Aggelopoulos and Meissl published a nice paper in the physiology literature a few years ago, showing the action spectrum of SCN responses to light *in vivo* in live rats. They found that scotopic light resulted in cell firing in the SCN. Even at intensities as low as 1×10^9 photons, which is two logs below our crypto-chrome sensitivity, the SCN fires — but it doesn't phase shift. The SCN is seeing light at fluences that don't affect it in terms of phase shifting. We don't know how this comes to pass. Perhaps PACAP isn't released appropriately, and PACAP is needed in addition to glutamate in order to affect phase shifting.

Menaker: You need to be a little careful there, because the SCN does more than phase shift. For instance, it may be that it is using that very dim light to suppress melatonin in the pineal. There is a difference in sensitivity.

Van Gelder: Again, how these different pigments play into the different output pathways is not clear. For example, in the pupillary light responsiveness we see a marked diminution in pupillary response comparing wild-type to *rd/rd* mice. We see no diminution in phase-shifting response for entrainment in these animals. This means that, for pupillary responses, 90% of the response is driven by the outer retina, whereas for circadian responses it may be that none or only a small fraction is normally mediated by the outer retina.

Foster: Simplistically, in entrainment the rods and cones provide transitional information at the sharp junction between a LD cycle, and the inner retinal opsins provide irradiance information. What was striking about that rodless-coneless mouse under dim LD cycles is that it was phase advanced and the onsets were bouncing all over the place. Something about the precision of entrainment is lost with the rods and cones under relatively low light conditions.

Rosbash: Are these all discontinuous, incubator LD transitions?

Foster: Yes. You have raised a very interesting question. We are finally doing the experiment we have been wanting to do for a long time, which is putting the rodless-coneless mouse under the natural photoperiod in a greenhouse. In the pupillary studies we find massive attenuation of sensitivity in the rodless-coneless mouse. If you look at the dynamics of the response, it is completely hidden underneath the rod/cone response. When we strip that away, then we start to see it. We have thought about this as a two-stage process in pupillary restriction. You go from a dim environment to a bright environment and the rods and cones provide the immediate transition information. But what they are very bad at doing is giving a sustained response to a sustained stimulus. If you want to maintain pupillary constriction under relatively bright light conditions, you need an irradiance detector system.

Menaker: There is old work by Kavanau in the 1960s (Kavanau 1962a,b) which has recently been repeated in hamster (Boulos et al 2002). This clearly shows that natural transitions in light/dark cycles increase the range of entrainment dramatically. The clock will entrain to T cycles much longer or shorter with natural transitions than it will with abrupt transitions. This has to mean something. It isn't clear what it means, but the complexity of the retinal input to the SCN must be functional at some level. Perhaps the function is in getting information about the rate of transition.

Foster: Plotting twilight is an incredibly difficult sensory task: the more receptor inputs you have the easier it is. What is amazing is that under natural conditions, some animals are able to anticipate twilight beautifully crisply.

Loros: In your rodless-coneless masking, is there any anticipation of lights on or lights off?

Foster: These animals are under an LD cycle, and then we turn the lights on for an hour into the D portion of the LD cycle and measure the suppression of activity. For both the rodless-coneless and the wild-type this suppresses activity more or less the same.

Young: Is PIAS3 ordinarily a suppressor of STAT?

Van Gelder: Yes, PIAS3 suppresses STAT3 *in vivo* and *in vitro*. It basically functions as a sumoyl ligase. It interacts with a large number of proteins that require sumoylation for their function.

Young: Could it be that the CRY interaction with PIAS3 has to do with regulation of the Jak/Stat pathway? Have you looked at light-dependent regulation of the pathway?

Van Gelder: We have not. We haven't looked at any of the downstream kinase pathways in this model. It is hard to isolate these cells because there are only about 500 per mouse retina.

Young: There is a PIAS in flies. It might be interesting to see if this interacts with fly CRY.

Ishida: Have you looked at the circadian oscillation of PIAS3 or WD repeat protein?

Van Gelder: Not yet.

Hardin: Thinking of the potential difference in function in the ganglion cells of CRY versus elsewhere, have you looked at the subcellular localization of the CRY in those cells?

Van Gelder: We haven't co-stained yet with melanopsin. It appears that in the vast majority of ganglion cells cryptochrome is predominantly nuclear, but we haven't done a double label specifically in those cells. That is a really good idea.

Hardin: Are these other interacting proteins known to be nuclear?

Van Gelder: There is no information on WDR9. It was an orphan that came out of the genome project, and is one of the WD40 repeat-containing proteins. PIAS can be nuclear or cytoplasmic, and some models suggest it sits on the nuclear membrane and helps things come in and out of the nucleus.

Hastings: Can you say anything about the time course of the PIAS interactions? One of the interesting things about the melanopsin-containing cells was the temporal dynamics of the responses to light. Do you see this mechanism being an acutely responsive mechanism, or is it just a constitutive mechanism where its presence or absence is required?

Van Gelder: There are two temporal constraints on any inner retinal photoreceptor. One is that it is a relatively slow integrator. It takes 30 seconds before cell firing is seen after the dim lights go on. Second, it does not appear to attenuate. Jocha Meijer has data showing that even after one or two hours of light exposure the SCN firing rate remains unchanged. There is no down regulation or attenuation. This is atypical of an opsin-based pigment. To address this we have had to go to fairly extreme lengths in the yeast two-hybrid, because light breaks apart the interaction. To data we have done nuclear run-on assays for LacZ and Ade3 (which are two of the two-hybrid markers) in liquid cultures, turning the lights on and seeing how rapidly the transcription rate falls off. In our preliminary experiments, within five minutes we see about three-quarters of the transcripts have gone and at 45 min there is no transcript left. Then if we put the yeast back in the dark, 45 min later we return to baseline levels of transcript being formed. We think the kinetics are within a five minute window and it will be difficult to narrow this down with this assay.

Rosbash: If one makes the assumption that in the rodless-coneless mouse there is no change in the remaining photoreceptors — you are just subtracting the contribution of rods and cones — when you do an action spectrum of the phase shifting in the rodless-coneless mice, are you saying you don't get something that looks like it could be 80–90% cryptochrome?

Foster: We see no evidence for the contribution of another pigment. The irradiance response curves are so amazingly univariate. What I would have predicted is that if there was a contribution from a flavoprotein-like response you would see it at those shorter wavelengths.

Van Gelder: Joe Takahashi, you did a beautiful action spectrum on hamsters for photic phase shifting and found univariate responses that were easily fittable to an action spectrum in a wild-type animal. But you did not see the contribution of the inner retinal photopigment in their action spectra initially.

Takahashi: This was work in Mike Menaker's lab in 1984. We did full fluence response curves at three wavelengths: 400, 500 and 600 nm. The slopes of those three curves were the same. The principle of univariance held for that data set. I think there is a species difference. Many years later we did another species of hamster and we found the same thing as Russell Foster. We never published this.

Foster: This is fascinating, because there is a weird photoreceptor complement in the golden hamster.

Takahashi: Hamsters are interesting because they do lack one of the cone pigments.

Van Gelder: Yet they are highly UV sensitive. They have a UV pigment, but it is just not an opsin.

Menaker: We need to worry about species differences and assay differences. This is obviously a much more complicated situation than anyone suspected in 1984. The thing about the hamster data that made us suspicious is that if you look at the action spectrum you could say it is rods, no problem.

Takahashi: Not really. It is too narrow. There are some really old action spectra from the Karolinska Institute that were done in rats. Under photopic conditions the action spectrum is much narrower than any opsin nomogram. This is common in rodents: it is because of the contribution of multiple pigments at high light intensities. This is what the hamster action spectrum looks like: it peaks near 500 nm, but it is narrower than an opsin nomogram.

Rosbash: Russell Van Gelder, it seems to me that you have pupillary data that are quite different from Russell Foster's.

Van Gelder: They are somewhat different. I should have plotted them against each other. We find a slightly different spectrum but we are dealing with a slightly different mouse: ours is an *rd/rd* mouse as opposed to the *rd/rd cl* mouse.

One possible difference may be strain-related. Strain-specific differences are very common in mouse work and could partially underlie the difference. Our methods were fairly similar. The peak was 470 nm, and our surrounding points were 450 nm and 490 nm. We have less than 50% sensitivity at the two surrounding points. We are confident that the peak is within that interval.

Foster: Finding where that λ_{max} is is impossible on the basis of the data you showed. You are saying it is around 470 nm. That is not far from 479 or 480 nm.

Van Gelder: I think it is the same peak.

Foster: Why, then are you using that evidence to suggest that it is a cryptochrome response?

Van Gelder: Because you are using that evidence to suggest it is an opsin response! The question is, is the fit on a Dartnall template adequate to implicate the photopigment? We have shown genetic evidence that 90% of the residual pupillary sensitivity in *rd/rd* mice is dependent on cryptochrome, and that peak sensitivity is not 505, 530 or 390 nm. It is cryptochrome dependent. I have reservations about fitting action spectrum data with Dartnall curves because you can't put reliable error bars on your action spectrum. The fit depends on your curve fitting of the fluence responses and how accurate this is. I have spoken with my statistician about this. It is complex to perform non-linear correlation coefficients and put an adequate interval in so you can derive a meaningful error bar. I know none of your actual action spectra have error bars. This depends on the statistics of the fit you use to derive the ID_{50}.

Foster: So what you are doing is throwing away 60–70 years of what the visual community has been basing its work on.

Menaker: Good idea!

Van Gelder: Not at all. We do not know the absorption spectrum of purified, post-translationally modified cryptochrome, and we certainly don't know it in context *in vivo*.

Foster: How, then, can you say that your action spectrum describes a cryptochrome?

Van Gelder: No, I am saying that our action spectrum defines the *genetic contribution* of cryptochrome to this function.

Takahashi: Each of you have made valid points, but you are still talking at cross purposes. It is important to plot your spectral sensitivity function on a log scale, because to fit a nomogram it is not the peak that is used: the peak is highly unreliable. When you plot on a linear scale it emphasises the peak, which is a dangerous thing to do. It is the long wavelength fall of on a log fit that really gives you the γ_{max}. This is where the 60 years of visual pigment physiology comes in. I agree that the logistic function we use to fit the 50% point (which we define as our threshold for sensitivity measurements) is also statistically difficult to capture. It is the best we can do, though.

Foster: With the rodless-coneless data, because those irradiance response curves were so parallel, we actually derived an action spectrum from both the 50% and through the range.

Takahashi: That's what you should do: select different thresholds. The spectra should be superimposable. Your question reminded me of an experiment Dwight Nelson and I did, which was to compare the threshold for the pineal inhibition response with phase shifting, which we did in the same animals. There is a 30-fold difference in sensitivity. What I forgot is that the slope is different. The pineal response has a much sharper slope, and the phase shifting is a little less than 1.

Van Gelder: Marty Zatz has shown that the chick pineal system has at least two components, one that is vitamin A-depletable and one that is not. This suggests multiple photopigments even in the pineal photo-sensation.

Foster: Joe Takahashi, we echoed your work with the mouse. What is fascinating about the dynamic range of melatonin suppression is that it is so narrow compared with the three or four log unit range for phase shifting.

Rosbash: Why isn't the fall off a reliable indicator of the complexity of the photoreceptors?

Van Gelder: It depends what you are fitting it to. For an opsin it gives you a fit, but your data points have error in them that is not captured in the action spectrum. I know of no statistical means to rigorously do the Dartnall fit taking into account the curve fit used on the fluence curves.

Foster: The real problem with wild-type data is to show that univariance doesn't hold. The data are noisy and there are subtle changes in slope.

Menaker: Since the pineal melatonin suppression has been brought up, let me tell you about some unpublished data that we have on albino hamsters. These are rather odd. If you keep these animals in constant bright light, they lose the ability to suppress pineal melatonin. Then they behave reproductively as if they were in constant darkness. In other words, the pineal escapes from light suppression. We thought originally that this might be due to some developmental effect on the retina, but it turns out that it is not. It must have something to do with bleaching of some particular pigment. I was going therefore to ask a rather naïve question: what do we know about reconstitution of cryptochrome photoreception?

Van Gelder: Essentially nothing. To my knowledge, no one has reconstituted mammalian cryptochrome *in vitro* in a system that can do anything other than an absorption spectrum, which as we have discussed is somewhat unreliable with a protein that has multiple partners and likely is post-translationally modified. The downstream signalling mechanism is not known. There are some interesting recent plant data showing light-dependent phosphorylation of cryptochrome, but these have not been replicated in the mammal yet. The bleaching of the pigment is completely unknown at present.

Menaker: Is anything known about the bleaching time-course of cryptochrome?

Kay: There are just the old redox-dependent spectra, I don't know of any studies on bleaching.

Cahill: We did an action spectrum for suppression of melatonin in *Xenopus* retina. We tried to deal with this variability issue. We got some estimates from our all-fit program of 95% confidence intervals which we did plot on the action spectrum. When we did this we were fairly disappointed with what we were able to exclude as possible photoreceptors. We were able to eliminate the 650 nm cones, but anything from 500 nm down we couldn't really throw out. When we plotted the actual points they fell very nicely on the green opsin. The peak could have been over a range of 40 nm.

Foster: It was because of this variability that we first started using the pupillary response. Having done phase shifts for most of my life it is quite extraordinary to see an error bar. The error bar falls within the data points. This is why we were so pleased with the action spectrum, because it shows so little variation we can be confident of the fit.

Van Gelder: I disagree. The measured pupillary response depends on the exact second that you measure the pupil size, because there is second-to-second variation in the pupil size that does produce somewhat of an error bar.

Foster: We haven't seen that.

Dunlap: One of the experiments done in a microbial system to prove the involvement of flavins is to feed the experimental organism flavin analogues and thereby move the peak of the action spectrum in a predictable way. Is this feasible in mice?

Van Gelder: We could inject them in the eye where toxicity might not be such a problem. That is a good idea.

Foster: I want to go back to the triple knockout. You showed that the *rd/rd;mCry1$^{-/-}$;mCry2$^{-/-}$* doesn't show a pupillary response.

Van Gelder: No, it shows a response to very bright light: 10^{14} photons per cm^2 per cell. That is quite bright: it's a halogen dissecting light at maximal brightness run through a 470 nm filter.

Foster: So there is some residual photosensitivity in the triple knockout. Do you ascribe this to residual cones or to the novel opsin?

Van Gelder: I think it is the novel opsin on the basis that the action spectrum in those animals does not peak anywhere near a green or blue cone in a mouse.

Schibler: I have a question on the yeast two-hybrid system. Doesn't this argue that cryptochrome falls apart and it must be involved in directing light signalling? As far as I know there is no light-sensitive molecule in yeast.

Van Gelder: Yes, it would be an odd argument to suggest that yeast have been waiting for 2 billion years for a cryptochrome to be introduced to trigger a whole latent phototransduction cascade! I think it does suggest strongly that cryptochrome itself has photopigment properties.

Rosbash: When we saw cryptochrome disappear within a few minutes of light exposure, we spent some time trying to make mutants to change the action spectrum. We failed.

Van Gelder: Phase shifting is a difficult assay because of the inherent variability and measurement difficulties. The pupillary response should provide a more robust platform for doing this. Our plan is to reintroduce different cryptochromes into the eye by this viral rescue technique and then see by site-directed mutagenesis whether we can actually show a shift in the action spectrum.

Kay: It is hard to get changes in spectra. In Cashmore's lab they have something like 40 alleles of *Cry1*. They had cases where they had lost the light-harvesting chromophore and they saw the spectrum narrow down. But if I recall correctly they couldn't get subtle shifts.

Van Gelder: At this point even these sorts of data would be useful in excluding models.

Kay: It was useful for them to assign it, but one would want something a little more satisfying.

References

Boulos Z, Macchi MM, Terman M 2002 Twilights widen the range of photic entrainment in hamsters. J Biol Rhythms 17:353–363

Kavanau JL 1962a Twilight transitions and biological rhythmicity. Nature 194:346–348

Kavanau JL 1962b Activity pattern on regimes employing artificial twilight transitions. Experientia 18:382–384

General discussion I

Menaker: What is the ideal set of experiments to resolve the problems we have been discussing so far? It has got to the point where the data are not really in conflict, but we aren't in agreement.

Rosbash: Let's define the problems first. It is not clear to me what these are.

Van Gelder: For me, the key question is what are the key inner retinal photopigments?

Rosbash: Russell Foster, do you take issue with the fact that it is cryptochrome plus one or more opsins?

Foster: I can't exclude a cryptochrome, but on the basis of our data we find no evidence for it. The experiments can be done. When Russell Van Gelder re-plots the data that he has on a log scale, and perhaps gets some more points, then we will have the resolution to say one way or another.

Kay: You could put a melanopsin knockout on top of the rodless-coneless and double *Cry* mutants.

Foster: The difficulty will be sorting out an element of photosensory pigment from an element that is required for the response. Indeed, the way that people have separated out RGR (this photoisomerase) from a classical photosensory pigment is to see whether it will activate a phototransduction cascade.

Rosbash: Let's consider the following scenario. You send a grant proposal into a NIH panel on cryptochrome, and you want US$200 000 per year for five years to show whether cryptochrome really is a light-sensitive molecule as opposed to a player in the phototransduction cascade. My guess is that you would have trouble getting the money and they would argue that this is a done deal. Your scepticism isn't unique, but you'd have trouble finding a lot of company.

Foster: I suppose this is because I come from a photopigment-type background. When you talk about these sorts of issues with photobiologists they have been dealing with these sort of issues for much longer, and they use much stricter criteria for defining a photopigment. You need to match action spectra to absorption spectra, and look at spectral elements.

Van Gelder: In the absence of accurate absorption spectrum data the argument is moot. We have no idea what a cryptochrome actually does *in vivo,* so matching spectra does not apply to a pigment where you don't know what the *in situ* spectrum is. Many *in vivo* modifications could alter the absorption spectrum, including the redox state of the cell, bound protein co-factors to cryptochrome, and masking functions of other retinal pigments.

Foster: That is right. But you are arguing that we can't get an action spectrum because we don't know what the absorption spectrum is, and then you are presenting action spectra and saying that it is probably cryptochrome.

Van Gelder: The reason for showing the action spectra is to demonstrate the genetic influence of the cryptochrome mutation—that is, what is the spectrum of the photopigment that is cryptochrome-dependent? I can't say at this point that cryptochrome is *the* photopigment until we have more biochemical data. But we have well established now that cryptochrome is functioning on this pathway by analysis of the photic sensitivity of the mutants that we have studied. We are left with the situation where 90% of pupillary responsiveness in a Rd mouse is attributable to cryptochrome function because it disappears when we take the two cryptochromes out.

Young: Perhaps the only remaining door left is what opsin does cryptochrome control, if it does that?

Van Gelder: This brings us back to the vitamin A question, where we have shown in the $Rbp^{-/-}Cry1Cry2$ knockouts that they have essentially no suprachiasmatic nucleus (SCN) photoresponses that we can detect. We are thus able to fully deplete a *Cry* mutant to eliminate all photoresponses. We therefore presume that there is no hidden photopigment in those animals. We thus have to just explain the difference between the *Rd-Cry* mutant and *Rbp-Cry* mutant, and the only difference is that there is 10% photosensitivity left, which I presume is due to a novel opsin.

Foster: But you can't completely vitamin A-deplete the mice. In the Dowling studies, after years of vitamin deprivation in mice he basically said that if the mouse isn't dead it isn't fully vitamin A-deprived.

Van Gelder: We deprive them enough. The old studies used global vitamin A depletion, which depleted retinals, retinols and retinoic acid. Survival may be a retinoic acid versus a retinal issue. The point is, how do you explain the complete loss of SCN photoresponse in *Rbp* (vitamin A-depleted) *Cry1Cry2* triple mutants?

Foster: You just see an attenuation of a response.

Van Gelder: No, it has gone. We see no SCN gene induction by light.

Foster: I can't explain it.

Menaker: It comes down to a question of how much of the photic response is dependent on which of the photoreceptors. The only interesting answer to that is what happens in the real world.

Van Gelder: In the broader world there are clinical implications of these questions. For example, we have done a clinical study on blind children. We put wrist actographs on children and assorted them by the type of eye disease. We broke them into children with glaucoma (where the pressure destroys the retinal ganglion cells but leaves the outer retina intact) and children who have hereditary retinal degeneration (such as retinitis pigmentosa). We studied about 25 blind and 12 normal children for two weeks in a set environment. Our outcome measure was daytime napping and stability of wake onset. We found that the relative risk of excessive daytime sleepiness for the children who had retinal ganglion cell disease was 20-fold over control children and 10-fold over children with outer retinal degeneration. These children had an average of over 30 minutes a day of napping. I don't think the only interest in this is what happens in a normal wild-type animal.

Menaker: That's the real world. The comment that I made was that the interesting thing is what happens in the real world. The other comment is that we know virtually nothing about how these things are worked out in diurnal animals. There is a crying need for someone to get away from all these nocturnal rodents and figure out what is going on in an animal whose retina has evolved to deal with bright light.

Weitz: What about general experiments to dissociate the function of the clock in the eye from potential photoreception? We know that there is a circadian clock in the eye, and we know from fish and birds that this clock controls visual sensitivity and retinal physiology. In birds, for example, circadian clock control accounts for an order of magnitude difference in visual sensitivity. When one examines photoreceptive behaviour in a *Cry1/Cry2* double mutant, in principle one is confounding the removal of two potential visual pigments with something that, by disrupting the clock in the eye, might set the eye at a particular state of visual sensitivity. Has anyone looked in *Per1/Per2* double mutants or *Bmal* mutants to see whether disrupting the clock in the eye itself has consequences for photic sensitivity? This is a potential confounding factor.

Van Gelder: We are currently making the *Rd/Bmal* mice to answer that question. Rae Silver has some results relating to this. She has looked for clock gene expression in the melanopsin-positive ganglion cells. Remarkably, she does not find *Per* expression in those cells.

Weitz: That is not relevant to my question. The problem is that there is a clock in the mammalian eye. It doesn't matter where the clock is or what cells it is in, there is a clock in the retina that controls visual sensitivity by means yet to be clarified. We know that the clock is not in the melanopsin-expressing ganglion cells, but that doesn't matter for the issue that I am raising.

Menaker: It is also true that the clock in the eye affects more than the visual sensitivity. It even affects the circadian wheel-running behaviour. That is an important point to consider.

Schibler: Paolo Sassone-Corsi, you have these cell lines that are light responsive. Does the RNAi work in zebra fish?

Sassone-Corsi: We are just checking those cells. In zebra fish we are using RNA to try to block the various CRYs that these cells contain.

Schibler: Have you overexpressed them to see whether the cells become more sensitive?

Sassone-Corsi: That's another possibility. These cells can be really nicely infected by adenovirus-based vectors. We could also try introducing additional signalling molecules that these cells don't normally contain.

Light signalling in *Cryptochrome*-deficient mice

Xavier Bonnefont, Henk Albus*, Johanna H. Meijer* and
Gijsbertus T. J. van der Horst[1]

*MGC, Department of Cell Biology and Genetics, Erasmus MC, PO Box 1738, 3000 DR
Rotterdam and *Department of Neurophysiology, LUMC, Wassenaarseweg 62,
Rijksuniversiteit Leiden, Leiden, The Netherlands*

Abstract. The mammalian master clock driving circadian rhythmicity in physiology,
metabolism, and behaviour resides within the suprachiasmatic nuclei (SCN) of the
anterior hypothalamus and is composed of intertwined negative and positive
autoregulatory transcription-translation feedback loops. The *Cryptochrome 1* and *2* gene
products act in the negative feedback loop and are indispensable for molecular core
oscillator function, as evident from the arrhythmic wheel running behaviour and
absence of cyclic clock gene expression in *mCry1/mCry2* double mutant mice in constant
darkness. Recently, we have measured real-time multi-unit electrode activity recordings
in hypothalamic slices from *mCry*-deficient mice kept in constant darkness and observed a
complete lack of circadian oscillations in firing patterns. This proves that CRY proteins,
and thus an intact circadian clock, are prerequisite for circadian rhythmicity in membrane
excitability in SCN neurons. Strikingly, when *mCry*-deficient mice are housed in normal
light–dark cycles, a single non-circadian peak in neuronal activity can be detected in SCN
slices prepared two hours after the beginning of the day. This light-induced increase in
electric activity of the SCN suggests that deletion of the *mCry* genes converts the core
oscillator in an hour-glass-like timekeeper and may explain why in normal day–night
cycles *mCry*-deficient mice show apparently normal behaviour.

*2003 Molecular clocks and light signalling. Wiley, Chichester (Novartis Foundation Symposium
253) p 56–72*

Circadian clock-controlled rhythms provide most organisms with an orchestrated
temporal programme that allows for appropriate timing of physiology (i.e. blood
pressure, hormonal levels) and behaviour (i.e. alertness, sleep–wake cycle). The
mammalian central circadian pacemaker resides in the suprachiasmatic nucleus
(SCN) of the brain (Weaver 1998). At the molecular level, the core oscillator
driving the mammalian clock consists of interconnected autoregulatory

---

[1]This paper was presented at the symposium by Gijsbertus T. J. van der Horst to whom all
correspondence should be addressed.

transcription–(post)translation based feedback loops in which a set of clock genes negatively and positively affect each others and their own transcription level (reviewed by Reppert & Weaver 2002).

Molecular clocks do not oscillate with an exact 24 h periodicity but are entrained to solar day–night rhythms by light. Although responsible photoreceptor cells must reside in the eye, rods and cones and their visual pigments appear not involved as *Retinal-degenerate* (*Rd*) mice and rodless-coneless (*rdcl*) mice show a normal response to phase-shifting light stimuli (Foster 1998, Freedman et al 1999). Recent work suggests that the responsible photopigment is melanopsin, present in a subset of retinal ganglion cells with projections into the SCN (Berson et al 2002, Hattar et al 2002), although a redundant role for other photopigments is not excluded.

In vivo, the SCN is thought to send nervous and humoral output signals to other areas of the brain as well as to peripheral organs. Circadian behaviour is driven by SCN-derived electrical rhythms, which show high firing rates during daytime and low activity during the night (Green & Gillette 1982, Meijer et al 1997). Circadian activity of the master clock in the SCN is self-sustaining, as is evident from the persistence of oscillations of electrical activity in SCN slices or dispersed SCN neurons (Welsh et al 1995). In circadian mutant animals with an accelerated, retarded or damped clock (as in *tau* hamsters, heterozygous and homozygous *clock* mutant mice respectively), the electrical discharge rhythm of dissociated SCN neurons closely relates to the free running period of locomotor activity (Liu et al 1997, Herzog et al 1998), indicating that the molecular clockwork, directly or indirectly, governs parameters of the neuronal cell membrane to generate circadian rhythms in membrane potential and discharge rate. However, information on the electrophysiological behaviour of the SCN in the absence of a circadian core oscillator has been lacking. The introduction of gene-targeting techniques has allowed specific inactivation of one or more clock genes in the mouse, which in some cases — i.e. *mCry1/mCry2* double knockout mice (van der Horst et al 1999, Vitaterna et al 1999) and *Bmal1(Mop3)* knockout mice (Bunger et al 2000) — resulted in a loss of core oscillator function. The present paper describes the electrophysiological properties of the SCN of *Cryptochrome*-deficient mice, carrying an inactivated circadian core oscillator (Albus et al 2002).

Mammalian cryptochromes and the circadian clock

Photolyases are ingenious DNA repair enzymes that remove UV-induced DNA damage in a single-step process requiring light energy captured by blue light-harvesting chromophores. Despite strong evolutionary conservation, placental mammals lack photolyase activity (reviewed by Yasui & Eker 1998). Instead, they express two photolyase-like proteins (designated CRY1 and CRY2) that contain a photolyase core domain and a C-terminal extension, and strongly resemble plant *cry*ptochrome blue-light receptor proteins (Todo et al 1996, van der Spek et al 1996, Kobayashi et al 1998). Subsequent analysis of *mCry1* and *mCry2* single and double knockout mice—generated to study the potential circadian photoreceptor function of the mammalian CRY proteins as anticipated on the basis of the role of plant cryptochromes and the high expression levels of *mCry1* and *mCry2* mRNA in the retina (Miyamoto & Sancar 1998)—revealed a complex role for these proteins in the circadian timing system (Thresher et al 1998, van der Horst et al 1999, Vitaterna et al 1999).

In constant darkness *mCry1* and *mCry2* deficient mice display accelerated and retarded biological clocks, respectively (Thresher et al 1998, Okamura et al 1999, van der Horst et al 1999, Vitaterna et al 1999). This suggests that mCRY1 and mCRY2 have an antagonistic clock-adjusting function. Strikingly, a complete lack of both free-running behavioural rhythmicity and cyclic *mPer1* and *mPer2* expression is seen in *mCry1/mCry2* double mutant mice kept in constant darkness (Okamura et al 1999, van der Horst et al 1999, Vitaterna et al 1999). Thus, the mCRY proteins are essential for maintenance of circadian rhythmicity. An important clue to our understanding of the role of CRY proteins in the mammalian core oscillator was obtained by Reppert and co-workers, who provided evidence that CRY proteins strongly inhibit CLOCK/BMAL1-driven transcription of E-box containing genes, and thus are true clock genes that act in the centre of the negative feedback loop (Kume et al 1999, Shearman et al 2000).

Mice lacking mCRY1 or mCRY2 still entrain to light–dark cycles, which may point to redundancy in the expected photoreceptor function of these proteins (Thresher et al 1998, van der Horst et al 1999, Vitaterna et al 1999). The arrhythmic behaviour of *mCry1/mCry2* double knockout mice in constant darkness precludes analysis of circadian photoreception by classical phase shift experiments. However, since phase advancing or phase delaying light stimuli are known to induce c-*fos*, *mPer1* and *mPer2* mRNA in the SCN (Shigeyoshi et al 1997), the effect of brief light pulses on gene expression in the SCN of *mCry1/mCry2* double knockout these mice was investigated. An acute light pulse is still able to induce expression of *mPer2* (Okamura et al 1999, Vitaterna et al 1999), whereas induction of *mPer1* was observed in one study (Okamura et al 1999) and appeared lacking in another (Vitaterna et al 1999). This suggests that CRY proteins are dispensable for a functional light-transducing pathway. Yet, redundancy between CRY proteins and other photoreceptors can not be excluded, as light induction of c-*fos* gene expression was markedly reduced in rodless Cry-deficient mice (Selby et al 2000). Thus, the role of *mCry* as (circadian) photoreceptor remains a matter of debate (Cermakian & Sassone-Corsi 2002), particularly with the recent discovery of a melanopsin-containing subset of retinal ganglion cells that project into the SCN.

Recently, we have investigated the electrical behaviour of the SCN of *mCry1/mCry2* double knockout mice, using multi-unit electrode activity (MUA) recordings on cultured SCN slices (Albus et al 2002), and obtained data on the effect of light on the SCN at the level of a physiological output parameter rather than gene expression.

Electrophysiology of the SCN of *mCry*-deficient mice

SCN slices from wild type mice previously kept in constant darkness (DD) for at least one week and sacrificed at circadian time 2 (CT2, as deduced from behavioural activity patterns) exhibit a clear circadian rhythm in electrical activity (Fig. 1A; Albus et al 2002). Maximal discharge rates are observed during the subjective day and low firing rates at the subjective night. This pattern can be recorded over two consecutive days, with peak levels reached at CT 5.42 ± 1.18 h and 2.23 ± 1.58 h ($n = 7$, mean $\pm$ SD) on the first and second day of recording, respectively. When wild-type mice are kept under 12 h light–12 h dark cycles (LD) and sacrificed at the beginning of the light phase, around *zeitgeber* time 2 (ZT2), peaks in firing frequency were observed at CT 6.72 ± 0.91 h and 3.78 ± 0.76 h ($n = 9$, mean $\pm$ SD) on days one and two, respectively (Fig. 1B; Albus et al 2002). Thus, pre-exposure to light does not change significantly the pattern of electrical activity of cultured SCN slices from wild-type mice.

In contrast, SCN slices from *mCry1/mCry2* double knockout mice, previously housed under DD conditions and thus showing arrhythmic behaviour, display an initial high level of firing activity, with spike frequencies rapidly decreasing along the first hours of recording, stabilising to a plateau at mid-to-low level, and after about 20–30 h dropping to low levels (Fig. 2A; Albus et al 2002). This indicates that mCRY proteins are dispensable for the ability of neuronal cells to fire spontaneous action potentials. The electrical activity pattern in *mCry*-deficient SCN slices markedly differs from the MUA profile obtained from wild type slices in that any (circadian) pattern is lacking. From these experiments it can be concluded that mCRY proteins, and thus an intact circadian clockwork, are prerequisite for circadian electrical activity in SCN neurons.

SCN slices from *mCry1/mCry2* double knockout mice, housed under LD conditions and sacrificed at ZT2, display an initial decrease in discharge rate similar to that observed in slices from animals kept in DD (Fig. 2B; Albus et al 2002). Interestingly, at the end of the subjective day, action potential frequencies start to show a temporary increase, reaching a maximum at mid-subjective night (ZT 15.34 ± 4.59 h, $n = 5$). Prolonged recordings revealed that this peak in electrical activity does not reappear in the next 24 h of recording, and thus is unlikely to originate from an intrinsic circadian property of the mutant SCN but rather might be the result of exposure of the animals to light prior to isolation of the

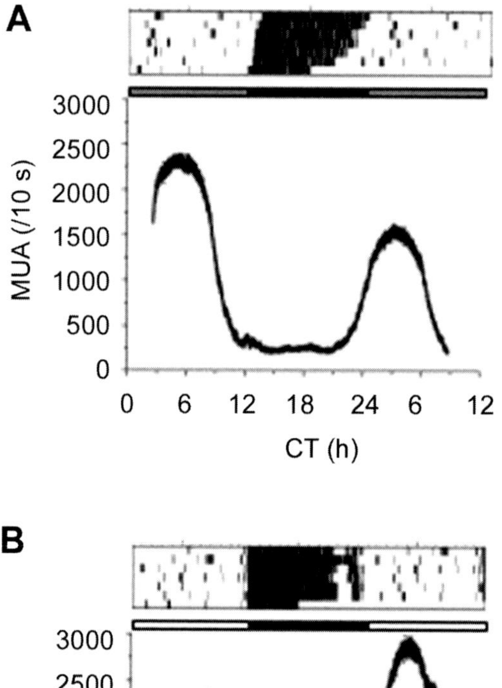

FIG. 1. Firing-rate patterns in SCN slices from wild-type mice kept in constant darkness (DD) or in 12 h light–12 h dark (LD) cycles. The running wheel activity patterns of the mice over the last 7 days prior to slice preparation is indicated above the records. (A) Circadian rhythm in multiunit activity (MUA) in the SCN of a wild-type mouse, previously kept under DD conditions and isolated at CT2; (B) Circadian rhythm in MUA activity in the SCN of a wild-type mouse, previously kept under LD conditions, and isolated at CT2. Black/grey and black/white bars above each plot indicate the subjective night/day and night/day respectively. CT, circadian time. (Reprinted from Albus et al 2002, with permission from Elsevier Science.)

SCN (i.e. between ZT0 and ZT2). If so, SCN slices from *mCry1/mCry2* double knockout mice sacrificed in the dark period may not show a peak in electrical activity. Indeed, *mCry1/mCry2* double knockout SCN slices isolated at ZT14 do not show a peak in electrical activity (Fig. 2C; Albus et al 2002) and in fact show a MUA pattern very similar to that of animals kept in DD (Fig. 2B; Albus et al 2002). This finding strongly suggests that the peak in firing rates is caused by exposure of the animal to light.

Further evidence for light-mediated increase in neuronal electric activity in the SCN of *mCry1/mCry2* mice was obtained from electrophysiological experiments with aged (412 months) animals. Whereas young *mCry1/mCry2* mice are behaviourally inactive during the light phase of a light–dark cycle, probably due to masking (van der Horst et al 1999, Mrosovsky 2001), animals start to become arrhythmic upon ageing. SCN slices taken from old animals at ZT2 do no longer show a peak in firing activity (X. Bonnefont, H. Albus, G.T.J. van der Horst and J.H. Meijer, unpublished data). This effect likely originates from a retinal defect as SCN neurons still respond to NMDA treatment with an increased firing rate.

Taken together, these data indicate that light can cause an increase in neuronal activity in the SCN of *mCry*-deficient mice, and that the absence of CRY proteins does not prevent light signalling from the retina to the SCN. Moreover, the ability of the *mCry*-deficient SCN to respond to light exposure electrophysiologically is lost upon ageing.

Conclusions

So far, data on the electrophysiological properties of circadian arrhythmic mammals have been obtained from studies with homozygous *Clock* mutant mice (Liu et al 1997, Herzog et al 1998, Nakamura et al 2002). These animals differ from *mCry1/mCry2* double knockout mice (van der Horst et al 1999, Vitaterna et al 1999) and *Mop3(Bmal1)* knockout animals (Bunger et al 2000) in that they maintain free-running circadian behaviour for at least some days (Vitaterna et al 1994), and as such may not be considered as complete circadian oscillator knockout. Analysis of multi-unit electrode activity in SCN slices of *mCry1/mCry2*-deficient mice (Albus et al 2002) has revealed three important properties of an arrhythmic SCN.

First, the SCN of *mCry1^{-/-}mCry2^{-/-}* mice appears electrically arrhythmic, independent of whether animals had been kept under a light–dark cycle or in constant darkness. This finding provides direct genetic evidence that mCRY proteins are indispensable for circadian rhythmicity of electrical activity in SCN neurons. Thus, an intact circadian clockwork appears prerequisite for circadian electrical activity in SCN neurons. This finding seems somewhat contradictory to a recent study of Nakamura et al (2002), who demonstrated electrical rhythmicity in SCN slices from arrhythmic homozygous *Clock* mutant mice. However, this

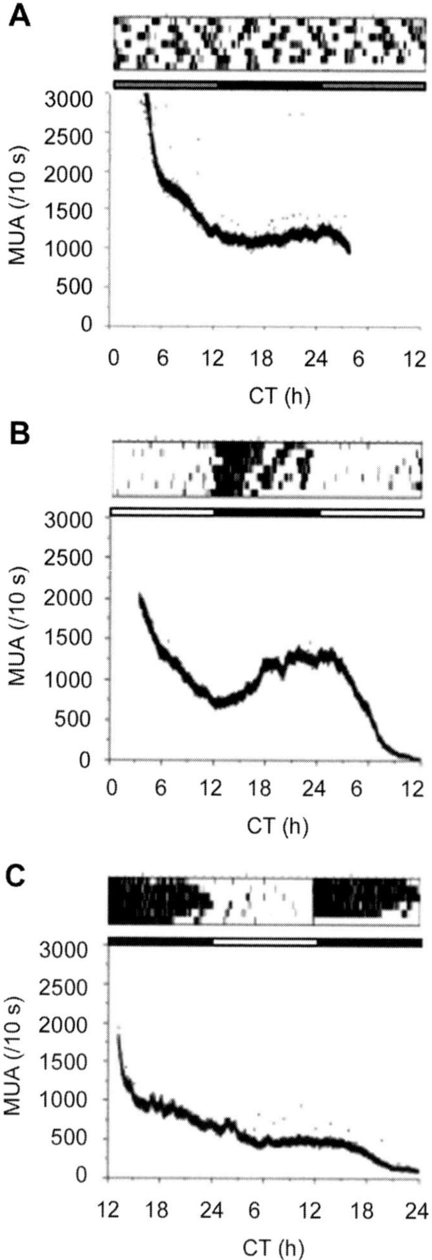

discrepancy is well explained by the ability of *Clock* mutant to pursue rhythmic behaviour for a few days before falling into arrhythmicity (Vitaterna et al 1994). Indeed, dispersed SCN neurons from homozygous *clock* mutant mice show a near-24 h rhythm in electrical activity for 2–3 days before turning into arrhythmicity (Herzog et al 1998).

Second, the SCN of young *mCry1−/− mCry2−/−* mice can still respond to light with a temporary increase in electrical activity. This finding is in line with previous observations that phase-resetting light pulses are still capable of inducing *mPer1* and *mPer2* gene expression in the SCN of *mCry*-deficient mice (Okamura et al 1999, Vitaterna et all 1999), likely via cAMP responsive elements (CREs) that respond to activated cAMP and mitogen activated protein kinase (MAPK) signalling pathways in a core oscillator-independent manner (Travnickova-Bendova et al 2002).

In this context it is interesting to note that *mCry1/mCry2*-deficient mice show relatively normal behaviour when housed under normal LD cycles (van der Horst et al 1999). Animals are inactive during the day and show large periods of wheel-running activity immediately after onset of the night, which contrasts the very short bouts of activity and rest observed in constant darkness (compare activity patterns Figs 1B, 2B and 2C). This suggests that periodic exposure to light, at least to some extent, can normalize behaviour (and presumably other clock-controlled output processes) in *mCry1/mCry2*-deficient mice. The mPER2 protein has been proposed to directly or indirectly act as an activator of the positive feedback loop of the molecular clockwork (Shearman et al 2000). Recently, the nuclear orphan receptor protein REV-ERBα has been shown to suppress *Bmal1* and *Clock* gene expression (Preitner et al 2002). As *Rev-erbα* expression itself is under negative circadian control by mPER2, this finding provides an attractive molecular explanation for the positive effect of mPER2 on *Bmal1* transcription. In view of these findings, and given our observation that nuclear entry of the mPER2 protein does not depend on mCRY proteins (Yagita et al 2002), we hypothesize that in *mCry1/mCry2* knockout mice the environmental LD cycle can act as a periodic trigger that kick-starts a crippled molecular clockwork, and as a result increased SCN neuronal activity and other clock-controlled output processes, by induction of *mPer* gene expression and

---

FIG. 2. Firing-rate patterns in SCN slices from *mCry1/mCry2* double knockout mice kept in constant darkness (DD) or in 12 h light–12 h dark (LD) cycles. Running wheel activity patterns prior to slice preparation are indicated above the records. (A) Circadian electrical activity in an SCN slice of a *mCry1/mCry2* double knockout mouse, kept under DD conditions; (B) Circadian electrical activity in the SCN of a *mCry1/mCry2* double knockout mouse, kept under LD conditions and sacrificed at ZT2; (C) Circadian electrical activity in the SCN of a *mCry1/mCry2* double knockout mouse, kept under LD conditions and sacrificed at ZT14. (Reprinted from Albus et al 2002, with permission from Elsevier Science.)

subsequent derepression of *Bmal1* transcription. This non-self-sustaining oscillator will block at the stage where mCRY proteins are required for silencing of CLOCK/BMAL1-driven transcription and will not restart in the absence of a new stimulus (as in SCN slices). This scenario, in which the *mCry*-deficient SCN in mice exposed to regular LD cycles works as an 'hour-glass' rather than a 'clock', is supported by the observation that *mPer2* mRNA levels at ZT6 are twofold increased compared to ZT18 when *mCry*-deficient mice are kept in LD cycles (Vitaterna et al 1999).

Interesting questions emerging are whether non-photic stimuli are also capable of triggering the proposed 'hour-glass' clock in the SCN of *mCry*-deficient mice and whether such SCN (or a transplanted wild-type SCN) can drive 'hour-glass' type peripheral oscillators via humoral factors. In addition, as peripheral clocks can phase-uncouple from the SCN clock by enforced feeding schedules or injections with glucocorticoids (Balsalobre et al 2000, Damiola et al 2000), it will be interesting to investigate whether such treatment can induce clock gene expression and clock-output processes in 'hour-glass' type peripheral oscillators.

Third, light-mediated induction of electrical activity in SCN neurons is lost in old *mCry1/mCry2* double knockout mice that have become arrhythmic under LD conditions. This suggests that in these animals an important photoreceptor cell or protein is lost upon ageing. Since aged *mCry1/mCry2* mice still contain rods and cones and NMDA can increase electrical activity in SCN slices obtained from these animals, the observed age-dependent loss of light-signalling into the SCN is likely to be attributed to either the loss of a photoreceptor protein other than cryptochromes and classical (rod and cone) opsins, or an age-dependent defect in signal transduction via the retinohypothalamic tract. Obvious candidates in this respect are the melanopsin-containing retinal ganglion cells projecting to the SCN (Berson et al 2002). These cells contain PACAP (Hannibal et al 2002), a neurotransmitter known to act on the SCN through its PAC1 (Hannibal et al 2001) and VPAC2 (Shen et al 2000, Harmar et al 2002) receptors. The VPAC2 seems indeed to have a crucial role in SCN photoentrainment since in its absence *mPer* genes are not induced by a nocturnal illumination (Harmar et al 2002). We are currently investigating whether aged *mCry1/mCry2* mice have lost melanopsin-containing ganglion cells or whether these cells display reduced expression of neurotransmitters.

Acknowledgements

This work has been supported by a Marie Curie Fellowship of the European Community program 'Improving Human Research Potential and the Socio-economic Knowledge Base' under contract number HPMF-CT-2000-00950 to X.B., a NWO–Hungarian co-operation grant to J.H.M. and a Spinoza Premium of the Netherlands Organisation for Scientific Research (NWO).

References

Albus H, Bonnefont X, Chaves I et al 2002 Cryptochrome-deficient mice lack circadian electrical activity in the suprachiasmatic nuclei. Curr Biol 12:1130–1133

Balsalobre A, Brown SA, Marcacci L et al 2000 Resetting of circadian time in peripheral tissues by glucocorticoid signaling. Science 289:2344–2347

Berson DM, Dunn FA, Takao M 2002 Phototransduction by retinal ganglion cells that set the circadian clock. Science 295:1070–1073

Bunger MK, Wilsbacher LD, Moran SM et al 2000 Mop3 is an essential component of the master circadian pacemaker in mammals. Cell 103:1009–1017

Cermakian N, Sassone-Corsi P 2002 Environmental stimulus perception and control of circadian clocks. Curr Opin Neurobiol 12:359–365

Damiola F, Le Minh N, Preitner N, Kornmann B, Fleury-Olela F, Schibler U 2000 Restricted feeding uncouples circadian oscillators in peripheral tissues from the central pacemaker in the suprachiasmatic nucleus. Genes Dev 14:2950–2961

Foster RG 1998 Shedding light on the biological clock. Neuron 20:829–832

Freedman MS, Lucas RJ, Soni B et al 1999 Regulation of mammalian circadian behavior by non-rod, non-cone ocular photoreceptors. Science 284:502–504

Green DJ, Gillette R 1982 Circadian rhythm of firing rate recorded from single cells in the rat suprachiasmatic brain slice. Brain Res 245:198–200

Hannibal J, Jamen F, Nielsen HS, Journot L, Brabet P, Fahrenkrug J 2001 Dissociation between light-induced phase shift of the circadian rhythm and clock gene expression in mice lacking the pituitary adenylate cyclase activating polypeptide type 1 receptor. J Neurosci 21:4883–4890

Hannibal J, Hindersson P, Knudsen SM, Georg B, Fahrenkrug J 2002 The photopigment melanopsin is exclusively present in pituitary adenylate cyclase-activating polypeptide-containing retinal ganglion cells of the retinohypothalamic tract. J Neurosci 22:RC191

Harmar AJ, Marston HM, Shen S et al 2002 The VPAC(2) receptor is essential for circadian function in the mouse suprachiasmatic nuclei. Cell 109:497–508

Hattar S, Liao HW, Takao M, Berson DM, Yau KW 2002 Melanopsin-containing retinal ganglion cells: architecture, projections, and intrinsic photosensitivity. Science 295:1065–1070

Herzog ED, Takahashi JS, Block GD 1998 Clock controls circadian period in isolated suprachiasmatic nucleus neurons. Nat Neurosci 1:708–71

Kobayashi K, Kanno S, Smit B, van der Horst GTJ, Takao M, Yasui A 1998 Characterization of photolyase/blue-light receptor homologs in mouse and human cells. Nucleic Acids Res 26:5086–5092

Kume K, Zylka MJ, Sriram S et al 1999 mCRY1 and mCRY2 are essential components of the negative limb of the circadian clock feedback loop. Cell 98:193–205

Liu C, Weaver DR, Strogatz SH, Reppert SM 1997 Cellular construction of a circadian clock: period determination in the suprachiasmatic nuclei. Cell 91:855–860

Meijer JH, Schaap J, Watanabe K, Albus H 1997 Multiunit activity recordings in the suprachiasmatic nuclei: *in vivo* versus *in vitro* models. Brain Res 753:322–327

Miyamoto Y, Sancar A 1998 Vitamin B2-based blue-light photoreceptors in the retinohypothalamic tract as the photoactive pigments for setting the circadian clock in mammals. Proc Natl Acad Sci USA 95:6097–6102

Mrosovsky N 2001 Further characterization of the phenotype of mCry1/mCry2-deficient mice. Chronobiol Int 18:613–625

Nakamura W, Honma S, Shirakawa T, Honma K 2002 Clock mutation lengthens the circadian period without damping rhythms in individual SCN neurons. Nat Neurosci 5:399–400

Okamura H, Miyake S, Sumi Y et al 1999 Photic induction of mPer1 and mPer2 in Cry-deficient mice lacking a biological clock. Science 286:2531–2534

Preitner N, Damiola F, Lopez-Molina L et al 2002 The orphan nuclear receptor REV-ERBalpha controls circadian transcription within the positive limb of the mammalian circadian oscillator. Cell 110:251–260

Reppert SM, Weaver DR 2002 Coordination of circadian timing in mammals. Nature 418:935–941

Selby CP, Thompson C, Schmitz TM, Van Gelder RN, Sancar A 2000 Functional redundancy of cryptochromes and classical photoreceptors for nonvisual ocular photoreception in mice. Proc Natl Acad Sci USA 97:14697–14702

Shearman LP, Sriram S, Weaver DR et al 2000 Interacting molecular loops in the mammalian circadian clock. Science 288:1013–1019

Shen S, Spratt C, Sheward W J et al 2000 Overexpression of the human VPAC2 receptor in the suprachiasmatic nucleus alters the circadian phenotype of mice. Proc Natl Acad Sci USA 97:11575–11580

Shigeyoshi Y, Taguchi K, Yamamoto S et al 1997 Light-induced resetting of a mammalian circadian clock is associated with rapid induction of the mPer1 transcript. Cell 91:1043–1053

Thresher R J, Vitaterna MH, Miyamoto Y et al 1998 Role of mouse cryptochrome blue-light photoreceptor in circadian photoresponses. Science 282:1490–1494

Todo T, Ryo H, Yamamoto K et al 1996 Similarity among the *Drosophila* (6-4)photolyase, a human photolyase homolog, and the DNA photolyase-blue-light photoreceptor family. Science 272:109–112

Travnickova-Bendova Z, Cermakian N, Reppert SM, Sassone-Corsi P 2002 Bimodal regulation of mPeriod promoters by CREB-dependent signaling and CLOCK/ BMAL1 activity. Proc Natl Acad Sci USA 99:7728–7733

van der Horst GTJ, Muijtjens M, Kobayashi K et al 1999 Cry1 and Cry2 are essential for maintenance of circadian rhythms. Nature 398:627–630

van der Spek P J, Kobayashi K, Bootsma D, Takao M, Eker APM, Yasui A 1996 Cloning, tissue expression, and mapping of a human photolyase homolog with similarity to plant blue-light receptors. Genomics 37:177–182

Vitaterna MH, King DP, Chang AM et al 1994 Mutagenesis and mapping of a mouse gene, Clock, essential for circadian behavior. Science 264:719–725

Vitaterna MH, Selby CP, Todo T et al 1999 Differential regulation of mammalian period genes and circadian rhythmicity by cryptochromes 1 and 2. Proc Natl Acad Sci USA 96:12114–12119

Weaver DR 1998 The suprachiasmatic nucleus: a 25-year retrospective. J Biol Rhythms 13:100–112

Welsh DK, Logothetis DE, Meister M, Reppert SM 1995 Individual neurons dissociated from rat suprachiasmatic nucleus express independently phased circadian firing rhythms. Neuron 14:697–706

Yagita K, Tamanini F, Yasuda M, Hoeijmakers JHJ, van der Horst GTJ, Okamura H 2002 Nucleocytoplasmic shuttling and mCRY-dependent inhibition of ubiquitylation of the mPER2 clock protein. EMBO J 21:1301–1314

Yasui A, Eker APM 1998 DNA photolyases. In: Nickoloff J A, Hoekstra MF (eds) DNA damage and repair. vol 2. Humana Press Inc, Totowa, New Jersey, p 9–31

DISCUSSION

Menaker: Are you really convinced that there is a one-to-one relationship between electrical activity in the SCN and behaviour? You kind of assumed this, but I don't think the evidence is very good. Work from Rae Silver's lab in which the SCN transplants are encapsulated, and therefore can't make electrical connections, suggests that it may not be so simple (Silver et al 1996, Le Sauteur & Silver 1999). In addition Chuck Weitz has shown that SCN secretions can drive behaviour.

Rosbash: I don't see why those are incompatible.

Menaker: It is a question of how directly the electrical activity has an effect. It is only correlative at the moment.

Weitz: Most people would assume that for secreted factors, even those acting at a distance in paracrine fashion, release is likely to be gated by electrical activity of the neuron. Most likely electrical activity of SCN neurons is important for the release of conventional synaptic factors or hypothetical paracrine factors.

van der Horst: I also assumed that there was this direct connection between electrical activity in SCN neurons and behavioural activity, which for nocturnal animals means low electric activity being associated with the behaviourally active period.

Stanewsky: There is one aspect of the *Cry* double knockout behaviour that has always puzzled me. It never looked to me that it was just masking, because they seem to anticipate the LD transition. This would fit perfectly with your assumption that it is an hour-glass mechanism.

van der Horst: Nicholas Mrosovsky has observed normal masking properties in our Cry double knockout mice and published a paper on this (Mrosovsky 2001). He indeed noticed that the animals show 'pre-dark' activity and assumed that there was a remnant damped oscillator. I disagreed with him, because this suggests that in a way the oscillator should be intact, and proposed the presence of an 'hour-glass' type of timekeeper acting through a crippled oscillator that gets kick-started every 24 h by dark–light or light–dark transitions. The light-induced electrical activity in the *Cry* double knockout SCN fits with such a model and could well be responsible for the suppressed behavioural activity in the light period and long period of wheel-running activity in darkness, including pre-dark running activity.

Van Gelder: I have some data which address this. We have looked very carefully for transients in the *Cry* double mutants by phase shifting them 6 hours to 12 hours. You would think that a weak clock that was being remasked would show a couple of transients, and we never saw evidence for this. If there is a residual oscillator, it is not enough to generate one or two cycles in the absence of the cryptochromes. I also wanted to comment on the ageing effects that you see. We looked very carefully for these as well in our mice. I should point out that Bert's mice and ours were independently derived and involve a different knock-in construct on a different genetic background. We tested our mice to 18 months and never saw a dimunition of masking effects in 100 lux light. I don't know where on the luminance curve your mice are.

van der Horst: I don't know the exact light levels. We are using regular light conditions in our animal facility, which it is somewhere around 300 lux.

Van Gelder: It is probably brighter than what we are using. We did see age-dependent loss of masking in some of those triple mutants that still showed masking activity, but we never saw it in the double mutants. I have some records of some mice that literally died on the wheel at 18 months, where you can see that they are pretty well masked in their last two or three days. I have a more general question. Is the terminology of masking versus clock-dependency still adequate? It seems to me that there are at least two forms of masking that we have to talk about. One is the SCN-independent form. Nicolas Mrosovsky and others have shown that SCN-lesioned animals will still show masking activity. This clearly doesn't require the pacemaking clock component. But there is also signalling where light can drive gene expression in the SCN in the absence of free-running rhythms. Presumably, that gene expression has output consequences so that the SCN can still drive an output signal. There are separate mechanisms for light affecting behaviour outside the SCN and light signalling through the SCN.

van der Horst: We really need to determine clock gene expression patterns in the SCN during the LD cycle to determine whether there is an hour-glass timekeeper that might contribute to suppression of wheel-running behaviour during the day.

Rosbash: There is masking and there are real oscillators, damped oscillators and then one might consider this at least as a candidate for a super-damped oscillator. The experiment that was done in flies to address this was to stick a timing mutant into that kind of 'arrhythmic' background and look at whether *Per* or some other clock component still contributes to the hourglass or hyperdamping feature of the rhythm. If the peak of electrical activity or behaviour now shifts in those mutants in a fairly predictable way, you would argue that the hourglass and oscillator are sharing clock core components.

van der Horst: That's a good point. We should check this behaviour in various double mutant combinations.

Van Gelder: The problem is, of course, that the *Cry* double mutant is arrhythmic to start with.

Rosbash: There are two things that can be measured: he could ask whether the behaviour shifts in any way with respect to the light–dark cycle, and now he also has an electrical peak. The question is, if you throw in to the animal something that shifts the period, would either of these change?

Loros: I think that is an excellent experiment. One of the things about the animal work is that you don't have the availability of period length mutants analogous to *Frq1* and *Frq7* in *Neurospora*. They are very informative in telling you if you are looking at a completely different regulatory system: whether the residual light response is completely unconnected to the *Cry*-based circadian system or not.

Sassone-Corsi: I'm very excited by the experiment that involved transplanting the normal SCN into the double *Cry* knockout mutants (Sujino et al 2003). This could tell us a lot about the relationship between the SCN masking and the entrainment of the other tissues.

Menaker: I disagree. The interpretation of that experiment is up for discussion.

Weitz: This would mean that we could ignore peripheral oscillators after all!

Menaker: It has nothing to do with peripheral oscillators.

Young: When you showed the data on the SCN transplant you used the word 'entrainment'. Did you actually see evidence for LD entrainment?

van der Horst: That was just a slip.

Young: So you are seeing an emergence of rhythmicity. You don't see any evidence for a light response.

van der Horst: Animals have not been tested in LD cycles yet. As for the emergence of behavioural rhythmicity in wild-type SCN-transplanted *Cry*-deficient animals, I think that analogous to the model of a light-inducible hour-glass-like timekeeper in the SCN of *Cry* double knockout mice, a transplanted SCN may induce hour-glass peripheral oscillators in the brain and periphery of double knockouts. These rhythmically kick-started peripheral oscillators should then drive regular periodicity in behaviour. I do not expect this behaviour to respond to light–dark cycles.

Weitz: In the experiment that you mention, there is another thing you could do, which would address this beautifully. In that transplanted *Cry1/Cry2* knockout, you could look at whether there are rhythms in the liver. This would be the experiment to do. If there are, this would imply that this could somehow drive crippled oscillators in the periphery. There is nothing about the locomotor assay that requires an oscillator outside the SCN. This actually proves that at least for the emergence of some rhythmicity, no oscillator outside the SCN is required.

Sassone-Corsi: What we don't know is what happens in peripheral oscillators.

Menaker: That result is no different, in principle, from an SCN-lesioned animal that gets a transplant and rhythmicity is restored. In other words, the knockout is destroying the rhythmic function of the animal's own SCN, and it is being replaced with a transplant.

Weitz: It is also destroying the rhythmic function of every other potential oscillator known.

Menaker: The outcome is the same.

Weitz: But the context is not at all the same.

Hastings: Taking all the data together, we can make a strong prediction. Presumably there is a crippled liver oscillator, but because there is now a functional SCN giving a functional activity cycle on a daily basis, you will get that single pulse per day which the crippled liver can use. So those animals will have a rhythmic liver, but each day they will have to receive the acute pulse which in fibroblast culture is sufficient to give it the required stimulation.

Schibler: Bert Van der Horst did an experiment where he showed that he couldn't induce circadian rhythms in *Cry* double knockout mouse embryonic fibroblasts. This is not surprising because they probably need the same components than in the clock. My strong prediction would be that everything is flat in the periphery.

Hastings: The functional SCN makes sure that each day there is a feeding cycle, for example. Each day the liver will receive a macronutrient stimulus. This would be sufficient to give a pattern to the gene expression, similar to reacting to the serum shock that in culture would cause an acute induction of *Per* but not a second wave of induction.

Menaker: It is clear that it is a good experiment.

Van Gelder: A clarification. The restored transplant is a wild-type SCN into a cry double knockout. Is the *Cry* double-mutant SCN-lesioned?

van der Horst: Yes.

Van Gelder: Have you tried transplanting into a non-lesioned *Cry* double mutant?

van der Horst: That has not been done.

Van Gelder: That would give you insight into the output pathways of the intact SCN.

Sehgal: Chuck Weitz said something about the SCN being sufficient to drive activity rhythms. What about the cycling clock proteins in the motor cortex?

Weitz: This experiment proves that they are not necessary for the emergence or rhythmicity.

Sehgal: How do you know that the implanted SCN is not also transmitting signals to the motor cortex?

Weitz: It just says that the other places don't have to have a functional clock. It is true that they could entrain rhythms. This is what everyone wants to know.

Sehgal: The question that interests me is whether the motor cortex is required, and if it is, does this tell us something?

Hastings: Notwithstanding Michael Menaker's data on the luciferase assays on brain tissues, it is possible for various brain regions to show only one or two cycles. The accepted idea is that what is happening in the motor cortex is a reflection of activity that is being caused by the SCN. It is not an intrinsic oscillation of the motor cortex that would be sustained for any time.

Sehgal: What I am getting at is whether cycling clock proteins are required in the motor cortex for activity rhythms to be generated.

Hastings: I suspect not, but in the motor cortex and striatum of these grafted mice we will probably see a cycle of gene expression there as a consequence of the activity cycle.

Weitz: But would a transplanted SCN have access to the output pathways that then control peripheral tissues? If it requires specific connections and certain neural outputs, then transplanted animals may not have access to these.

Hastings: The expectation is that the feeding rhythm would be sufficient to drive these events.

Sassone-Corsi: We know that the transplanted SCN doesn't receive normal input from the retina. Is it possible that the transplanted SCN is not communicating with the rest of peripheral tissues in the same way? I am not sure that the experiment of Bert van der Horst will tell us whether the transplanted SCN is able to entrain a non-functioning peripheral clock.

Cermakian: We have done almost the reverse experiments with fibroblast implants from *Clock*$^{c/c}$ mutants whose clock is quite impaired. When we put these fibroblast implants into mice whose SCN is functional, the SCN was not able to entrain them. So I don't think that a grafted SCN will be able to entrain peripheral tissues of these *Cry* double mutants.

van der Horst: It might depend on the type of mutation and its effect on the core oscillator. Assuming that the *Cry* double knockout mice have an hour-glass timekeeper, it would not surprise me when fibroblast implants from these animals turn out to display to some extent cyclic clock gene expression driven by periodic stimuli from the intact SCN.

Sassone-Corsi: I agree. What we don't know is how much the liver, for example, is affected in the double *Cry* knockout.

Rosbash: The dogma would have you believe that it would be transcriptionally dead for circadian gene expression.

Sassone-Corsi: It is one thing to graft an implant in a mouse where the SCN is still wired normally with the rest of the animal; it is quite another to transplant a SCN that is not able to wire itself with the rest of the animal.

Weitz: If you transplant SCNs into animals that already have intact SCNs, the grafts often don't work.

Menaker: That's not true. Most often, we see two bouts of activity which have different periods.

Weitz: Didn't that require a partial lesion in the animals first?

Menaker: That's true.

Ishida: We have the same data. If we feed *Clock* homozygote mutant mice during the daytime, the rhythmic expression of *Clock* is completely entrained as indicated by *Per2* and *Bmal1* expression patterns in the absence of clock protein.

Rosbash: The *Clock* mutant still has some clock protein. But that's a pertinent result, for sure.

Menaker: This is making clear how profound our ignorance is of the coupling pathways!

Loros: Where is the technology for producing mosaic mice? This would enable us to generate mice that were *Cry*-deficient in the liver but which had a wild-type SCN, for instance.

Weitz: We are all working on it, don't worry!

Takahashi: Everyone is working on tissue-specific knockouts and inducible expression. But no one is there yet. The hardest part is getting SCN drivers; the liver and the periphery are already done. Not much work has been done with genetic mosaics. The problem with mosaics in mice is that they are very fine grained. You can get patches, but you have to do cross-species work.

Rosbash: It would be hard to address this in the fly mosaics because each fly is a one-off.

Weitz: Conditional knockouts are the way to go.

References

LeSauter J, Silver R 1999 Localization of a suprachiasmatic nucleus subregion regulating locomotor rhythmicity. J Neurosci 19:5574–5585

Mrosovsky N 2001 Further characterization of the phenotype of mCry1/mCry2-deficient mice. Chronobiol Int 18:613–625

Silver R, LeSauter J, Tresco PA, Lehman MN 1996 A diffusible coupling signal from the transplanted suprachiasmatic nucleus controlling circadian locomotor rhythms. Nature 382:810–813

Sujino M, Masumoto K, Yamaguchi S, van der Horst GTJ, Okamura H, Inoye SI 2003 Suprachiasmatic nucleus grafts restore circadian behavioural rhythms of genetically arrhythmic mice. Curr Biol 13:664–668

Circadian light input in plants, flies and mammals

Satchidananda Panda*†, John B. Hogenesch†, Steve A. Kay*†[1]

*Department of Cell Biology, The Scripps Research Foundation, 10550, North Torrey Pines Road, La Jolla, CA, 92037 and †Genomics Institute of the Novartis Research Foundation, 10675, John J Hopkins Drive, San Diego, CA 92121, USA

Abstract. The rotation of our planet results in daily changes in light and darkness, as well as seasons with characteristic photoperiods. Adaptation to these daily and seasonal changes in light properties (and associated changes in the environment) is important to the sustained survival of higher life forms on our planet. Many organisms use their intrinsic circadian oscillator or clock to orchestrate daily rhythms in behaviour and physiology to adapt to diurnal changes. Some higher organisms use the same oscillator to monitor day length in selecting the appropriate season for reproductive behaviour. Organisms have developed irradiance measurement mechanisms to ignore photic noise (lightning, moonlight), and use the light of dusk and dawn for circadian photoentrainment. They have also devised multiple photoreceptors and signalling cascades to buffer against changes in the spectral composition of natural light. The interaction of the clock with ambient light is, therefore, quite intricate.

2003 Molecular clocks and light signalling. Wiley, Chichester (Novartis Foundation Symposium 253) p 73–88

In the last decade, molecular genetic analysis of the circadian system in many organisms has elucidated critical aspects of clock function. One common feature is a transcriptional translational feedback loop, where transcriptional activators regulate expression of repressor proteins that (directly or indirectly) inhibit their own transcription. This results in at least one clock component having an oscillation in its mRNA and protein levels with the period length of their oscillation read out as the circadian time of the system (Harmer et al 2001). To maintain synchrony with the changing environment, the phase of oscillation of the cycling component(s) needs to be adjusted on a daily basis. This is mostly achieved at twilight by resetting the phase of a cycling clock component usually to its daytime level. These daily phase adjustments are essentially an adaptive

---

[1]This paper was presented at the symposium by Steve A. Kay to whom all correspondence should be addressed.

response to ensure that the organism times its physiology appropriately and accounts for changing day length over the seasons.

There are two separate, but equally important, components of light input to the clock: photoperception and the signalling cascade that ultimately results in resetting of the oscillator. Molecular and genetic analyses have been employed to characterize the light-input pathway to the circadian oscillator in plants, flies and mammals. These studies have resulted in the identification of several circadian photoreceptors, and the characterization of their signalling has led to an increasing understanding of how organisms integrate light signals from the environment, transmit that information to the clock, and modify their physiology accordingly. This review will focus on describing both light-perception mechanisms as well as the integration of this information in the core clock in the model organisms *Arabidopsis thaliana*, *Drosophila melanogaster* and *Mus musculus*.

Light input in plants

In photosynthetic plants, light plays a crucial role during growth and development. Plants use an array of light-perception mechanisms to measure the spectral composition, irradiance (or light intensity), and direction of incident light to better modulate growth and optimize physiology. For example, young seedlings grown under canopy cover measure the relative ratios of red and far-red light in order to accelerate their hypocotyl or internode growth appropriately. A genetic deficiency in the photoperception mechanism can mimic the absence of light and result in a long hypocotyl phenotype under normal lighting conditions (Chory 1993). Genetic analysis of this photomorphogenesis in plants initially identified two major classes of photoreceptors — red/far-red absorbing phytochromes and blue light absorbing cryptochromes (Fig. 1) (Quail 2002). These phytochromes (PHYs) and cryptochromes (CRYs) have either been shown to bind to chromophores or bear sequence features indicative of chromophore binding. Their characterization has led to the identification of the first circadian photoreceptors in higher organisms.

Under subsaturating levels of constant red or blue light, *Arabidopsis* seedlings exhibit an inverse correlation between their free-running period length and the fluence rate (or light intensity), which constitutes a fluence rate response curve (FRC). Generation of FRCs for *Arabidopsis* seedlings deficient in one or more photoreceptors has elucidated the circadian light-input role of individual photoreceptors (Mas et al 2000, Somers et al 1998, Yanovsky et al 2000). Spectral properties of these photoreceptors and their downstream signalling components generate exquisite plasticity in circadian light input in plants. PHYA and CRY1, light-unstable but highly sensitive photoreceptors, act as low fluence-rate

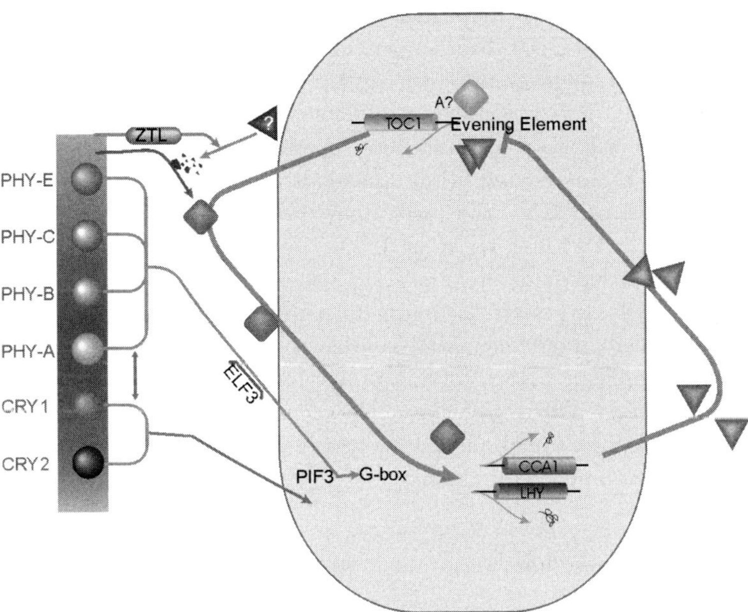

FIG. 1.   Circadian light input in *Arabidopsis*. Phytochromes and cryptochromes function as the respective circadian photoreceptors in the red and blue spectra of incident light. Upon photoreception phytochromes may activate the transcription factor PIF3 that binds to G-box *cis*-acting element and induces transcription of CCA1. CCA1 and LHY are two myb-domain containing transcription factors that are rhythmically expressed with a peak in early subjective morning, and repress transcription of TOC1 via binding to evening element in its promoter. In addition to its role in the core oscillator, TOC1 may also mediate some red light signalling to the oscillator. This may be mediated via its interaction with PIF3. A clock output ELF3, which is mostly cytoplasmic, negates photic input to the oscillator and thereby fine tunes the light sensitivity of the oscillator. ZTL protein may add another level of light input by its proposed function in light-induced degradation of a clock component.

photoreceptors and may function at dawn to adjust the oscillator. At high irradiance during the day, light-stable but less sensitive photoreceptors, CRY2 and PHYB, may adjust the pace of the oscillator. Finally, at night the accumulation of light-unstable but highly sensitive photoreceptors PHYA and CRY1 also raises the risk of light resetting by photic noise (moonlight, lightning). To buffer the oscillator against light resetting by photic noise, the clock-regulated transcriptional repressor, ELF3, peaks and modulates PHY and CRY signalling to the clock (Covington et al 2001).

Light resetting of the circadian oscillator is at least partly mediated by light-induced transcription of a cycling clock component (Wang & Tobin 1998) (Fig. 1). CCA1 and LHY are two myb domain-containing transcription factors, and exhibit a rhythm in their abundance with peak expression during the morning. Recently, Alabadi et al (2001) have suggested that they function as transcriptional repressors of TOC1, a pseudo response regulator thought to comprise the positive activating arm of the *Arabidopsis* circadian clock, thus constituting a basic transcriptional/translational feedback loop. A phytochrome-interacting protein, PIF3, is a bHLH transcription factor and binds to its target binding site in the promoter regions of CCA1, thus activating their transcription in response to light (Halliday et al 1999, Ni et al 1999). Interestingly, light input to the oscillator may also have a different target in the core oscillator, TOC1, as genetic analysis of TOC1 suggests that it functions in red light input to the clock. Furthermore, over-expression as well as under-expression of TOC1 in seedlings leads to arrhythmicity under conditions of constant light and darkness—a classic phenotype of a disrupted clock. However, seedlings expressing reduced levels of TOC1 exhibit a red light specific period alteration, which is indicative of its role in red light input to the clock.

Light input in *Drosophila*

In flies, the master circadian oscillator resides in a well-defined group of lateral neurons, and generates most of the overt rhythms in physiology including locomotor activity. Under normal light–dark (LD) conditions, the locomotor activity profile of adult flies exhibits two peaks, one tracking lights-on (dawn) and the other tracking lights-off (dusk). With a shift in the LD regime, the locomotor activity rhythm re-entrains within a day or two. Genetic analyses have shown that at least three input mechanisms exist originating from the compound eyes as well as from a cell autonomous deep-brain photoreceptor (Fig. 2). These systems function in tandem in the entrainment of the master oscillator.

Evidence for the involvement of ocular photoreception came from the characterization of the *norpA* mutation in flies, which renders them eyeless and devoid of ocellar function. Flies harbouring this mutation take a longer time to re-entrain their activity rhythms to a shifted LD cycle (Helfrich-Forster et al 2001). However, the observation that they still entrain suggested involvement of a second system in circadian photoperception.

A second system contributing to light entrainment was elucidated by the identification of cryptochrome, a clock component. Characterization of a fly mutant, *cry^{baby}*, showed that they also take a longer time to re-entrain their activity rhythms to changes in the light regime. Evidence that cryptochrome is a photoreceptor comes in part from the observation that the *cry^{baby}* mutation occurs in a well-conserved amino acid in the flavin-binding domain of the cryptochrome protein (dCRY). Strikingly, under constant darkness, a brief pulse of light does not phase shift the activity rhythm of *cry^{baby}* flies at all. Instead, these flies continue to

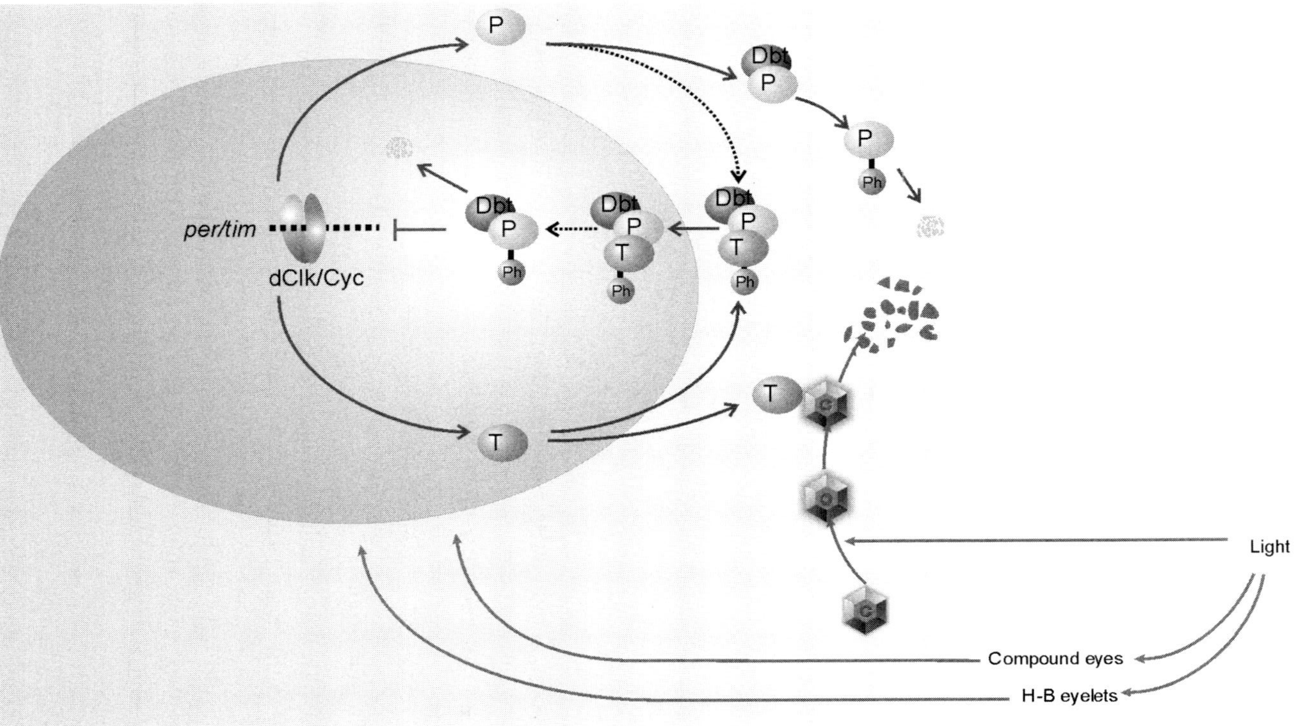

FIG. 2.   Circadian light input in *Drosophila*. At least three different circadian photoperception pathways have been described, out of which the cell autonomous *Cry*-mediated cascade is well characterized. dCRY, upon light activation binds to dTIM and facilitates its degradation via the proteasome pathway. TIM degradation, subsequently leads to degradation of PER, which is unstable in monomeric form. Degradation of the repressor components that are usually low during the subjective day resets the oscillator. The molecular target(s) of light input mediated by the compound eyes or the H–B eyelet has (have) yet to be identified.

exhibit unchanged activity rhythms. Furthermore, under constant light, which otherwise causes arrhythmia in the wild-type flies, the *cry^baby* flies retained rhythmic locomotor activity (Stanewsky et al 1998). Although these studies indicated that both cryptochrome and ocular photoreceptors play overlapping roles in circadian photoentrainment, the residual entrainment in the absence of either one to changed LD regimes (more than a brief pulse of light) suggested additional mechanisms for light input to the clock.

This third photic-input pathway was implicated by the observation that *cry^baby norpA* double mutants also entrain and re-entrain to LD regimes, although they take even longer to do so. Characterization of *glass* mutant flies, where both ocular and extraocular eyes in the Hofbauer–Buchner (H–B) eyelets are missing, have suggested the location, but not identity, of photoreceptors constituting this third system. In these experiments, *glass^60j cry^baby* mutants were found to completely lose their ability to photoentrain (Helfrich-Forster et al 2001). Because *cry^baby norpA* double mutants retain some residual photoentrainment, while *glass^60j cry^baby* double mutants do not, the H–B eyelets have been suggested to harbour this third photic input pathway.

The nature of the photoreceptor and the mechanism of light-induced signal transduction from the compound eyes and H–B structures to the master oscillator resident in the lateral neurons are still unknown. However, the role of dCRY in the clock has been characterized (Fig. 2). Light activated dCRY binds to dTIM and targets it for degradation (Ceriani et al 1999, Lin et al 2001, Naidoo et al 1999). Since the monomeric dPER is unstable in absence of dTIM, dPER degradation results. A light pulse, therefore, degrades the dTIM/dPER repressor complex, resulting in activation of the positive arm of the fly clock, the dCLOCK/ dCYCLE complex, and underlying the molecular basis of circadian entrainment in flies.

Light input in mammals

The location of the master circadian oscillator in mammals is in the suprachiasmatic nucleus (SCN) of the hypothalamus, making it highly unlikely that a deep brain photoreceptor would serve as a cell autonomous circadian photoreceptor like dCRY (how would light reach the SCN of an elephant?). Instead, ablation and enucleation studies have shown that ocular photoreceptors signal via the retinohypothalamic tract (RHT) to the SCN, resulting in appropriate changes to oscillator components and concomitant photoentrainment (Fig. 3) (Pando & Sassone-Corsi 2001). We are just beginning to understand the complexity of circadian light input in mammals.

Several candidate photoreceptors have been suggested as mediating circadian photoentrainment including visual photoreceptors, cryptochromes and

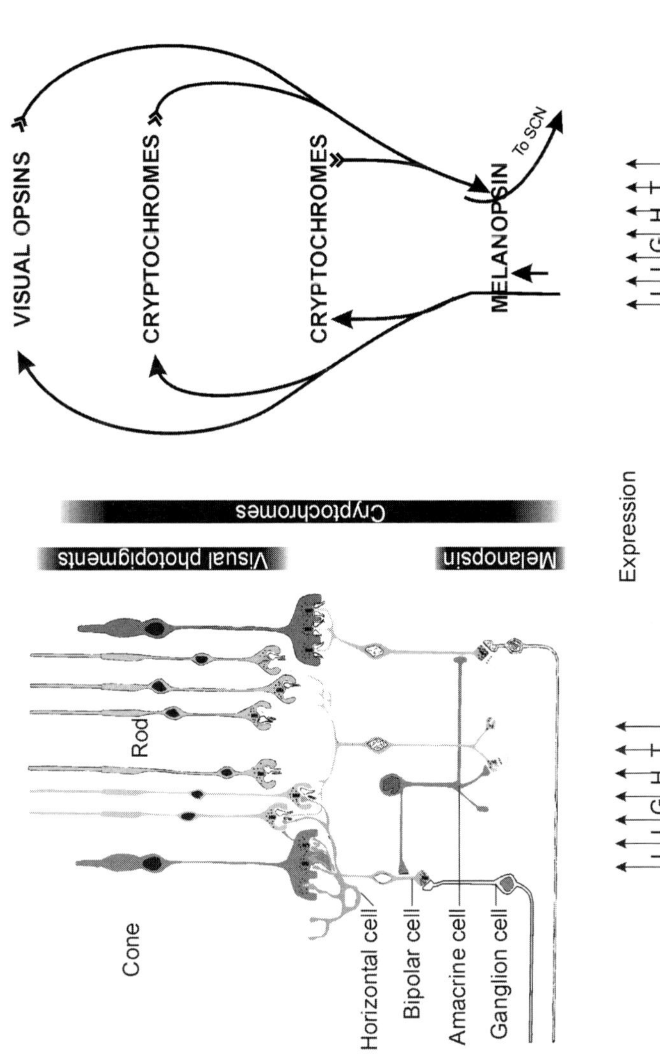

FIG. 3.   Circadian light input in mammals. Ocular photoreceptors mediate light entrainment of the master circadian oscillator in the SCN. Different cell types of the retina and the expression of visual photoreceptors and of CRYs are shown. Retinal light input to the SCN is transmitted via a small group of retinal ganglion cells, most of which exclusively express a putative photopigment melanopsin. CRYs and visual photoreceptors participate in some transcriptional changes in the SCN in response to light. However, loss of visual photoreceptors does not have any significant effect on light entrainment. Loss of melanopsin attenuates light induced phase shifts. The residual photic entrainment in such mice may be mediated by CRYs, visual photoreceptors or even a yet unknown photoreceptor.

melanopsin. Arguing against the involvement of visual photoreceptors (rods and cones) in photoentrainment in mammals are the observations that some visually blind people and mice (such as *rd* mice) display apparently normal responses to light (Czeisler et al 1995, Freedman et al 1999). Evaluation of the role of cryptochrome in photoentrainment has proven more complicated. In mammals, mCRYs are cytoplasmic (and possibly nuclear proteins) that are expressed in both the outer and inner nuclear layer (ONL and INL), as well as in the ganglion cell layer (GCL) of the retina (Sancar 2000). In addition, mCRYs are expressed in the SCN, and subsequent characterization of their biochemical function has shown that they can act as potent repressors of the core oscillator components, CLOCK and MOP3/BMAL (Griffin et al 1999).

The recent demonstration of light perception by melanopsin-containing retinal ganglion cells (RGCs) that directly project to the SCN has suggested melanopsin as an attractive candidate for a circadian photoreceptor (Berson et al 2002, Gooley et al 2001, Hannibal et al 2002, Hattar et al 2002, Provencio et al 1998, 2000). To test that hypothesis, several groups including ours have generated melanopsin (*Opn4*) null mice. To test any involvement of melanopsin in the core oscillator, $Opn4^{-/-}$ mice were entrained to a 12:12 LD cycle and subsequently released into constant darkness. Monitoring of their locomotor activity patterns revealed a normal tau of approximately 24 h, showing that melanopsin was not required for the normal functioning of the oscillator (unlike the cryptochromes). In addition, acute-light suppression of activity, or masking, was still intact in $Opn4^{-/-}$ animals. To characterize photoentrainment in melanopsin null mice, animals kept in DD were exposed to brief pulses of monochromatic light. Because electrophysiological responses of melanopsin positive RGCs have an action spectrum with a peak at 480 nm (Berson et al 2002), we assayed phase shift in activity rhythm of $Opn4^{-/-}$ animals in response to brief light pulses of this wavelength. These experiments revealed a requirement of melanopsin for normal responses to light. Importantly however, some residual photoentrainment was still apparent, implying that other photoreceptive systems may also be contributing. Although the experiments recounted above rule out the requirement of visual photoreceptors and cryptochromes for circadian photoentrainment, they do not rule out a situation where melanopsin may be epistatic to them. In that regard, generation of $Opn4^{-/-}$, *rd*, *Cry1*, *Cry2* animals and the various permutations will be of acute interest.

But how is light information transmitted to the oscillator in mammals? Unfortunately, the understanding of this process is only rudimentary at this time (Fig. 3). Both glutaminergic neurotransmission as well as pituitary adenylate cyclase-activating peptide (PACAP) signalling have been implicated in mediating the signal from retinal photoreceptors to the clock (Hannibal 2002). These signals result in induction of immediate early genes such as c-*Fos*, *Jun*, *Zif268* and many others, as well as *mPer1* and *mPer2*. This induction may be mediated by MAPK pathway- and CREB-mediated transcription (Obrietan et al 1998). Antisense against *mPer1* seems to block photoentrainment (Akiyama et al 1999). However, the *mPer1* knockout animal had apparently normal responses to light (Zheng et al 2001). How upregulation of the *Per* genes relates to modulation of the positive-activating arm of the mammalian clock is at this point unknown.

Conclusion

Investigation of circadian photoreception in plants, flies, and mammals is beginning to reveal some underlying themes. In none of the organisms, has the loss of a single photoreceptor or signalling intermediate completely abolished photic entrainment. Instead, multiple photoreceptive systems with overlapping absorption spectra may ensure photoreception under fluctuating spectral qualities in nature, and at the same time may offer a buffer against malfunction of any single light entrainment pathway. In that regard, future research may show that these multiple mechanisms impinge on the central oscillators via the same, or different, signal transduction pathways. It is slowly becoming apparent that multiple oscillator components may be targets of light resetting. These complex, highly redundant, systems highlight the critical importance of an organism's anticipation and adaptation to changes in their natural environment.

References

Akiyama M, Kouzu Y, Takahashi S et al 1999 Inhibition of light- or glutamate-induced mPer1 expression represses the phase shifts into the mouse circadian locomotor and suprachiasmatic firing rhythms. J Neurosci 19:1115–1121
Alabadi D, Oyama T, Yanovsky MJ, Harmon FG, Mas P, Kay SA 2001 Reciprocal regulation between TOC1 and LHY/CCA1 within the *Arabidopsis* circadian clock. Science 293:880–883
Berson DM, Dunn FA, Takao M 2002 Phototransduction by retinal ganglion cells that set the circadian clock. Science 295:1070–1073
Ceriani MF, Darlington TK, Staknis D et al 1999 Light-dependent sequestration of TIMELESS by CRYPTOCHROME. Science 285:553–556
Chory J 1993 Out of darkness: mutants reveal pathways controlling light-regulated development in plants. Trends Genet 9:167–172
Covington MF, Panda S, Liu XL, Strayer CA, Wagner DR, Kay SA 2001 ELF3 modulates resetting of the circadian clock in *Arabidopsis*. Plant Cell 13:1305–1315
Czeisler CA, Shanahan TL, Klerman EB et al 1995 Suppression of melatonin secretion in some blind patients by exposure to bright light. N Engl J Med 332:6–11
Freedman MS, Lucas RJ, Soni B et al 1999 Regulation of mammalian circadian behavior by non-rod, non-cone, ocular photoreceptors. Science 284:502–504
Gooley JJ, Lu J, Chou TC, Scammell TE, Saper CB 2001 Melanopsin in cells of origin of the retinohypothalamic tract. Nat Neurosci 4:1165
Griffin EA Jr, Staknis D, Weitz CJ 1999 Light-independent role of CRY1 and CRY2 in the mammalian circadian clock. Science 286:768–771

Halliday K J, Hudson M, Ni M, Qin M, Quail PH 1999 poc1: an *Arabidopsis* mutant perturbed in phytochrome signaling because of a T DNA insertion in the promoter of PIF3, a gene encoding a phytochrome-interacting bHLH protein. Proc Natl Acad Sci USA 96:5832–5837

Hannibal J 2002 Neurotransmitters of the retino-hypothalamic tract. Cell Tissue Res 309:73–88

Hannibal J, Hindersson P, Knudsen SM, Georg B, Fahrenkrug J 2002 The photopigment melanopsin is exclusively present in pituitary adenylate cyclase-activating polypeptide-containing retinal ganglion cells of the retinohypothalamic tract. J Neurosci 22:RC191

Harmer SL, Panda S, Kay SA 2001 Molecular bases of circadian rhythms. Annu Rev Cell Dev Biol 17:215–253

Hattar S, Liao HW, Takao M, Berson DM, Yau KW 2002 Melanopsin-containing retinal ganglion cells: architecture, projections, and intrinsic photosensitivity. Science 295: 1065–1070

Helfrich-Forster C, Winter C, Hofbauer A, Hall JC, Stanewsky R 2001 The circadian clock of fruit flies is blind after elimination of all known photoreceptors. Neuron 30:249–261

Lin FJ, Song W, Meyer-Bernstein E, Naidoo N, Sehgal A 2001 Photic signaling by cryptochrome in the *Drosophila* circadian system. Mol Cell Biol 21:7287–7294

Mas P, Devlin PF, Panda S, Kay SA 2000 Functional interaction of phytochrome B and cryptochrome 2. Nature 408:207–211

Naidoo N, Song W, Hunter-Ensor M, Sehgal A 1999 A role for the proteasome in the light response of the timeless clock protein. Science 285:1737–1741

Ni M, Tepperman JM, Quail PH 1999 Binding of phytochrome B to its nuclear signalling partner PIF3 is reversibly induced by light. Nature 400:781–784

Obrietan K, Impey S, Storm DR 1998 Light and circadian rhythmicity regulate MAP kinase activation in the suprachiasmatic nuclei. Nat Neurosci 1:693–700

Pando MP, Sassone-Corsi P 2001 Signaling to the mammalian circadian clocks: in pursuit of the primary mammalian circadian photoreceptor. Sci STKE 107:RE16

Provencio I, Jiang G, De Grip WJ, Hayes WP, Rollag MD 1998 Melanopsin: an opsin in melanophores, brain, and eye. Proc Natl Acad Sci USA 95:340–345

Provencio I, Rodriguez IR, Jiang G, Hayes WP, Moreira EF, Rollag MD 2000 A novel human opsin in the inner retina. J Neurosci 20:600–605

Quail PH 2002 Photosensory perception and signalling in plant cells: new paradigms? Curr Opin Cell Biol 14:180–188

Sancar A 2000 Cryptochrome: the second photoactive pigment in the eye and its role in circadian photoreception. Annu Rev Biochem 69:31–67

Somers DE, Devlin PF, Kay SA 1998 Phytochromes and cryptochromes in the entrainment of the *Arabidopsis* circadian clock. Science 282:1488–1490

Stanewsky R, Kaneko M, Emery P et al 1998 The cryb mutation identifies cryptochrome as a circadian photoreceptor in *Drosophila*. Cell 95:681–692

Wang ZY, Tobin EM 1998 Constitutive expression of the CIRCADIAN CLOCK ASSOCIATED 1 CCA1 gene disrupts circadian rhythms and suppresses its own expression. Cell 93:1207–1217

Yanovsky MJ, Mazzella MA, Casal JJ 2000 A quadruple photoreceptor mutant still keeps track of time. Curr Biol 10:1013–1015

Zheng B, Albrecht U, Kaasik K et al 2001 Nonredundant roles of the mPer1 and mPer2 genes in the mammalian circadian clock. Cell 105:683–694

DISCUSSION

[Editor's note: the first section of this discussion relates to data presented by Steve Kay in his oral presentation on the role of an exterior coincidence mechanism involving CONSTANS

and FLOWERING LOCUS T in regulating flowering timing in Arabidopsis (see Yanovsky & Kay 2002).]

Weitz: I have a question about the phase-specific features of CONSTANS (CO) function in *Arabidopsis*. Do you think this is circadian phase specific, or is it light specific? Is there some independent light pathway that regulates it?

Kay: It is external coincidence. Both are required. If you activate phytochrome A (*phyA*) and cryptochrome 2 (*cry2*) in wild-type plants at the end of the day in short days this has no effect on flowering because the phase of CO expression is such that it is at too low a level. It really seems to be entrainment regulating CO waveform. This is where Bunning in the 1940s was brilliant in proposing how this could work. This story will get even prettier when we fulfil the prediction that this light is leading to some post-translational modulation of CO protein or potentially a CO partner.

Takahashi: You have an interesting model for testing Pittendrigh's internal versus external coincidence idea. You can ask whether light is necessary by using T cycles driven by temperature. This is what Pittendrigh always wanted to do. You just have to drive for 21 or 28 h with temperature, change the phase and if it is internal coincidence you will get photoperiodic induction, if it is external you won't.

Kay: We have driven this before with temperature cycles, but not with the goal of measuring flowering time. We have no evidence for internal coincidence.

Rosbash: Does a negative response negate the hypothesis?

Takahashi: If he shows the molecular correlates, yes. This is the difference: you have something tangible to measure inside.

Kay: Exactly. That is what Marcelo Yanovsky is doing now. At the protein level we can see FLOWERING LOCUS T (FT) cycling. CO is incredibly low.

Dunlap: Over the course of the light/dark cycle light is acting through CO to chronically induce FT. So in long day you are getting more and more, until finally you get enough.

Kay: That is correct.

Dunlap: But it is happening at the light–dark transition, not the other way round. This is why I have a hard time understanding why this is called an 'acute effect'. Does the induction happen when the light is turned off?

Kay: The acute induction experiments were done under conditions to define the role of *phyA* and *cry2* in the light dependency of FT expression. This should be distinguished from plants growing under LD. When light is still there at the end of the day, because CO is now high enough, you are essentially acutely booting up FT levels. It is acute relative to the prior part of the day. In other words, FT comes rocketing up as long as CO is high.

Dunlap: What about the 10 min before the lights went off? Why didn't it acutely induce then? So it's not 'acute induction' by the light–dark transition that you mean, but just general light induction of FT during extended lights-on. Why wouldn't this happen during short light periods?

Kay: Because CO wouldn't have been high enough in short days. The light induction is occurring in the late part of the day; not at the transition.

Menaker: Can you do T cycles with short light pulses? That should answer this question.

Kay: Yes.

Dunlap: This is where I was sort of going. It is not the resonance experiments but the dark induction experiments where you can keep plants in constant darkness and give them short pulses of light at circadian intervals, and get photoperiodic induction.

Kay: That is true for absolute photoperiodic plants. But unfortunately *Arabidopsis* is a facultative long-day plant and it seems to require several rounds of signalling to do that. It is not clear to me that this type of resonance experiment is going to hold up for *Arabidopsis*.

Loros: Do you think that there is another player perhaps acting as a repressor on FT that CO might interact with?

Kay: This is more George Coupland's domain. He doesn't know exactly how CO induces FT expression. He has CO oestrogen receptor in cell culture, and if you give it oestrogen it translocates to the nucleus and you see FT come on. There is no evidence that CO binds DNA. No one knows what the partner is or whether this is a derepressor and so on.

Hastings: There are great parallels between this model and melatonin in mammals. The clock sets up the duration of expression of melatonin in the night, whereas CO is a day marker. The presence of light turns melatonin off, whereas light has a positive effect on CO. Apart from the signs being positive and negative, the properties seem similar. Would you predict that if there is constitutive over-expression of CO you would turn the plant into a permanent long-day response? Would it flower all the time no matter how short the days are?

Kay: George has done this experiment and it does. This hasn't been published yet though.

Ishida: You mentioned that *phyA* and *cry2* are important for photoperiodicity in long day plants. What about short day plants?

Kay: This comes back to the problem that we have great model systems for genetics which are not always the best photoperiodic responders. All I can say in answer to your question is that research on rice, which is a short-day plant, has really taken off. We know that CO cycles in rice and it is a night-time gene. All the same players seem to be there in rice. Where the polarity change occurs is unknown. There is no obvious key there in terms of RNA levels of CO and FT.

Sassone-Corsi: You mentioned that *phyB* interacts with *cry2* in the nuclear speckles. Is there colocalization?

Kay: This is a separate issue that we haven't yet tied into the photoperiodic story. We know that some of those speckles appear to be the proteasomes.

Sassone-Corsi: Does the position of those speckles change during the cycle?

Kay: There is a lot going on. It seems that anyone who studies signalling ends up studying a gigantic protein complex. I agree that the way to study this is to probe the composition of these protein complexes in a time-dependent way. We are lucky at Scripps to have John Yates who can do mass spectrometry profiling. We are now co-immunoprecipitating at different times of the day. These experiments are useful, but it is important to combine them with imaging. We found a couple of new phase-specific transcription factors like this, for example. Time-dependent analysis of protein complex formation will be a crucial technique.

Van Gelder: If I understand the TOC1 knockdown story correctly, you basically get an arrhythmic read-out except when you put the plants into blue light where you restore a *tau*. Is that correct?

Kay: We don't understand all of this. In constant white light the TOC1 null is a short period with robust rhythmicity. This is like the reference allele we found many years ago. In constant red light and DD it is arrhythmic. The key here is that it is arrhythmic, CAB goes really low and CCR2 goes really high. In constant darkness it is also arrhythmic but at a median level. I think what you are seeing in constant red light is that in the absence of TOC1 the light is pushing the clock into arrhythmicity. This is reminiscent of what we found with another gene called *ELF3*, where ourselves and Andrew Millar showed that the transcriptional repressor, ELF3, is required to stop light signals from hitting the clock at the end of the day. This actually gates phototransduction pathways. If you remove ELF3, again you get light-dependent arrhythmia. I suspect we will soon find ELF3 as being part of this competing protein complex between TOC1, phytocrome, PIF3 and ELF3. I can't explain all of these different phenotypes in the same way that you guys can't explain *Cry* single knockouts.

Van Gelder: This is reminiscent of mammalian work in which rhythmicity is restored to DD-arrhythmic *Clock* mutants and *Per* mutants in LL conditions.

Takahashi: These are Sergei Daan's results. I don't think this should be called a restoration because the original phenotype in clocks has a long period. The duration after which the mouse loses rhythmicity is variable, indicating that there are genetic modifiers. In a Bl6 background you can have a mouse that has a 28 h rhythm for three months in rare cases. The *Clock* phenotype has a variable interval before it goes arrhythmic. I would rather call this enhancing than restoration.

Kay: The *Arabidopsis* clock has a light-sensitive phase. If you don't gate that light-sensitive phase the clock stops. This is what I think is happening in the *Toc1* mutant.

Loros: It is important to point out that the *Clock* mutant in mouse is a partially functional mutation. When used in experiments people often interpret their results as if the mutation was a null, because they assume that 'mutant' equals null or lack of function.

Takahashi: I agree. May people use the incorrect notation $Clk^{-/-}$.

Loros: This is a continuing problem in the literature.

Takahashi: In our original phenotypic description of clock where we say that the rhythm disappears after about two weeks, this is the average of a group of animals on that particular genetic background.

Weitz: Didn't a single light pulse restore rhythmicity to the clock mutants?

Takahashi: Just the transition of going from DD to LL or LL to DD will induce rhythmicity.

Kay: It is confusing. There is presumably some average between the red-light effect and the blue-light effect which means that in white light you end up with short periods.

Takahashi: How short is the period in blue light?

Kay: It is 6 h.

Loros: Have you looked at temperature compensation of this 6 h rhythm?

Kay: Are you kidding?

Loros: No. You could see whether, with changing temperatures, you could move the period length towards the circadian range.

Kay: We have almost never done a compensation experiment. But you are right, this is worth doing. For us the priority is more the biochemistry of TOC1 and PIF3, competing with phytochrome.

Menaker: Would it be fair to say that there are analogies but not homologies among clock mechanisms in plants and other major groups of organisms?

Kay: Yes. It is clear that it is analogous. I wanted to mention the PAS proteins that we have because these connect so nicely to *Neurospora*, and are extremely homologous to the white collar phototropin-type PAS/LOV domain. There is that domain that is well conserved between *Neurospora* and plants.

Takahashi: What about cryptochrome?

Kay: We have to be very careful about cryptochrome. Cashmore has made the observation that he thinks the cryptochromes are more similar to their own photolyases than to each other, and has used this observation to suggest that cryptochromes have each arisen independently. It is not surprising that cryptochromes could have arisen independently and still be involved in clocks. One of the most stressful stimuli a cell can be exposed to is light. You can imagine one of the first clocks being built around light dependency of DNA repair. It is wrong to think of cryptochromes as a conserved element of clocks because they could have arisen independently.

Loros: When we look at evolution we have to consider it molecule by molecule. In a review you recently wrote with Mike Young (Young & Kay 2001), you discussed the independent evolution of fungal, plant and animal clocks. You said that although we have PAS domains that are similar, even the similarity of White Collar 1 and BMAL1 is limited. I beg to differ on that. The similarity between those two proteins extends throughout the entire BMAL protein and it is quite even over the entire protein. This clearly predicts some BMAL and White Collar 1 ancestor.

Kay: The problem is that the PAS domain is a good way of mediating protein–protein interactions, and therefore could have evolved convergently.

Loros: Certainly, but if you get rid of the PAS domains in White Collar 1, BMAL1 is still going to come up in a sequence-based homology search like BLAST.

Kay: I don't believe the HERG channel evolved from PYP, so Libby Getzoff and I fell out of our chairs after we did the threading and prediction of the PAS structure when McKinnon published the HERG channel and his crystal structure was exactly the PYP PAS domain module. This is absolutely convergent evolution. He only mentioned it as a PAS domain channel in passing. No one had really pulled out HERG as being a member of the PAS family.

Rosbash: You could back off the specifics of PAS domains in BMAL and White Collar. The more general question is whether we are in a position to say with confidence that these things did evolve multiple times, or are there very strong molecular connectors between systems? Ravi Allada and I have recently published a paper on CK2, a kinase and a new clock mutant (Lin et al 2002). CK2 is implicated in plant clocks, *Neurosopora* clocks and now in animal clocks. My feeling is that the jury is still out as to whether we have hit on the key common proteins or whether it is independent evolution.

Dunlap: I was writing this up for a textbook (Dunlap 2003) and the one point I could make for sure is that there are very few chronobiologists working now who were actually present 900 million years ago when the divergence happened, so it will all be speculation. What we can do is point out similarities and the differences, of which there are many. The similarities are strongest among the animals and next between animals and fungi which are, phylogenetically, closest relatives based on all including the newest phylogenies from Rogers (Simpson & Roger 2002). Both fungi and animals use heterodimers of PAS proteins as transcriptional activators, and we've shown that the sequences of WC1, and human BMAL1 are quite similar (Lee et al 2000), having a BLAST score better than 10^{-5}. They're not just a little similar, they are quite similar and, moreover, have the same role in the same aspect of metabolism, building a clock. It seems too much to have happened by accident.

Golden: We have to think about recruitment of molecules to provide a function. This is becoming obvious with respect to the particular folds that are available to be used for performing a particular function. There will be particular things that need to happen with respect to the clock machinery in terms of keeping time. As we find out more about the structure of the proteins things will clarify. One reason I am thinking more about recruitment of various molecules is that we are getting structural information from the cyanobacterial clock. We know that some very typical folds are present. In the cyanobacterial circadian system we have at least two cases of a *pseudo*-receiver domain, which is also present in TOC1, but this doesn't meant that *Arabidopsis* got it from its cyanobacterial forebear. A pseudo-receiver is present in the N-terminus of KaiA. But this can't be seen from looking at the sequence: you have to get the structure. Receivers have been studied for years in the bacterial field as signal transduction phosphotransfer modules, and this is not what we see these receivers doing. Folds of proteins are multifunctional. There is so much plasticity, and a limited number of folds. These organisms have had a long time to recruit something that works. Structure will help us a lot in answering these questions.

Young: I seem to remember that several years ago Wally Gilbert calculated how many protein domains one could start with and build all current proteins, and he came up with a figure of 600. If we believe numbers like this, everything is wide open.

References

Dunlap JC 2003 Molecular biology of circadian pacemaker systems. In: Dunlap JC, Loros JJ, Decoursey P (eds) Chronobiology: biological timing. Sinauer Assoc, Sunderland MA, p 210–251

Lee K, Loros JJ, Dunlap JC 2000 Interconnected feedback loops in the *Neurospora* circadian system. Science 289:107–110 [Erratum in Science 290:277]

Lin JM, Kilman VL, Keegan K et al 2002 A role for casein kinase 2α in the *Drosophila* circadian clock. Nature 420:816–820

Simpson A, Roger AJ 2002 Eukaryotic evolution: getting to the root of the problem. Curr Biol 12:R691–R693

Yanovsky MJ, Kay SA 2002 Molecular basis of seasonal time measurement in *Arabidopsis*. Nature 419:308–312

Young MW, Kay SA 2001 Time zones: a comparative genetics of circadian clocks. Nat Rev Genet 2:702–715

Orphan nuclear receptors, molecular clockwork, and the entrainment of peripheral oscillators

Nicolas Preitner*†, Steven Brown*, Juergen Ripperger*, Nguyet Le-Minh*‡,
Francesca Damiola*¶ and Ueli Schibler*[1]

*Department of Molecular Biology and NCCR Frontiers of Genetics, Sciences II, University of
Geneva, 30, Quai Ernest Ansermet, CH-1211 Geneva, Switzerland, †Department of Cell
Biology and Program in Neuroscience, Harvard Medical School, 240 Longwood Avenue, Boston,
MA 02115, ‡Institute for Molecular and Cellular Biology, 30 Medical Drive, Singapore
117609 and ¶Centre de génétique moléculaire et cellulaire, UMR CNRS 5534, Université
Claude Bernard Lyon I, 43, boulevard du 11 Novembre 1918, 69622 Villeurbanne Cedex,
France

Abstract. Here we summarize our work on two aspects of circadian timing: the roles of
orphan nuclear receptors in the molecular clockwork, and phase entrainment of
peripheral oscillators. With reference to the former, studies on *cis*-acting regulatory
elements within the *Bmal1* promoter revealed that REV-ERBα, an orphan nuclear
receptor provides a link between the positive and negative limbs of the molecular
oscillator. Specifically, REV-ERBα controls the cyclic transcription of *Bmal1* and *Clock*,
the positive limb components. In turn, the circadian expression of *Rev-Erbα* itself is
driven directly by the molecular oscillator: it is activated by BMAL1 and CLOCK, and
repressed by PERIOD1/2 and CRYPTOCHROME1/2 proteins (the negative limb
members). With regard to phase entrainment, it was initially believed that only the
suprachiasmatic nucleus (SCN) was capable of generating circadian rhythms. However,
circadian oscillators have recently been discovered in many peripheral tissues. In the
absence of a functional SCN pacemaker, these peripheral clocks dampen after a few
days. Hence, the SCN must periodically synchronize these subsidiary timekeepers. It
may accomplish this task mostly through an indirect route: namely, by setting the time
of feeding. In addition to feeding cycles, body temperature rhythms and cyclically
secreted hormones might also serve as *zeitgebers* for peripheral clocks.

*2003 Molecular clocks and light signalling. Wiley, Chichester (Novartis Foundation Symposium
253) p 89–101*

---

[1]This paper was presented at the symposium by Ueli Schibler, to whom correspondence should
be addressed.

Our laboratory started to work on circadian rhythms by serendipity, while studying the liver-specific transcription of the serum albumin gene. We isolated a cDNA copy for a transcription factor that we dubbed DBP (for albumin promoter D-element Binding Protein). DBP, a basic leucine zipper (bZip) transcription factor, is the founding member of the PAR (proline-acidic amino acid rich)-domain bZip transcription factors, a small subfamily of bZip proteins consisting of DBP, TEF and HLF. It turned out that DBP protein and mRNA accumulation undergo circadian cycles with amplitudes in excess of one hundred-fold (Wuarin & Schibler 1990).

Initially, we assumed that cyclic *Dbp* expression in peripheral organs was the direct consequence of rhythmic hormone levels in the blood. Thus, we imagined that deciphering the relevant hormones and their responsive elements within cis-acting regulatory *Dbp* elements would reveal the mechanism controlling circadian *Dbp* transcription. In fact, the mechanism turned out to be quite different: *Dbp* expression cycles are governed not by extracellular hormones, but rather by cell-autonomous molecular oscillators present in most peripheral tissues (Balsalobre et al 1998, Ripperger et al 2000). These independent circadian clocks have a molecular makeup similar to those of SCN neurons (Yagita et al 2001). Nevertheless, in contrast to the master pacemaker residing in the SCN, peripheral cell oscillators dampen after a few days. Hence, peripheral clocks must be periodically entrained by the SCN (for review see Reppert & Weaver 2002), and these clocks in turn control oscillations of DBP transcription.

The search for *cis*-acting elements within the 5′-flanking region of *Dbp* revealed two evolutionary conserved DNA elements of the type RGGTCA (where R is A or G). Such elements are known to participate in the binding of a series of nuclear hormone and orphan receptors. The closer inspection of these two elements identified both of them asROREs, binding sequences for members of the two small orphan nuclear receptor families ROR (Retinoic acid receptor-related Orphan Receptor, comprising RORα, RORβ, and RORγ) and REV-ERB (comprising REV-ERBα and REV-ERBβ). While RORs are transcriptional activators, REV-ERBs act as transcriptional repressors (Forman et al 1994, Jetten et al 2001). Biochemical studies revealed three prominent protein:RORE complexes, one of which accumulated in a highly rhythmic fashion. The accumulation of this protein, identified as REV-ERBα, closely paralleled the phase of rhythmic *Dbp* transcription rates. Disappointingly, however, subsequent genetic experiments demonstrated that REV-ERBα has little if any influence on circadian *Dbp* transcription.

Even though the studies on ROREs within the *Dbp* promoter did not teach us much about *Dbp* transcription, we did uncover an important function of circadian REV-ERBα accumulation: this repressor governs the cyclic transcription of *Bmal1* and *Clock*, two central components of the molecular clock (Preitner et al 2002).

Orphan nuclear receptors and circadian rhythms

REV-ERBα: a link between the positive and negative limb of the rhythm generating feedback loop

The mammalian molecular oscillator, similar to that in *Drosophila* and *Neurospora*, is composed of two interconnected feedback loops, one within the negative limb and one within the positive limb. The members of the positive limb (CLOCK and BMAL1) activate transcription of the genes encoding the negative limb components (cryptochromes and period proteins). PER and CRY proteins then form heterotypic protein complexes that are translocated into the nucleus, and once these complexes reach a critical threshold level, they suppress the activity of CLOCK and BMAL1. As a consequence, the concentration of CRY and PER proteins falls below the threshold required for autorepression, and a new cycle of *Cry/Per* transcription can initiate (for review, see Reppert & Weaver 2002).

CRY and PER proteins not only repress the activity of their own genes, but they also stimulate the expression of BMAL1 and CLOCK. How can the CRY and PER repressors activate the transcription of these positive limb components? The most likely scenario would imply a hitherto unknown repressor whose gene is under the negative control of CRYs and PERs.

The inspection of the proximal *Bmal1* promoter sequence revealed that this extraordinarily well conserved sequence (identical over 170 nucleotides in mouse, rat and human) contains two RORE elements, binding sites for members of the REV-ERB and ROR orphan nuclear receptors. Protein–DNA binding studies with these RORE sequences and liver nuclear extracts identified three major proteins occupying these elements with high affinity: REV-ERBα, RORγ and RORα. The high-amplitude accumulation cycle of REV-ERBα was found to be in direct antiphase with *Bmal1* transcription rates, a situation that would be compatible with a role for REV-ERBα as a repressor of *Bmal1* transcription. This conjecture was directly confirmed in *Rev-erbα*-deficient mice, because *Bmal1* transcription and mRNA accumulation are constitutively high in these animals. REV-ERBα is not only a critical regulator of cyclic *Bmal1* transcription, but also governs the approximate twofold oscillation in *Clock* mRNA synthesis (Preitner et al 2002).

Subsequent studies revealed that the transcription of *Rev-erbα* is regulated by essentially the same mechanisms as those of *Per* and *Cry* genes: it is activated by BMAL1 and CLOCK and repressed by PER and CRY. Therefore, REV-ERBα directly connects two antiphasic feedback loops within the positive and negative limbs of the oscillator (Fig. 1).

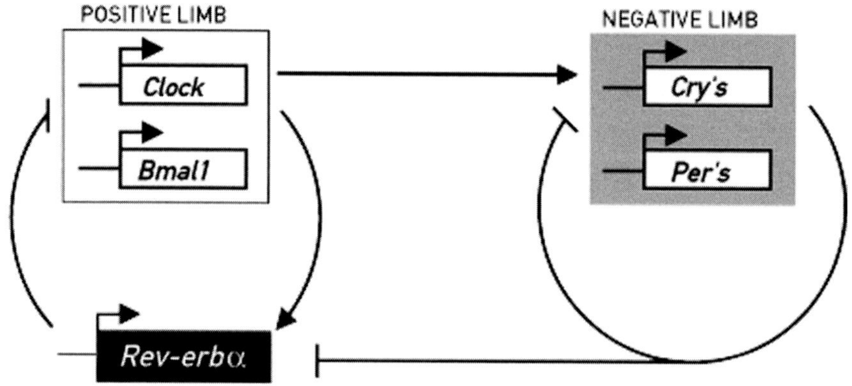

FIG. 1.   The orphan nuclear receptor REV-ERBα interconnects the positive and negative limbs of the molecular oscillator. The components of the negative limb, CRYs and PERs, repress transcription from their own genes by interfering with activation by CLOCK and BMAL1, the positive limb components. Circadian *Rev-erbα* transcription is regulated by the same mechanisms as for *Per* and *Cry* transcription, and REV-ERBα periodically represses *Bmal1* and to a lesser extent *Clock* transcription. This web of interactions leads to the anticyclic expression of positive and negative limb components (reproduced from Preitner et al 2002, with permission from Elsevier).

What are the activators of Bmal1 and Clock transcription?

As REV-ERB and ROR family members are respectively repressors and activators that bind to the same element, it is likely that REV-ERB and ROR proteins act antagonistically on the two RORE elements within the *Bmal1* promoter. In liver, both REV-ERBα and REV-ERBβ display a cyclic accumulation, although the latter is expressed at much lower levels than the former. Liver cells also express RORα and RORγ, two of the three ROR members (Preitner et al 2002). RORβ, the third member of the family, is only detectable in the brain (including the SCN). Lazar and colleagues have discovered that transcriptional repression by REV-ERBα requires two RORE elements, because two DNA-bound REV-ERBα molecules are required for recruitment of the corepressor NcoR1 (Zamir et al 1997). Our studies demonstrate that REV-ERBα efficiently represses transcription from the *Bmal1* promoter, in which the two direct RORE repeats are spaced by 25 nucleotides.

In contrast to the repression by REV-ERB proteins, transcriptional activation by members of the ROR orphan receptors may require only a single RORE (Sundvold & Lien 2001, Zamir et al 1997). We thus propose that the competitive binding of ROR and REV-ERB members to RORE elements accounts for the cyclic transcription of *Bmal1* (Fig. 2). ROR:coactivator complexes fill the two RORE elements at times when REV-ERB proteins are present at nadir levels, but are displaced by the cooperatively binding [REV-ERB]2:NcoR1 complexes

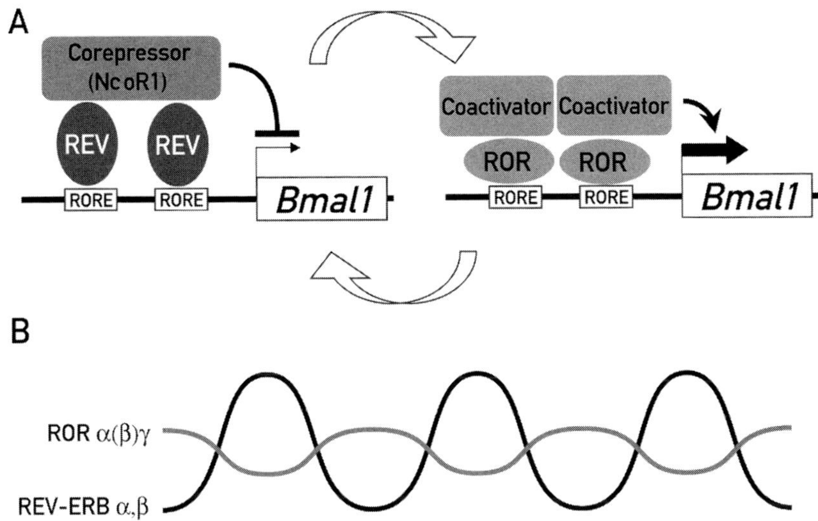

FIG. 2.   Speculative model on the competitive binding of REV-ERB repressors and ROR activators to the two ROREs within the *Bmal1* promoter. Two REV-ERB (REV) monomers, when bound to the two ROREs, recruit the corepressor NcoR1. This leads to cooperative repressor–corepressor binding on the *Bmal1* promoter. At nadir concentrations of REV-ERB repressors, the RORE elements are filled with ROR activators, which individually recruit coactivators (e.g. GRIP1 and PBP, see Atkins et al 1999). In peripheral tissues such as the liver, RORα and RORγ are expressed, while in the SCN (and other brain regions) all three ROR isoforms (α, β, γ) may accumulate. In liver, RORγ displays low-amplitude circadian accumulation, while RORα is expressed at similar levels throughout the day. In the SCN, all three ROR isoforms may accumulate in a circadian fashion (Ueda et al 2002).

when REV-ERB proteins reach high levels. Hence, we speculate that ROR members are also important transcriptional regulators of *Bmal1* (and, perhaps, *Clock*) transcription.

Behavioural phenotypes in Rev-Erbα-deficient mice

Rev-erbα-deficient mice still display a strongly rhythmic expression of most clock and clock-controlled genes. Likewise, the locomotor activity of these animals is strongly circadian under entrained conditions (light–dark cycles, LD), in constant darkness (DD), and in constant light (LL). Therefore, REV-ERBα, and consequently rhythmic *Bmal1/Clock* expression and the coupling of the positive with the negative limb, are not essential for rhythm generation. Nevertheless, REV-ERBα fulfils three important functions in the mammalian timing system: (1) it participates in the determination of the period length; (2) it increases the precision of the circadian system (by reducing the noise); and (3) it greatly constrains light-induced phase advances during the late night (see Fig. 3) (Preitner et al 2002).

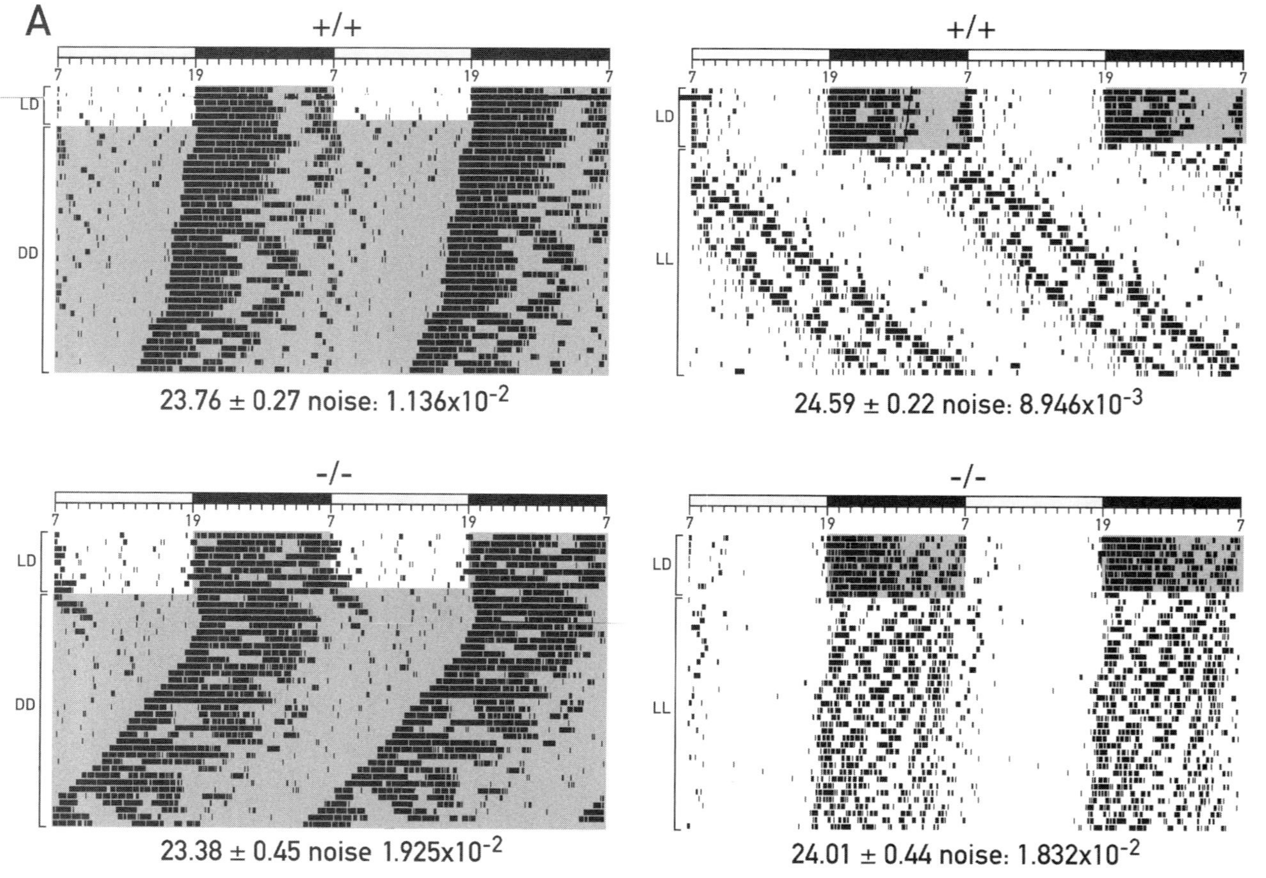
A
+/+
7
19
7
19
7
LD
DD
23.76 ± 0.27 noise: 1.136x10^{-2}
+/+
7
19
7
19
7
LD
LL
24.59 ± 0.22 noise: 8.946x10^{-3}
-/-
7
19
7
19
7
LD
DD
23.38 ± 0.45 noise 1.925x10^{-2}
-/-
7
19
7
19
7
LD
LL
24.01 ± 0.44 noise: 1.832x10^{-2}

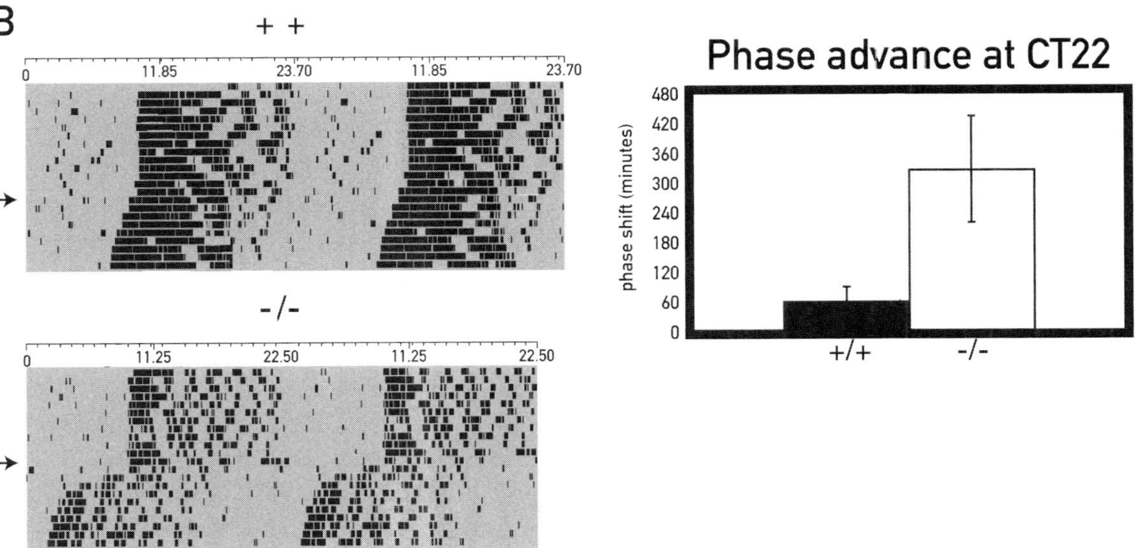

FIG. 3. Circadian phenotypes of *Rev-Erbα* deficient mice. A. *Rev-Erbα*-deficient mice (−/−) display circadian activity patterns with a shorter period length than those of wild-type mice (+/+) kept in constant darkness (DD) and constant light (LL). The mean period lengths (M) ± standard deviation (SD) and the noise levels (= SD/M) are shown below the panels. Note that the noise is about twice as high in *Rev-Erbα* [−/−] as compared to *Rev-Erbα* [+/+] mice! (B) Light pulses (500 lux during two hours) delivered to mice during the second half of the night (CT 23) provoke large phase shifts of five to six hours in *Rev-Erbα* [−/−] mice, but only small phase shifts of 60 min in *Rev-Erbα* [+/−] mice (adapted from Preitner et al 2002, with permission from Elsevier).

Entrainment of peripheral clocks

Feeding time is the dominant zeitgeber

How does the master pacemaker in the SCN maintain the amplitude and phase of oscillating gene expression in peripheral tissues? Our studies, along with those of the Menaker group, have demonstrated that feeding time appears to be a dominant *zeitgeber* for many if not most peripheral oscillators (Damiola et al 2000, Stokkan et al 2001). Thus, if mice or rats, which are nocturnal rodents, are fed exclusively during the day for seven to 10 days, the phase angle of circadian gene expression in liver, heart, kidney and pancreas changes by 180°. In contrast, feeding time has little if any influence on the phase of cyclic gene expression in the SCN, irrespective of whether the animals are kept in LD or DD. It is thus likely that the SCN entrains circadian oscillators primarily by driving rest–activity cycles, which in turn determine feeding time (Fig. 4). The chemical nature of the timing cues provoked by feeding and/or food processing are not yet known. Conceivably, they might include gastrointestinal hormones (for review see Rehfeld 1998), metabolites such as glucose (Hirota et al 2002), and changes in intracellular concentrations of reduced and oxidized nicotinamide adenine dinucleotides (Rutter et al 2001).

The role of glucocorticoid hormones

For a number of reasons, cyclically secreted glucocorticoid hormones have been considered as likely timing cues for the entrainment of peripheral clocks. However, our experiments with mice in which the glucocorticoid repressor gene was inactivated specifically in the liver demonstrated that glucocorticoid signalling is not essential for phase regulation of circadian clocks in peripheral tissues (Balsalobre et al 2000). However, glucocorticoid hormones do play an important role in the context of the food-dependent entrainment of peripheral clocks. In daytime-fed animals deficient of glucocorticoid signalling, the kinetics of phase inversion are much faster than in control mice (Le Minh et al 2001). Hence, in intact mice glucocorticoids counteract the feeding-induced uncoupling of peripheral clocks from the central oscillator (see Fig. 4).

Body temperature rhythms

In mammalian tissue culture cells, robust circadian gene expression can be entrained by 12 h temperature cycles with an amplitude of 4 °C (e.g. 37 °C versus 33 °C) (Brown et al 2002). We thus wondered whether physiological temperature rhythms, themselves circadian and with an amplitude of 1–4 °C in most mammals, could also sustain cyclic clock-gene transcription. To this end, we engineered a

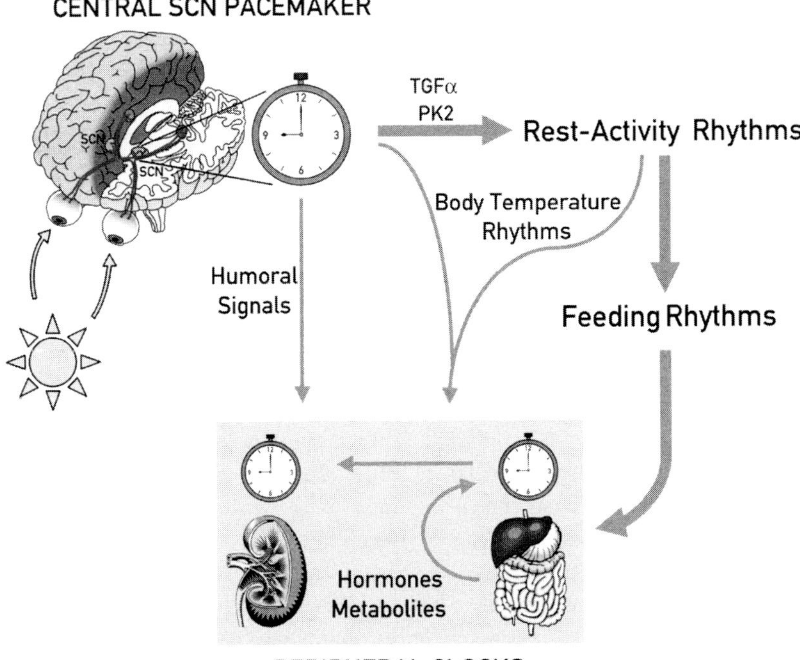

FIG. 4.   Phase entrainment of peripheral oscillators. The SCN, whose pacemaker is entrained by daily changes in light intensity, synchronizes peripheral clocks mostly by determining the phase of rest–activity cycles, and thus feeding time. Transforming growth factor alpha (TGFα) and prokineticin 2 (PK2) are strong candidates for diffusible signals involved in the timing of rest–activity cycles (Cheng et al 2002, Kramer et al 2001). The food-related signals synchronizing peripheral clocks have not yet been identified, but they may involve gastrointestinal hormones, changes in redox potential, or metabolites. Humoral signals whose cyclic secretion is controlled by the SCN, and body-temperature rhythms, may also contribute to the phase entrainment of peripheral clocks (reproduced from Schibler et al 2003, with permission from SAGE).

computer-driven incubator that is capable of reproducing physiological body temperature profiles recorded from the peritonea of mice with a precision of 0.1 °C in tissue culture dishes. These temperature oscillations, unlike step gradients, cannot entrain circadian gene expression *de novo*, but they can prolong the oscillation of already-entrained gene expression rhythms (Brown et al 2002).

The phase of gene expression in peripheral tissues of intact animals can also be inverted by temperature, by exposing animals to 'cold days' and 'hot nights' (the contrary of what is usually encountered in natural habitats) to reverse the pattern of body temperature in these animals. These experiments are, however, somewhat more difficult to interpret, since the external housing temperature influences feeding rhythms as well as body temperature. Nevertheless, the temperature-driven changes in the feeding behaviour that we observed were insufficient to account for the 180° phase-resetting behaviour of peripheral clocks in these animals. Therefore, we propose that body temperature cycles may also contribute to determining the steady-state phase of peripheral oscillators.

Conclusions

Our work on mammalian circadian oscillators was initially stimulated by studies on the circadian transcription factor DBP. Identifying binding sites for nuclear factors within the *Dbp* promoter region then led to research on the ROR and REV-ERB families of transcriptional regulatory proteins. The study of these nuclear orphan receptors in turn shed light on the genetic circuitry coupling the feedback loops within the negative and positive limbs of the circadian oscillator.

The discovery of peripheral oscillators in mammals has raised the question of how the SCN master pacemaker synchronizes these subsidiary clocks. Current evidence suggests that feeding time is the dominant *zeitgeber* for most peripheral oscillators. The molecular mechanisms involved in food-dependent phase resetting are not yet understood, and deciphering the signalling pathways involved will probably require many additional years of intense genetic and biochemical research.

Acknowledgements

We thank Nicolas Roggli for preparing the artwork. This work was supported by grants from the Swiss National Science Foundation (through an individual grant to U.S. and the NCCR program *Frontiers in Genetics*), the State of Geneva, the Louis-Jeantet Foundation for Medicine, and the Bonizzi-Theler Stiftung.

References

Atkins GB, Hu X, Guenther MG, Rachez C, Freedman LP, Lazar MA 1999 Coactivators for the orphan nuclear receptor RORα. Mol Endocrinol 13:1550–1557
Balsalobre A, Damiola F, Schibler U 1998 A serum shock induces circadian gene expression in mammalian tissue culture cells. Cell 93:929–937
Balsalobre A, Brown SA, Marcacci L et al 2000 Resetting of circadian time in peripheral tissues by glucocorticoid signaling. Science 289:2344–2347
Brown SA, Zumbrunn G, Fleury-Olela F, Preitner N, Schibler U 2002 Rhythms of mammalian body temperature can sustain peripheral circadian clocks. Curr Biol 12:1574–1583
Cheng MY, Bullock CM, Li C et al 2002 Prokineticin 2 transmits the behavioural circadian rhythm of the suprachiasmatic nucleus. Nature 417:405–410
Damiola F, Le Minh N, Preitner N, Kornmann B, Fleury-Olela F, Schibler U 2000 Restricted feeding uncouples circadian oscillators in peripheral tissues from the central pacemaker in the suprachiasmatic nucleus. Genes Dev 14:2950–2961

Forman BM, Chen J, Blumberg B et al 1994 Cross-talk among ROR alpha 1 and the Rev-erb
    family of orphan nuclear receptors. Mol Endocrinol 8:1253–1261
Hirota T, Okano T, Kokame K, Shirotani-Ikejima H, Miyata T, Fukada Y 2002 Glucose down-
    regulates Per1 and Per2 mRNA levels and induces circadian gene expression in cultured Rat-1
    fibroblasts. J Biol Chem 277:44244–44251
Jetten AM, Kurebayashi S, Ueda E 2001 The ROR nuclear orphan receptor subfamily: critical
    regulators of multiple biological processes. Prog Nucleic Acid Res Mol Biol 69:205–247
Kramer A, Yang FC, Snodgrass P et al 2001 Regulation of daily locomotor activity and sleep by
    hypothalamic EGF receptor signaling. Science 294:2511–2515
Le Minh N, Damiola F, Tronche F, Schutz G, Schibler U 2001 Glucocorticoid hormones inhibit
    food-induced phase-shifting of peripheral circadian oscillators. EMBO J 20:7128–7136
Preitner N, Damiola F, Lopez-Molina L et al 2002 The orphan nuclear receptor REV-ERBalpha
    controls circadian transcription within the positive limb of the mammalian circadian
    oscillator. Cell 110:251–260
Rehfeld JF 1998 The new biology of gastrointestinal hormones. Physiol Rev 78:1087–1108
Reppert SM, Weaver DR 2002 Coordination of circadian timing in mammals. Nature 418:
    935–941
Ripperger JA, Shearman LP, Reppert SM, Schibler U 2000 CLOCK, an essential pacemaker
    component, controls expression of the circadian transcription factor DBP. Genes Dev
    14:679–689
Rutter J, Reick M, Wu LC, McKnight SL 2001 Regulation of clock and NPAS2 DNA binding
    by the redox state of NAD cofactors. Science 293:510–514
Schibler U, Ripperger J, Brown SA 2003 Peripheral circadian oscillators in mammals: time and
    food. J Biol Rhythms 18:250–260
Stokkan KA, Yamazaki S, Tei H, Sakaki Y, Menaker M 2001 Entrainment of the circadian clock
    in the liver by feeding. Science 291:490–493
Sundvold H, Lien S 2001 Identification of a novel peroxisome proliferator-activated receptor
    (PPAR) gamma promoter in man and transactivation by the nuclear receptor RORalpha1.
    Biochem Biophys Res Commun 287:383–390
Ueda HR, Chen W, Adachi A et al 2002 A transcription factor response element for gene
    expression during circadian night. Nature 418:534–539.
Wuarin J, Schibler U 1990 Expression of the liver-enriched transcriptional activator protein
    DBP follows a stringent circadian rhythm. Cell 63:1257–1266
Yagita K, Tamanini F, van der Horst GT, Okamura H 2001 Molecular mechanisms of the
    biological clock in cultured fibroblasts. Science 292:278–281
Zamir I, Zhang J, Lazar MA 1997 Stoichiometric and steric principles governing repression by
    nuclear hormone receptors. Genes Dev 11:835–846

DISCUSSION

Sehgal: What happens with the rhythm phenotype in the *Dbp/Hlf/Tef* triple
knockouts?

Schibler: It is very interesting. In *Drosophila* there is a single orthologue of these
transcription factors, called *par domain protein 1*. Mutations in this gene have a very
strong phenotype. Mice with *Dbp* null alleles have a 30 min period shortening. In
contrast, mice with an *Hlf* or *Tef* mutation display a 30 min longer period than
wild-type mice. But the triple knockout mice have a wild-type period. Thus,

during the evolution of mammals, the activity of PAR bZip proteins has been tuned so that they do not influence the molecular oscillator as an ensemble.

Young: What do you think might be compensating when *Rev-Erbα* is absent?

Schibler: The rhythm of the protein accumulation of CLOCK is flat, and BMAL1 is only two or threefold higher than in wild-type mice. I think the *Rev-Erbα–Clock/Bmal1* feedback loop is important in setting up the concentration of CLOCK and BMAL1. With regard to the huge light pulse-induced phase advance in *Rev-Erbα* mutant animals, we see this only in mice kept in constant darkness for at least two weeks. It means that with time, in the absence of the *Rev-Erbα* control system, the oscillator gets much more sloppy. I heard from Shin Yamazaki in Michael Menaker's lab that they find something similar with the *tau* hamsters.

Menaker: The phase shift only happens after several weeks in constant darkness.

Schibler: So the entrainment system becomes different after the animals are left in DD for a long time.

Hardin: From your ROR EMSA experiments it still looked like it was rhythmic even though there was no band shift for REV-ERBα. Is this an ROR that is itself rhythmic?

Schibler: Absolutely. The ROR is not under the direct control of REV-ERBα, even though it has a similar phase as BMAL1.

Hardin: With regard to the mechanism here, are heterodimers forming between activators and repressors, or is it simply competition for the binding sites?

Schibler: REV-ERBα can bind as a monomer to hormone-binding half sites, plus an AT-rich region. Or it can bind as a dimer to a direct repeat-2 element of the type RGGTCANNRGGTCA, in which the NN is frequently CT. In this case, you only need the AT-rich sequence on the upstream half site. However, Mitch Lazar has shown that you can space two REV-ERBα molecules by quite a distance, and you need at least two monomers to bind the corepressor NcoR1. This cooperative binding of three proteins to two DNA elements will probably make the REV-ERBα-NcoR-DNA complex more stable than a complex composed of one ROR activator, one coactivator, and one DNA element.

Okamura: In the *Rev-Erbα* knockout mice, you showed that behavioural rhythms are sustained without a shorter period length. Do those animals show a flat *Cry* mRNA profile?

Schibler: No. The valley is much wider in the wild-type as opposed to the knockout animals, because it is only repressed when REV-ERBα is active during that narrow window.

Okamura: At the protein level, mCRY proteins show the robust rhythm very similar to mPER proteins. Do you speculate that the Cry transcription rhythm is not needed for the circadian rhythm?

Schibler: At least not a high-amplitude rhythm. I would assume that cyclic CRY accumulation is largely determined by protein–protein interactions with other cyclic proteins, such as PER1 and PER2.

Van Gelder: In some of your work you have shown that glucocorticoids are sufficient but not necessary for the entrainment of the peripheral oscillator for causing phase shifts. What happens to glucocorticoid rhythm in the starved animals or the voles? Is it still present but the liver ignores it, or is that output rhythm lost?

Schibler: I haven't looked at this, but it is a good experiment.

Rosbash: The voles remind me of strange animals that live where there is no light, or no change in light. People frequently ask me about this at seminars. Has anyone looked at these sorts of animals with modern molecular techniques?

Schibler: Urs Albrecht has worked on the Israelian blind mole.

Rosbash: Don't those animals have some sort of relationship with the light?

Schibler: As I remember from Albrecht's work, about 30% are night active and 70% are day active (or vice versa), and they can switch.

Rosbash: It wouldn't be surprising if they had all this circadian stuff going on.

Menaker: There are very few such situations that lend themselves to experiment.

Weitz: I am frequently asked about deep sea animals.

Hastings: Remember that many marine animals migrate up and down the water column, and this migration is under circadian control. Even though the deep sea is a continuously dark environment, the organisms that inhabit it will move vertically as a result of clock-controlled mechanisms to sample light as well. To study organisms completely devoid of light cues you would need to go to deep-sea hydrothermal vents.

Loros: There are amphibians or other vertebrates that live in caves that one could presumably take into the lab more easily.

Menaker: One of the best organisms to work on from this perspective is the blind cave fish. There are some that live in deep caves and never come out, but there is a lot of work on the evolution of the blind cave fish and fish with intermediate kinds of environments.

Rosbash: There are worm-like organisms inhabiting the edges of deep ocean hydrothermal vents. One of my ex-graduate students read in *National Geographic* about a researcher who collects these, and went on a trip with him. After collecting the worms, they freeze them immediately in liquid nitrogen and are now trying to crystallize proteins. I don't know why we couldn't look for clock genes in these deep-sea vent creatures. I would guess that expression is flat or they are not expressed.

General discussion II

Schibler: Bert van der Horst, did you ever try to food entrain a *Cry* double knockout? One suggestion has been that you wouldn't need negative factors if you could regulate the clock by redox.

van der Horst: No, we haven't tried yet. Given the fact that it has been shown that mutagenesis of the conserved tyrosines, thought responsible for intraprotein electron transport in the mCRY1 protein, does not prevent CRY1-mediated repression of clock *Bmal*-driven transcription it is unlikely that such redox mediated control involves CRY proteins. The idea was that you have CLOCK/ BMAL heterodimers bound to the E box promoter, and the redox status of the cell, more specifically the $NAD^+/NADH$ ratio, was influencing the binding state of BMAL. So indeed it would be a good idea to test whether limited availability of food, and as a result periodic changes in metabolism, may kick-start an hour-glass timekeeper.

Van Gelder: There is a confounder to that experiment, which is the food-entrainable oscillator that Fred Stephan has worked on. It is pretty mysterious. He sees in SCN-lesioned animals that food will create anticipation in a somewhat hour-glass way, but this is clearly anticipated at a 24 h level. We isolated the same oscillator in the same $math5^{-/-}$ mice that I talked about briefly. These mice don't have any retinal ganglion input to the SCN, so they don't entrain to light. They free run at about 24.5 h as opposed to 23.6 h. We wanted to bring these mice back into 24 h entrainment by some non-photic stimulus. We used wheel locking to restrict activity and try to entrain the clock back. We were unable to do this, although our heterozygotes entrained perfectly. The wild thing was that when we actually looked at their drinking activity as our surrogate measure of activity, very clearly these mice had an enormous anticipatory response. They would anticipate the wheel lock in their drinking activity and start drinking two hours before the wheel was released to them, yet their free-running period remained at 24.5 h. I think there are these sub-oscillators that are going to confound any attempt to try to do a central entrainment in something like the *Cry* delta mutant by food.

Menaker: It is not only the food anticipatory oscillator that may be extra-SCN, but there is also a methamphetamine-induced oscillator that has yet to be identified or explained. It may be the same one, but we don't know.

Rosbash: Is the liver of a SCN-lesioned animal entrainable with food?

Menaker: I don't know.

Okamura: Shibata's group shows that SCN-lesioned animals are food entrainable, and CLOCK in the liver is oscillating (Hara et al 2001).

Menaker: How does this apply to the food-anticipatory activity, which is a locomotor response? This is the question.

Young: I have a question concerning the *Cry* double knockouts and the degeneration of the masking over long periods of time. You were looking for changes in the retina as evidence of degeneration of photoreceptive machinery. But isn't it also possible that the degeneration is outside the retina? What can an isolated eye do? Do you get a pupillary response in an isolated eye, or does that require hooking to the central system?

Van Gelder: It depends on the species. In the non-mammals such as the frog, the iris is directly photosensitive, but no one has demonstrated this in mice. There are reports of hooded rats and hamsters with intrinsic pupillary light responses.

Foster: We took a lot of trouble showing that there wasn't intrinsic photosensitivity of the pupil by measuring in the non-illuminated eye.

Van Gelder: And the *math5*$^{-/-}$ mutants, which lack retinal ganglion cells, also lack a pupillary response.

Young: Have you looked anywhere outside the eye for evidence that there might be degeneration?

van der Horst: We only looked in the retina for changes in morphology because on the basis of data obtained by the Sancar group with cryptochrome-deficient rodless–coneless mice (Selby et al 2000), suggesting redundancy between cryptochromes and classical opsins, we anticipated that aged *Cry*-deficient animals might have lost rods and cones. This is not the case, but we haven't looked yet at ganglion cells.

Rosbash: In the double *Cry* knockouts, what are the levels of the clock proteins such as PER, relative to a wild-type oscillatory cycle?

van der Horst: Per1 and *Per2* mRNA levels are high in the periphery. At the protein level, we observed continuously high levels of nuclear mPER1 protein in the liver.

Rosbash: I was thinking of this in terms of the transcription versus protein only issue: it is a pretty good argument that we might have expected that the kinases would allow it to continue to cycle. In other words, if there is a way to establish an oscillator without the transcriptional circuitry (kinases and post-transcriptional regulation only), one might have expected persistent protein cycling.

Hastings: In the SCN of the *Cry* double-knockout mice, the PER2 protein is destabilized. There is no PER2 protein, even though there is mRNA. In Steve's knockouts, if you knockout some of the PERs then you destabilize the CRYs. There is a reciprocal co-stabilization taking place in the SCN.

Schibler: This is in the nuclei only.

Hastings: In the SCN the only staining we see is nuclear.

Rosbash: It could be just like the liver except you can't do the biochemistry.

Hastings: It may well be.

Rosbash: The liver results are all based on biochemistry — they are based on Western blots. The SCN results are based on histochemistry.

Loros: Steve Kay has a melanopsin knockout but he hasn't put it in a rodless–coneless background. What do you think you would see if you did this?

Van Gelder: That is obviously a critical experiment, and it would be a cosmic joke of tremendous proportions if a gene with the expression of melanopsin pattern didn't have a reasonable phenotype when it is appropriately unmasked in a retinal-degenerate background. Whether that is going to be a bright-light or dim-light effect is hard to say.

Kay: We saw our effect of the knockout in the presence of the other photoreceptors.

Van Gelder: One of the very interesting things I didn't mention — because we have only seen it sporadically — is that occasionally we get phase inversions in our *Cry* double-mutant animals kept under relatively dim light–dark conditions, where we actually see them become diurnally active. We have seen this in about 30% of our mice.

Foster: This is seen in normal mice if you have an LD cycle that is very dim.

Van Gelder: The masking is presumably contributing to this where positive masking is dominant over negative masking. One possibility is that you may see some phase inversion if cryptochrome is involved more in one masking pathway and a melanopsin-dependent process is involved in the other masking pathway. You may actually see a diurnal mouse come out when you take the rods and cones out.

Takahashi: If you have a mouse that is clockless it is much more susceptible to disturbances. In clock mutants that are arrhythmic, if you are not careful the noise in the animal facility will actually drive a diurnal day-active activity pattern. We see this in *Cry* double knockouts too. They are arrhythmic, so they are highly susceptible to being disturbed. This leads to this diurnal activity pattern.

Van Gelder: We don't see this in our DD *Cry* mutants, and we often see it in yoked controls who are in the same chamber under the same lighting conditions. There is the question of masking inputs. What are the photopigments there? Russell Foster and Nicholas Mrosovsky published work showing that masking is preserved in rodless-coneless animals, but there is a differential effect on positive and negative masking.

Foster: The data are noisy, but in general positive masking was attenuated, if not abolished, in the rodless–coneless background. But it is messy.

Menaker: Joe Takahashi, the question was what happens with the rodless–coneless melanopsin knockouts. I guess you wouldn't expect them to be arrhythmic. In order to have some fun with this I am going to stick my neck out and say that those animals will not respond to light. That is my prediction. They will not entrain or phase shift.

Foster: If melanopsin is the photopigment, I agree. That triple knockout will abolish everything. If it is acting as a photoisomerase then what you will see is some residual light responses with very long latencies, where the thing is struggling to regenerate chromophore.

Weitz: Since when is there no possibility of another photopigment?

Menaker: There is, without doubt.

Van Gelder: Genetically you cannot demonstrate that melanopsin is the photopigment by that result. Photopigment function must be demonstrated through biochemistry, not genetics. The genetic result just says that a particular protein is involved somewhere in the pathway in some way.

Weitz: Phyllis Robinson, I believe, reported reconstitution of melanopsin and showed by roughly the same criteria used for rhodopsin that it binds a chromophore.

Van Gelder: The λ_{max} of reconstituted melanopsin was 424 nm, so there is a problem; this doesn't match the OP_{479} spectrum.

Weitz: I am not saying that its properties must account for the full action spectrum of phase-shifting — it need not be the sole photopigment. However, it is a novel mammalian opsin that has apparently been shown to behave like a photopigment. I don't care about the lambda max in this context.

Menaker: Are you satisfied with those experiments? I have heard people carp about them.

Weitz: I don't know them in detail.

Foster: I don't want to discuss the details of the experiments, but it is possible to show that a bleachable pigment isn't sufficient. You can get this with RGR. What you have to show is that this pigment can activate a transduction cascade in some way. This is the way to tease apart the photoisomerase as distinct from the photosensory molecule.

Rosbash: Changing topic a little, I am not a great believer in redundancy except as an oversimplified, imprecise term. In yeast genetics over the last 10–15 years, the more work is done and the better the assays get, the more things turn out to have distinct functions. Ralph Stanewsky and I have a slightly different view on the parallel fly cryptochrome/opsin issue. My take is that cryptochrome is 90–95% of it, and in LD entrainment with bright lights and incubators you get virtually no phase shifting in the *Cry* knockouts. It is the circadian photoreceptor molecule. I view the distribution as having *Cry* as the leading player, and then maybe there is some opsin contribution that is not redundant. Plants have always been unique for all sorts of reasons. But I am sceptical that the plant view is going to eventually explain the mammalian world. My guess is that one molecule does the vast majority of entrainment and phase shifting, and then something else will have some other physiological task, rather than being redundant.

Menaker: There is another way to look at this. If you have multiple photoreceptors, it doesn't mean that they are redundant. It may mean that they have different but equally important functions.

Van Gelder: An example of this is rods and cones. You could say at a gross level that they are redundant for vision, because you can see with either, to a certain extent. It depends on the exact conditions: the rods are good for dim light, the cones better for bright light.

Menaker: If you knock out one molecule, can you delay or advance phase shifts? That's the question.

Van Gelder: The light world out there is very complex. You have to deal with a number of different conditions and you have also to deal with the intrinsic properties of the proteins and how they work for or against a particular function. For example, known opsins do adapt over time. Most G-protein-coupled receptors show adaptation with continual stimulation. This may not be an advantageous thing for integrating long light pulses for determining phase shifting.

Rosbash: What do you mean by long light pulses?

Van Gelder: Like a 12 h light cycle: non-parametric versus parametric entrainment. There you may require a different property in your photopigment that, for example, flavin might be able to provide. On the other hand the pupillary argument is a good one: why don't pupils come back open again in constant bright light? I see this clinically all the time—I shine a bright light in someone's eye for a long time and watch the pupil close and it never comes back up again. But I know that the rods and cones are adapting. Another pigment may be responsible for keeping the pupil constricted under continuous light conditions so that the outer retina is not bleached.

Rosbash: Which specific assays would you recommend?

Van Gelder: If you use a gross assay like entrainment you may say that they are redundant in that they provide adequate information for the clock to entrain or not entrain.

Rosbash: I was just pointing out that in fly experiments, despite the fact that the fly lives in this same complex external world, essentially everything can be wiped out with one mutant.

Menaker: You have to be very careful here: flies do not live in the same world.

Rosbash: I'm referring here to phase shifting.

Stanewsky: Phase shifting is an artificial phenomenon which an organism usually never does. What the *Cry* mutant is blind for is something that the fly never experiences in its life.

Rosbash: I would argue that it is even more extreme. A mutant is much more likely to be insensitive to the three minute change in day length that can occur daily at some latitudes than to a much stronger pulse. I would bet anything that those mutants are blind to the natural daily change in daylight.

Stanewsky: They do entrain nicely to normal LD cycles and they can also do phase shifts of 6 h. I don't think the *Cry* mutant will have a problem in nature.

Young: I thought part of the question was that if you take *Cry* away in a simpler system like a fly, what is left behind is opsin based. Is the photic information that ultimately has an impact on the phase of the rhythm interpreted by something you would call photoentrainment? Is there a path to the clock, or is it affecting a very different pathway? It comes back around, and you might not be dealing with photoentrainment. It is a misnomer to think of these as redundant systems. I would ask the same question in the mammal. When you see what looks like redundancy, are you really influencing something like activity through one of these pathways, completely independently of the molecular mechanisms that we are thinking about.

Van Gelder: It is the subtlety that we don't understand. For example, *Cry1* and *Cry2* are redundant by every criteria we have applied, although Urs Albrecht has now seen specific interactions between *Per2* and *Cry1*. We see identical pupillary responses in a *Cry1 Rd* versus a *Cry2 Rd*. Our data are a little different than Bert van der Horst's on the single alleles: we just see period variability in both, not a consistent long or short period. Basically the two cryptochromes seem to be roughly interchangeable for each other if you have two copies of any of them by our assays. Yet there has been a selective pressure throughout mammalian evolution to keep both of these expressed. They must be doing something subtly different.

Weitz: Well, they produce different circadian periods when individually knocked out.

Van Gelder: It is a very subtle effect. What we find in our *Cry1* single copies is that their periods are very unstable.

Rosbash: The redundancy argument is better for the *Crys* than it is for *Cry* versus melanopsin, because there are lots of examples of pairs of genes in all sorts of organisms for which it is very hard to find different phenotypes.

Menaker: We can't forget that organisms are exquisitely tuned to their particular environments. A fly does not have the same environment as a rat, even when they are both out in the real world. Even hamsters which live in the desert haven't got the same light environment as voles or rats. The response to the external light conditions will be finely tuned. This will require a lot of subtlety which we probably don't see much of.

Lee: Steve Kay, in the melanopsin knockout mice, have you looked at the light induction of *Per*? Are these animals normal in this respect?

Kay: We have pulled SCNs from these mice and this experiment is planned.

Rosbash: Russell Van Gelder, when you do retrograde tracings from the SCN, do you light up the *Cry*-containing cells?

Van Gelder: Yes, but the *Cry* cells are a superset of that family. Those cells light up plus many other cells.

Rosbash: What about the melanopsin-containing retrogradely labelled cells, compared with the cryptochrome-containing cells?

Van Gelder: Other people have studied this. Gary Pickard has done extensive tracing with this virus and he has shown that the melanopsin-containing cells are only a subset of the cells that are retrogradely labelled by virus. We don't know what is in those other cells. It could be a problem with the virus. As far as we can tell, we have yet to see retinal ganglion cells that are cryptochrome negative by histology. Cryptochrome is nearly ubiquitously expressed.

Weitz: A large number of ganglion cells in the retina are probably clock cells and have cryptochrome on that basis alone. Again, it is this issue of how to distinguish a photoreceptor role from a clock role.

Van Gelder: If you didn't find them there that would be a problem, but finding them there doesn't really implicate their function as photopigments at this point.

Menaker: However, as I understand it the melanopsin-containing cells are not clock cells.

Van Gelder: That is what Rae Silver says. They don't express *Per*.

Weitz: We have looked at most of the known clock proteins and find that there is no overlap at all between melanopsin immunoreactivity and immunoreactivity for various circadian clock proteins. It looks like the melanopsin-containing cells do not express any clock proteins.

Van Gelder: Unfortunately, the current sera for cryptochromes are not very good.

Weitz: The question about whether or not there are additional projections from the retina to the SCN that are melanopsin-negative is a key one. It looks like PACAP and melanopsin together account for most of the RHT.

Van Gelder: Hopefully the diphtheria toxin/melanopsin mouse will help answer this question.

Foster: For what it is worth, in the rodless–coneless mouse we see *Fos* induction in a subset of ganglion cells in response to light. About 50% of these are melanopsin expressing.

Van Gelder: The presumption is that there must be another photopigment in the other cells.

Foster: Whether or not *Fos* induction is a marker of a photoreceptor or not is an open question. It is a marker of a depolarized ganglion cell.

References

Hara R, Wan K, Wakamatsu H et al 2001 Restricted feeding entrains liver clock without participation of the suprachiasmatic nucleus. Genes Cells 6:269–278
Selby CP, Thompson C, Schmitz TM, Van Gelder RN, Sancar A 2000 Functional redundancy of cryptochromes and classical photoreceptors for nonvisual ocular photoreception in mice. Proc Natl Acad Sci USA 97:14697–14702

SCN: ringmaster of the circadian circus or conductor of the circadian orchestra?

Alec J. Davidson, Shin Yamazaki* and Michael Menaker[1]

*Department of Biology, University of Virginia, Charlottesville, VA 22904 and *Department of Biological Sciences, Box 1812-B, Vanderbilt University, Nashville, TN 37235, USA*

Abstract. The mammalian circadian system is composed of multiple circadian oscillators in both the brain and the periphery. Unravelling the organization of this system is a major challenge that the field is only beginning to take on. Clearly the suprachiasmatic nucleus of the hypothalamus (SCN) plays a key role and sits at or near the top of the organizational hierarchy, the details of which are largely unknown. The SCN has often been characterized as a 'master oscillator' that controls other oscillators downstream in the hierarchy, but there is little information about the nature of that control or how rigid or flexible it may be. Indeed, characterization of the SCN as 'master' may be exaggerated since other central circadian pacemakers are known to exist and the extent of feedback onto the SCN from other oscillators remains unexplored. We have tried to make some of the issues concerning the role of the SCN within the entire system more explicit using the somewhat fanciful metaphor referred to in the title.

2003 Molecular clocks and light signalling. Wiley, Chichester (Novartis Foundation Symposium 253) p 110–125

Without the ringmaster and his whip the circus grinds to a halt. Horses will not circle the ring and tigers will refuse to jump through flaming hoops. However, even in the absence of its conductor an orchestra may continue to play, each musician keeping his own time until they drift gradually out of synchrony and the music fails.

Which of these metaphorical roles does the suprachiasmatic nucleus of the hypothalamus (SCN) play as it influences rhythms in the rest of the body? The answer hinges upon the interpretation of the effects of SCN lesions and on the rhythmic behaviour of tissues and organs isolated from SCN influences.

Until recently most circadian researchers have considered the SCN to be a sort of ringmaster using a whip to drive rhythmicity throughout the organism. This view

---

[1]This paper was presented at the symposium by Michael Menaker, to whom correspondence should be addressed.

has been based upon a number of seminal studies. First, SCN lesions abolish behavioural and endocrine rhythmicity (Moore & Eichler 1972, Stephan & Zucker 1972). Second, transplanted SCN tissue can restore behavioural rhythms to an SCN-lesioned, arrhythmic host animal and the circadian period of the donor determines the resulting period of the host (Ralph et al 1990). Third, isolated slices (Gillette 1986, Shibata & Moore 1988) and neurons (Herzog et al 1998) from SCN, but not other brain areas, express circadian rhythms in electrical activity.

A somewhat different view is supported by a number of other studies. Behavioural rhythmicity persists in SCN-lesioned rodents under the influence of either periodically available food (Davidson & Stephan 1999, Stephan et al 1979) or continuous treatment with methamphetamine (Honma et al 1992). Rhythmicity of at least some circadian genes persists in isolated peripheral tissues for periods of time ranging from several days to two or three weeks (Yamazaki et al 2000, our unpublished results). Removal of the retina (which is known to contain its own circadian oscillator; Tosini & Menaker 1996) changes the distribution of free-running periods of locomotor activity and the duration of active time (α) (Yamazaki et al 2002a). To pursue our metaphor, these results suggest that circadian signals may be relayed to the periphery by a baton rather than a whip; the SCN may indeed be the conductor of an orchestra composed of dozens, if not thousands of potentially independent oscillators.

Clocks without hands

SCN slices and neurons continue to exhibit electrical rhythmicity *in vitro*, but other variables oscillate in cultured SCN as well. The *Period 1* gene, a member of the transcriptional–translational feedback loop that comprises the core molecular oscillator in mammals (Reppert & Weaver 2002), has provided a valuable tool with which to assay rhythmicity in isolated tissues from rodents. The m*Per1–luciferase* (*Per-luc*) rat bears a reporter construct that allows for dynamic recording of gene expression *in vitro* (Yamazaki et al 2000). Consequently, SCN can be cultured from this transgenic animal and the explant will express rhythms in bioluminescence that correlate with *Per1* mRNA expression (Fig. 1). The rhythms are remarkably self-sustained and the phase of the first complete cycle *in vitro* is consistent among animals and correlates with the prior phase of the intact animal. This technology, since it does not rely on electrical rhythms from neurons, has allowed us to assay rhythmicity in dozens of tissues outside the CNS (Davidson et al 2003, Stokkan et al 2001, Yamazaki et al 2000). As shown in Fig. 1C, these tissues are also nicely rhythmic *in vitro*. Other authors have also reported on rhythmicity outside the mammalian SCN (Balsalobre et al 1998, Tosini & Menaker 1996).

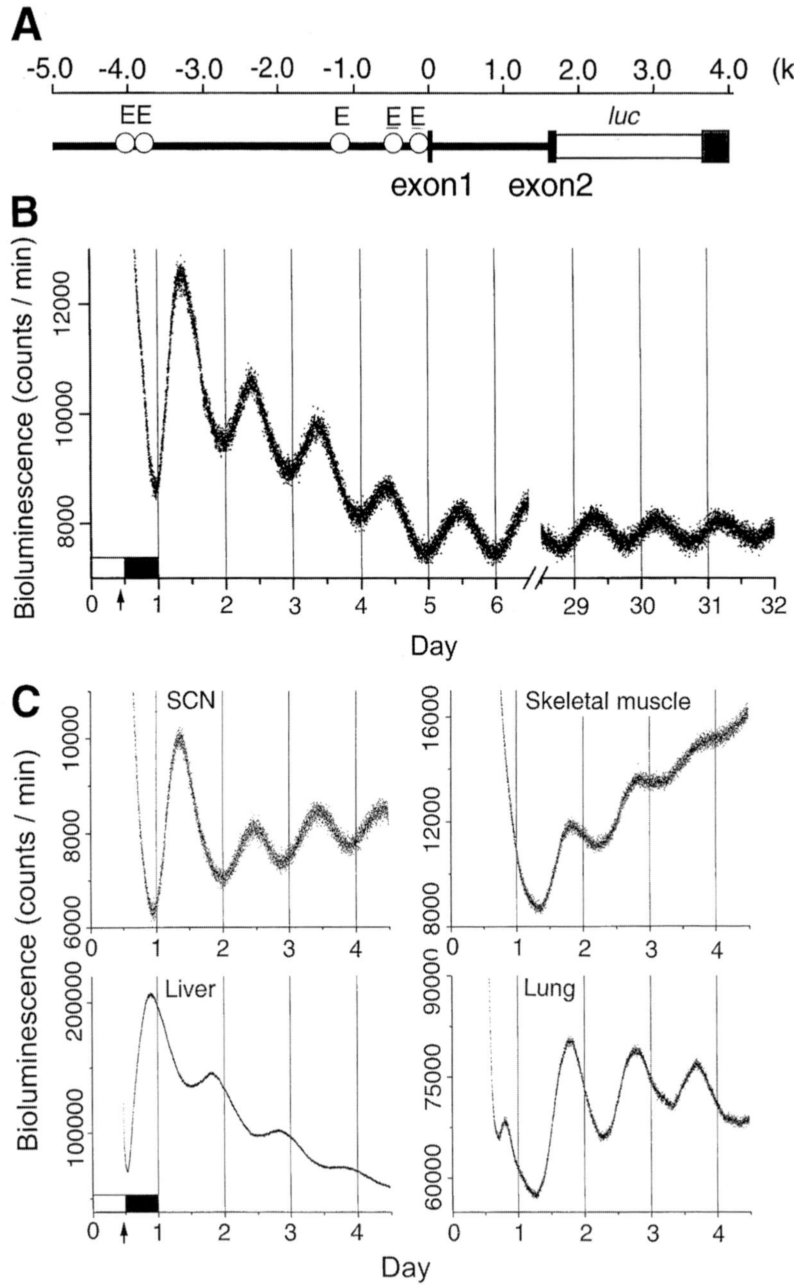
A
-5.0   -4.0   -3.0   -2.0   -1.0   0   1.0   2.0   3.0   4.0   (kb)
EE
E
E E
luc
exon1   exon2
B
Bioluminescence (counts / min)
12000
10000
8000
0   1   2   3   4   5   6   29   30   31   32
Day
C
Bioluminescence (counts / min)
SCN
10000
8000
6000
0   1   2   3   4
Skeletal muscle
16000
12000
8000
0   1   2   3   4
Liver
200000
100000
0   1   2   3   4
Lung
90000
75000
60000
0   1   2   3   4
Day

However, there is an apparently important difference between SCN and non-SCN rhythms *in vitro*: SCN explants remain rhythmic while other regions of the brain and peripheral tissues damp after a minimum of two, but a maximum of about 15 cycles (Fig. 1). The robustness and persistence of the rhythmicity depends on the tissue and does vary in individual cultures. It has been assumed from results like these that the SCN is a self-sustained pacemaker while peripheral clocks are damped oscillators, but it cannot be excluded that in isolated peripheral tissues, *Per1* may damp despite the persistent rhythmicity of other clock components or that the cultures lack some factor continuously present *in vivo* that is required for sustained rhythmicity of non-SCN tissue.

Although the *Per1-luc* rat has been a valuable model which has provided information on the widespread distribution of circadian oscillators (Davidson et al 2003, Stokkan et al 2001), the dysphasia-inducing effects of phase-shifting (Abe et al 2002, Yamazaki et al 2000) and the deleterious effects of ageing on the circadian system (Yamazaki et al 2002b), it has some important limitations. In using *Per1-luc* as a reporter, we are only measuring one component of a complex molecular clock mechanism; other components may have different kinetics. More importantly, we have not yet determined what role, if any, this molecular oscillation plays in either CNS or peripheral organ function. We have a clock, but are as of yet unaware of the nature of the clock's hands or of the mechanisms by which the two are coupled. Although many physiological functions show diurnal rhythmicity (for examples see the reviews on rhythmicity in the GI system by Lawrence Scheving [Scheving 2000] or on the cardiovascular system by Yi-Fang Guo and Phyllis Stein [Guo & Stein 2003]) we have yet to make functional connections between these potential hands and the clock whose molecular rhythmicity we measure in our cultures.

The whip or the baton?

Although the SCN is critical for behavioural and endocrine rhythmic output, there remains the fundamental question of which, if any, peripheral oscillators require

---

FIG. 1.   (A) Diagram of the mouse *Per1* transgene. Heavy line, *mPer1* fragment; open bar, luciferase fragment; shaded box, polyadenylation fragment; circles represent E boxes. (B) Representative circadian rhythm of bioluminescence from a cultured SCN explanted from a *Per1-luc* transgenic rat. Black and white bars show the animal's previous LD conditions. The explant was made just before lights-off (arrow), and luminescence was monitored immediately. The near 24 h rhythm peaked in the middle of the subjective day and persisted for 32 days *in vitro*, at which time the culture was removed from the assay. Rhythmicity persisted for more than 2 weeks in the seven SCN cultures that were maintained for this length of time; other SCN cultures were terminated after shorter times while still rhythmic. (C) Circadian rhythms expressed *in vitro* from several different tissues from the same animal. The tissues were explanted just before lights-off (arrow) (From Yamazaki et al 2000, with permission).

the SCN for their own rhythmic maintenance. If the SCN is ringmaster, then it drives peripheral oscillators; without the whip, peripheral rhythmicity disappears. Alternatively, the SCN as conductor provides phase information for otherwise independent oscillators in the periphery; it regulates phase and tempo and occasionally amplitude of specific groups in the orchestra.

The ringmaster role appears to be supported by several studies that, according to their authors' interpretations, 'prove' that peripheral clocks lose rhythmicity in the absence of the SCN (for example Furukawa et al 1999, Iijima et al 2002, Sakamoto et al 1998, Terazono et al 2003). However serious logical flaws in the design of all these experiments make that interpretation untenable. In every case, SCN lesions are followed by at least 2–3 weeks of behavioural monitoring to ensure arrhythmic behaviour. Animals are then placed in a light–dark regime (LD) and subsequently sacrificed in two or more groups according to clock, or *zeitgeber* time. Several subjects make up each time point, and separate groups comprise the circadian time series. If the SCN lesion did indeed abolish rhythmicity in peripheral organs, then one would expect the results that were in fact reported in these studies: controls show normal organ rhythms in LD, and SCN-lesioned groups are arrhythmic. However, there is another equally likely interpretation: suppose that peripheral organs were still rhythmic, but were free-running in the animal. With the conductor gone, the orchestra members are now playing each at his own tempo. Each animal, indeed each organ clock will have a slightly different *tau*; in the three or more weeks following the SCN ablation, the phases of the liver, for example, in each animal will drift relative to the livers of other animals in the group and become distributed around the clock — OUR clock. Control animals are also free-running during the locomotor screen, but the return to LD enables the SCN to reset the peripheral clocks so that all subjects are once again in-phase prior to tissue collection. The livers of the lesioned animals are not reset since the conductor is absent and we know that peripheral oscillators are not directly responsive to light (our unpublished observations). Tissue collection is done according to clock time and since there is no reason to assume that the livers collected at each timepoint, if they are rhythmic, are in phase with one another, the expected waveform is precisely the same as that predicted by the alternative explanation: flat. Therefore the result does not discriminate between the alternative hypotheses; the arrhythmic waveform could be the consequence of either arrhythmic tissue, or individually rhythmic tissues out-of-phase with one another.

It is important to recognize that for the moment this critical question remains unresolved and therefore ones choice of metaphor becomes a matter of taste. Ours clearly runs to SCN as conductor, using a variety of coupling signals of varying strengths to maintain adaptive synchrony among the many peripheral oscillators of which the organism is composed. In any system organized in this way one would

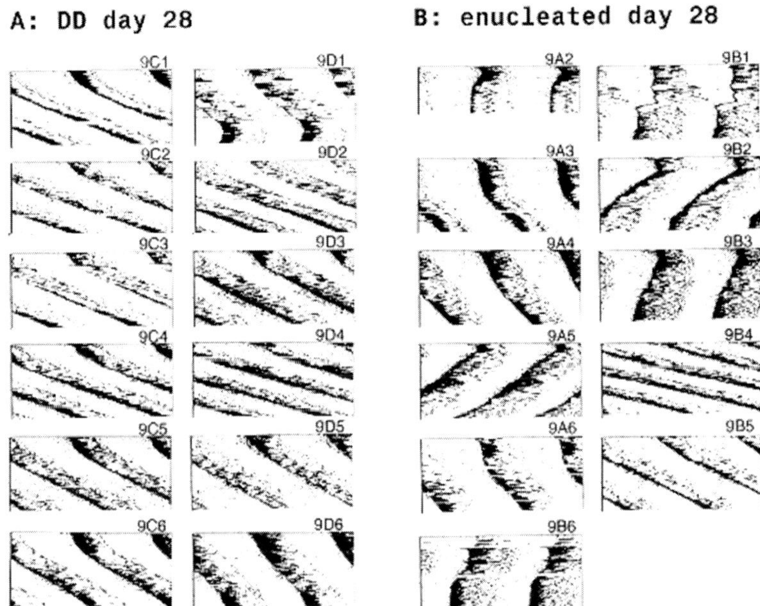

FIG. 2.   Wheel-running records of hamsters enucleated or placed in continual darkness (DD) at an age of 28 days. The enucleated group have a much broader range of free-running periods (Yamazaki et al 2002a, with permission from SAGE).

expect feedback loops of all sorts. Indeed there are indications that non-SCN oscillators feed back, although weakly, to the SCN. One example mentioned earlier is the role played by the retina (and perhaps its oscillator) in the fine-tuning of SCN output. Enucleated hamsters show a much broader range of circadian periods than do hamsters maintained in constant darkness (Yamazaki et al 2002a) (Fig. 2). A second example is the mutual coupling that occurs between the food-entrainable oscillator and the light-entrainable oscillator when rats are given a cycle of food availability with a period of 24 h, but are otherwise free-running in constant darkness (Stephan 1986). The SCN-dependent free-running rhythm of locomotor activity displays a variety of effects including shortening, lengthening, and even occasional synchronization with the rhythm of food-anticipatory activity when the two rhythms are at certain phase relationships with one another.

Hands without clocks

Although many tissues have been shown to be capable of rhythmicity in the absence of the SCN, there are several output rhythms, or 'hands' for which there is no anatomically identified pacemaker structure. The two premier examples are food-anticipatory activity (FAA) and chronic methamphetamine-induced rhythmicity. Both occur in the absence of a SCN (Davidson & Stephan 1999, Honma & Honma 1995). In some respects these rhythms appear similar and may share a common mechanism. For example methamphetamine-induced rhythms can be entrained with restricted feeding schedules (Honma et al 1992).

FAA, first described by Richter (1922), is the bout of heightened locomotor activity that precedes a daily timed meal. The literature has recently been reviewed by Stephan and therefore we will not give a detailed account here (Stephan 2001, 2002). It should be emphasized, however, that this phenomenon has been demonstrated in a wide variety of species and is likely to be an important tool used by animals to restrict their foraging behaviour to a time-domain that is both safe and productive. It should also be emphasized that the food-entrained oscillator (FEO) has so far eluded structural identification.

Although the hypothalamus has always been an obvious target for lesion studies aimed at identifying the FEO, ventromedial (Honma et al 1987, Mistlberger & Rechtschaffen 1984), paraventricular and lateral hypothalamic (Mistlberger & Rusak 1988) lesions failed to permanently abolish FAA. Hippocampal, amygdala and nucleus accumbens lesions were also ineffective (Mistlberger & Mumby 1992). Since lesion experiments failed to identify the FEO in the brain, attention turned to the periphery. In particular, the gastrointestinal (GI) system and the liver became prime candidates since they respond in many different ways to food inputs and contain clock genes that oscillate *in vitro*, removed from SCN influence.

Use of the *Per-luc* rat model made possible studies on the role of food in the entrainment of these and other peripheral clocks (Stokkan et al 2001). Rats were fed under restriction during the day. Organ explants from these rats were then cultured and the phases of the *Per-luc* rhythm were shifted by 12 hours in the liver (Fig. 3), and somewhat lesser amounts in stomach and colon (Fig. 4B). The SCN was completely unaffected by the feeding schedule.

This finding immediately prompted speculation by ourselves and others (Stephan 2002) that the FEO might indeed be located in the liver or GI tract. We sought to test this hypothesis by comparing the phases of the rhythms of FAA and *Per1* expression in the liver and GI system following a number of manipulations. First we entrained rats to daytime feeding, then allowed them to return to *ad libitum* meals. FAA emerged as expected during the food restriction, and then disappeared during *ad lib* conditions. After 10 days of *ad lib* feeding, the rats were completely food deprived for 2 days, during which time FAA re-emerged at its previous daytime phase (Fig. 4A). This result has been widely reported in the literature (Clarke & Coleman 1986, Coleman et al 1982, Rosenwasser et al 1984) and suggests that during *ad lib* feeding that follows a food-restriction treatment, the FEO continues to oscillate but its behavioural output (FAA) is masked.

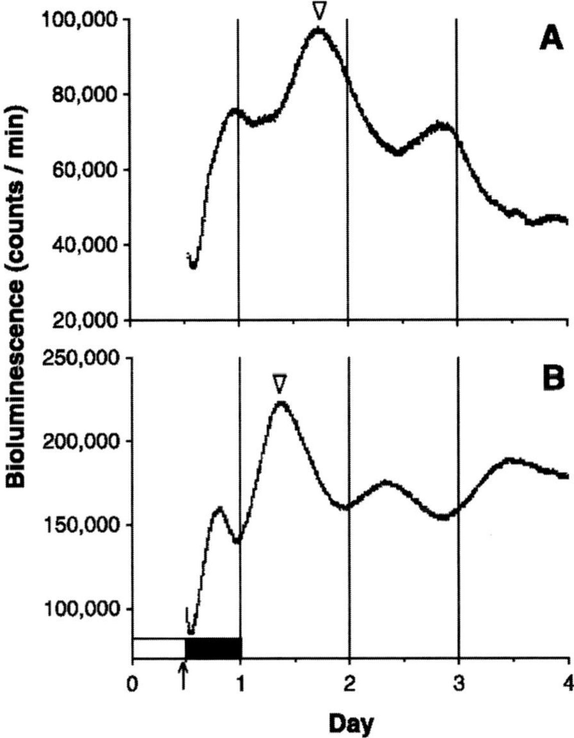

FIG. 3.   Rhythms of light emission by liver explants. Shown are raw data from (A) an *ad lib*-fed control animal and (B) an animal that had been exposed to a 4 h (daytime) restricted feeding regimen for 7 days. Both animals had been kept on the light cycle indicated by the white and black bars in (B), and both were killed and the tissues explanted at the time shown by the arrow. Because the pattern of light emission is quite variable during the first 12–14 h after explantation, we consider that the phase of the tissue *in vivo* is best reflected by the phase of the peak during the first full subjective day (12–36 h after explant). The phase of these peaks is consistent from animal to animal. Here, the phase chosen is indicated by the inverted triangles (From Stokkan et al 2001, with permission).

Subsequent deprivation then unmasks the behaviour revealing the FEO in its previous phase relationship to the (prior) restricted feeding. If the FEO were located in the liver or GI tract, then *Per1* rhythmicity should be phased with the unmasked FAA. Specifically, restricted feeding should reset the phase of *Per1* in the organs containing the FEO, and that phase should persist during the *ad lib* feeding and subsequent deprivation. However the experiment showed unequivocally that although the phase of *Per1* rhythmicity in the liver, stomach and colon did entrain to restricted feeding, it shifted back to its normal nocturnal position during *ad lib* feeding (Fig. 4B) and remained nocturnal during food deprivation *despite the recurrence of FAA during the day*. Since the kinetics of *Per1* in

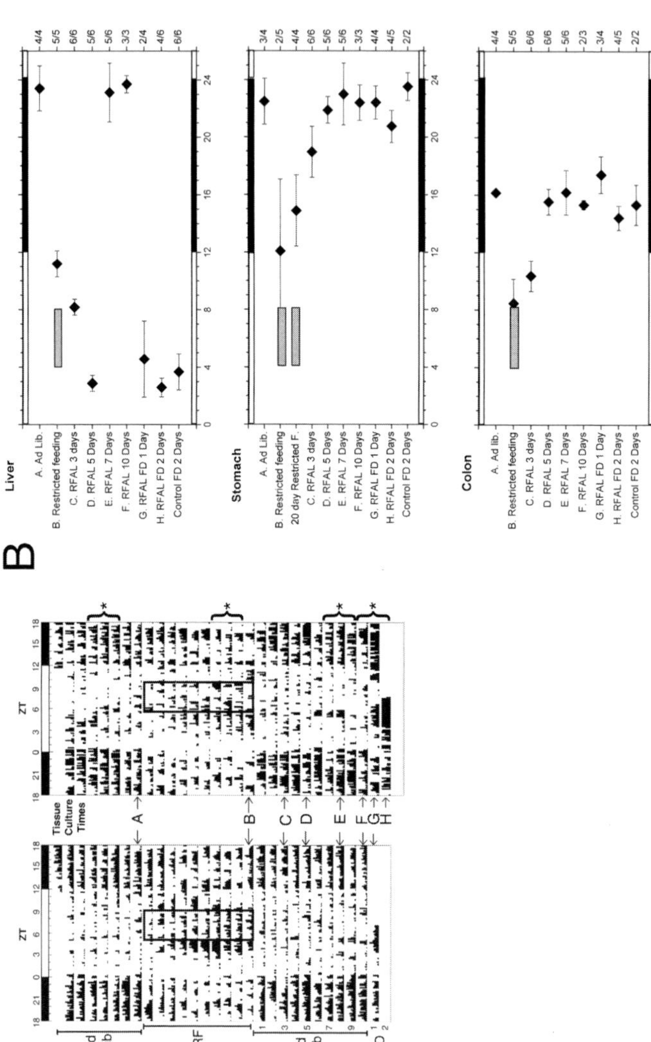

FIG. 4.   (A) Single-plotted event records from 2 rats. Cage activity counts per 10 min were derived from abdominal body temperature transmitter signal strength. Rats were *ad libitum*-fed powdered chow for at least 1 week, restricted fed (RF; ZT 0500–0900) for 10 days (depicted as a box), returned to *ad lib* for 10 days, then food deprived (FD) for either 1 or 2 days before tissue collection. The light–dark cycle is shown above the records. The letters between the records refer to days on which organ cultures were made from subsets of the rats in the experiment. (B) Phases of *Per-luc* expression for peripheral tissues. ZT and the light–dark cycle are shown on the x-axis. The data are group means ± SEM. The primary y-axis shows the experimental conditions for each group accompanied (where applicable) by letters that refer to culture times shown in Fig. 1. The secondary y-axis shows number of cultures attempted followed by number of cultures that contribute to the data shown. Discrepancies occur when cultured tissue is arrhythmic or has inconsistent peak phase from day to day. The grey boxes show mealtime where applicable. RFAL x, restricted feeding followed by *ad lib* for x days before culture; RFAL FD, 10 days RF, then 10 days *ad lib*, then food deprivation (1 or 2 days) (From Davidson et al 2003, with permission).

the GI tract and liver responded differently than did FAA in the same paradigm, it seems unlikely that FAA is timed by a *Per1*-based clock in any of these organs.

Conclusions

The many circadian oscillators in the periphery are likely to have important influences on organ function, and circadian organization at the systems-level almost certainly has significant effects on overall health. Exciting developments at the intersection between clocks and human health support this belief. For example, cancers grow faster in arrhythmic animals (Filipski et al 2002), acute cardiovascular disease has a significant rhythmic component (Guo & Stein 2003), and most of us have experienced first-hand the effects of jet lag and/or shift work.

Important questions that remain unanswered include the following. What physiological (as opposed to molecular) variables are rhythmic? How can description of the molecular clock in specific organs help us to understand organ function? How do such physiological rhythms interact with one another and feed back to the central or other peripheral clocks? And how does disruption of these rhythms, both on a single-organ and systems level, affect specific disease or general health? How can we correct these circadian dysfunctions prior to and during disease to allow for better health outcomes?

At the risk of overstretching our metaphor, we can summarize our view of circadian organization as follows. A living thing is like a symphony. The score (genetic program) specifies the notes; but equally importantly, the timing with which they must be played. The conductor (SCN) interprets the score, in particular its dynamic temporal aspects, and conveys his interpretation to the individual members of the orchestra. He uses a baton rather than a whip because musicians (peripheral oscillators) are independent interpreters in their own right and must be coaxed, not driven. The aesthetic quality of the performance (fitness) depends heavily upon how successfully the flow of information (coupling) regulates synchrony among the performers.

Acknowledgements

This work was supported by NIMH grants R01 MH062517 and R01 MH56647, and NSBRI grant NCC9-58-167.

References

Abe M, Herzog S, Yamazaki M et al 2002 Circadian rhythms in isolated brain regions. J Neurosci 22:350–356
Balsalobre A, Damiola F, Schibler U 1998 A serum shock induces circadian gene expression in mammalian tissue culture cells. Cell 93:929–937

Clarke JD, Coleman G J 1986 Persistent meal-associated rhythms in SCN-lesioned rats. Physiol Behav 36:105–113

Coleman G J, Harper S, Clarke JD, Armstrong S 1982 Evidence for a separate meal-associated oscillator in the rat. Physiol Behav 29:107–115

Davidson A J, Poole A, Yamazaki S, Menaker M 2003 Is the food-entrainable oscillator in the digestive system? Genes Brain Behav 2:1–8

Davidson A J, Stephan FK 1999 Feeding-entrained circadian rhythms in hypophysectomized rats with suprachiasmatic nucleus lesions. Am J Physiol 46:R1376–R1384

Filipski E, King M, Li X et al 2002 Host circadian clock as a control point in tumor progression. J Natl Cancer Inst 94:690–697

Furukawa T, Manabe S, Watanabe T et al 1999 Daily fluctuation of hepatic P450 mono-oxygenase activities in male rats is controlled by the suprachiasmatic nucleus but remains unaffected by adrenal hormones. Arch Toxicol 73:367–372

Gillette MU 1986 The suprachiasmatic nuclei. Circadian phase-shifts induced at the time of hypothalamic slice preparation are preserved *in vitro*. Brain Res 379:176–181

Guo YF, Stein K 2003 Circadian rhythm in the cardiovascular system. Chronocardiology. Am Heart J 145:779–786

Herzog ED, Takahashi JS, Block GD 1998 Clock controls circadian period in isolated suprachiasmatic nucleus neurons. Nat Neurosci 1:708–713

Honma S, Honma K 1995 Phase-dependent phase shift of methamphetamine-induced circadian rhythm by haloperidol in SCN-lesioned rats. Brain Res 674:283–290

Honma S, Honma K, Nagasaka T, Hiroshige T 1987 The ventromedial hypothalamic nucleus is not essential for the prefeeding corticosterone peak in rats under restricted daily feeding. Physiol Behav 39:211–215

Honma S, Kanematsu N, Honma K 1992 Entrainment of methamphetamine-induced locomotor activity rhythm to feeding cycles in SCN-lesioned rats. Physiol Behav 52:843–850

Iijima M, Nikaido T, Akiyama M, Moriya T, Shibata S 2002 Methamphetamine-induced suprachiasmatic nucleus-independent circadian rhythms of activity and mPer gene expression in the striatum of the mouse. Eur J Neurosci 16:921–929

Mistlberger RE, Mumby DG 1992 The limbic system and food-anticipatory circadian rhythms in the rat. Ablation and dopamine blocking studies. Behav Brain Res 47:159–168

Mistlberger RE, Rechtschaffen A 1984 Recovery of anticipatory activity to restricted feeding in rats with ventromedial hypothalamic lesions. Physiol Behav 33:227–235

Mistlberger RE, Rusak B 1988 Food-anticipatory circadian rhythms in rats with paraventricular and lateral hypothalamic ablations. J Biol Rhythms 3:277–291

Moore RY, Eichler B 1972 Loss of a circadian adrenal corticosterone rhythm following suprachiasmatic lesions in the rat. Brain Res 42:201–206

Ralph MR, Foster RG, Davis FC, Menaker M 1990 Transplanted suprachiasmatic nucleus determines circadian period. Science 247:975–978

Reppert SM, Weaver DR 2002 Coordination of circadian timing in mammals. Nature 418: 935–941

Richter C 1922 A behavioristic study of the rat. Comp Psychol Mono 1:1–55

Rosenwasser AM, Pelchat R J, Adler NT 1984 Memory for feeding time. Possible dependence on coupled circadian oscillators. Physiol Behav 32:25–30

Sakamoto K, Nagase T, Fukui H et al 1998 Multitissue circadian expression of rat period homolog (rPer2) mRNA is governed by the mammalian circadian clock the suprachiasmatic nucleus in the brain. J Biol Chem 273:27039–27042

Scheving L A 2000 Biological clocks and the digestive system. Gastroenterology 119: 536–549

Shibata S, Moore RY 1988 Electrical and metabolic activity of suprachiasmatic nucleus neurons in hamster hypothalamic slices. Brain Res 438:374–378

Stephan FK 1986 Interaction between light- and feeding-entrainable circadian rhythms in the rat. Physiol Behav 38:127–133
Stephan FK 2001 Food entrainable oscillators in mammals. In: Turek TJFW, Moore RY (eds) Handbook of behavioral neurobiology 12. Circadian clocks. Kluwer Academic/Plenum Publishers, New York, p 223–241
Stephan FK 2002 The "other" circadian system. Food as a Zeitgeber. J Biol Rhythms 17:284–292
Stephan FK, Swann JM, Sisk CL 1979 Anticipation of 24-hr feeding schedules in rats with lesions of the suprachiasmatic nucleus. Behav Neural Biol 25:346–363
Stephan FK, Zucker I 1972 Circadian rhythms in drinking behavior and locomotor activity of rats are eliminated by hypothalamic lesions. Proc Natl Acad Sci USA 69:1583–1586
Stokkan KA, Yamazaki S, Tei H, Sakaki Y, Menaker M 2001 Entrainment of the circadian clock in the liver by feeding. Science 291:490–493
Terazono H, Mutoh T, Yamaguchi S et al 2003 Adrenergic regulation of clock gene expression in mouse liver. Proc Natl Acad Sci USA 100:6795–6800
Tosini G, Menaker M 1996 Circadian rhythms in cultured mammalian retina. Science 272: 419–421
Yamazaki S, Numano R, Abe M et al 2000 Resetting central and peripheral circadian oscillators in transgenic rats. Science 288:682–685
Yamazaki S, AlonesV, Menaker M 2002a Interaction of the retina with suprachiasmatic pacemakers in the control of circadian behavior. J Biol Rhythms 17:315–329
Yamazaki S, Straume M, Tei H, Sakaki Y, Menaker M, Block GD 2002b Effects of aging on central and peripheral mammalian clocks. Proc Natl Acad Sci U S A 99:10801–10806

DISCUSSION

Rosbash: I was going to ask a question about the food-anticipatory activity, but then I realized that the question was probably equally valid for the eye lesion experiments that you mentioned at the end. Why is there a separate oscillator? Is that a unique interpretation to those results?

Menaker: It is a unique interpretation for the food-entrainable oscillator. It has to be extra-SCN, because it persists with SCN-lesioned animals. It is not a unique interpretation of the enucleation experiments.

Rosbash: Let's separate the two and focus on the food-entrainable activity. You have an SCN-lesioned animal from the outset — that is an animal with a very early SCN lesion — and then you end up setting up oscillations by virtue of this restricted feeding. These are the only locomotor activity rhythms that one sees in that lesioned animal, because the normal activity wouldn't be there because the SCN isn't there. Is that correct?

Menaker: That is one observation.

Rosbash: It is a key observation. Without that background observation, if you just had what you presented in the absence of the lesions you would know that it is not just some secondary output from a normal oscillator, so the food is setting up an oscillator somewhere. The results of the SCN lesion and the non-lesion are the same. This food anticipatory activity is essentially indistinguishable whether the animal is lesioned or not lesioned.

Menaker: I think that's correct. There is a big literature on this, which hasn't received the attention it deserves.

Rosbash: Now you are interpreting the eye in more or less the same way?

Menaker: The interpretation of the eye experiment is much more problematic. The way I would like to interpret this is that there is an oscillator in the eye, and an oscillator in the SCN, and they are involved in a conversation. The consequence of that conversation is the value of the free-running period.

Rosbash: It was clear that this was how you interpreted it.

Menaker: That is not the only interpretation, for sure.

Weitz: I have a question concerning the assumption underlying your interpretation of the phase of the liver and GI tract oscillators, with respect to the food-anticipatory behaviour. I interpret the literature exactly the same way you do: that is, there is a food-entrainable oscillator that drives food-anticipatory activity, and when *ad lib* feeding returns, the expression of this activity is masked and disappears virtually immediately upon *ad lib* activity. Then a very long time later, in some cases many weeks later, as soon as food deprivation occurs the anticipatory behaviour pops up again at exactly the same phase. The interpretation, as I understand it, is that the oscillator is running persistently at that phase, and the expression of anticipatory activity has been masked by *ad lib* feeding. This interpretation requires the belief that this oscillator is virtually uncoupled from the SCN — its phase persists for a long time and ignores the SCN. This disagrees with all the luciferase data you have presented from whatever tissue you have looked at. All the other oscillators were coupled to some extent to the SCN.

Menaker: I don't think that's fair. It certainly presents a challenge. For instance, the liver oscillator is coupled to the SCN perhaps only through feeding behaviour. If the SCN is controlling feeding behaviour and the liver is following that, then this is a kind of coupling, but it doesn't argue against the other interpretation.

Weitz: During *ad libitum* feeding such an oscillator would then return to its original phase relationship with the SCN.

Menaker: That's what we showed.

Weitz: But this is not true for the oscillator underlying food-anticipatory activity. It doesn't behave in the way that you have described.

Menaker: That is the basis for our interpretation that the liver doesn't contain the oscillator that controls the behaviour.

Schibler: I noticed that the peak of the phase in the liver of your adult rats is towards the end of the night. If you look at the induction of mRNA, the peak is actually at the beginning of the night. That's a huge difference. The question is, what is the half-life of luciferase, and could it be responsible for that difference? But the half-life in cells is about 2 h, and it must be similar if you look at the difference between protein accumulation and mRNA accumulation. It cannot be the luciferase itself. Can it be the mRNA? You didn't make an effort to put destabilizing elements into the 3′ region of your mRNA. This could account for one or two hours but not more. Completely counterintuitively, if you have a fixed function of transcription and look at the difference between the 1 h half-life and 24 h half-lives of the products, this makes only a 4 h difference. The question is, in the promoter regions that you are using, are there some elements that may be important for phase? This does not change your interpretations with regard to changes in the phase. But it would be interesting to follow the different tissues just at the mRNA level.

Menaker: That's true, but we haven't done that. The other thing we must remember is that in our hands the peak of *Per* expression in the liver varies a good deal from one experiment to another. This has to do with how the rats are responding to the immediate situation that they are in. We need to start measuring feeding behaviour, because they feed at different times under somewhat different circumstances. This changes with age, too. The right way to do your experiment is to get the *Per* expression and the RNA from the same animals.

Stanewsky: Is the eliciting of this food-anticipatory activity dependent on how long the rats were kept on the restricted feeding schedule? Is it all connected with a learning process?

Menaker: It develops after 2–3 d. It can be seen immediately, but gets stronger. The length of time past a week or so doesn't seem to have much influence on the phase of the tissue. The question of whether it is a learning process depends on how you define learning.

Stanewsky: Would it make sense to look in brain structures associated with learning?

Menaker: Perhaps. There is a lot of brain to look at.

Hastings: With regard to the neural substrate of the food-entrainable oscillator, there is evidence pointing towards the ventromedial hypothalamus. Your recent paper looking at luciferase emissions from different brain regions (Abe et al 2002) showed a weak cyclicity in mediobasal hypothalamus in antiphase to the SCN. I am just wondering what you think the ventromedial hypothalamus rhythm might look like if you took it from the food-restricted rat. Could you soup up the endogenous cyclicity there, and also set its phase by the time of the phase of restricted feeding?

Menaker: That is the kind of experiment that Alec is planning to do. Obviously, if you find the right part of the brain, it should be souped up and it should be phase-controlled.

Hastings: And it should persist in animals with SCN lesions.

Menaker: That's right.

Dunlap: At the superficial level the food-entrainable oscillator looks sort of like *zeitgedachtnis* in bees (Renner 1960). They can be trained to feed at a certain time and they will remember this. Even if you interrupt this for some time, they will remember it for weeks. Have you thought about similarities with this?

Menaker: I don't think we understand *zeitgedachtnis* in bees. It seems very analogous though.

Rosbash: What is the evidence that this food-anticipatory activity is circadian, as opposed to an hour-glass timer from the previous feeding cycle?

Menaker: It free runs for a month. It reappears in a defined phase.

Rosbash: The defined phase is triggered by the food reappearing?

Menaker: No, the way you get it back is to deprive the animal of food.

Hastings: Are the rats not on a LD cycle throughout that time? Could it not be an interval relative to lights-on that is being remembered?

Menaker: It works either way.

Loros: In free run, what is the period like?

Menaker: It is a circadian period.

Loros: In a SCN-lesioned animal, if you still have this going on, what is the period? Is it the same period?

Menaker: I don't think it can be that precisely defined, but it is circadian because you can do if for 7 d and it is near the phase that it was when you masked it. If you do it for 14 d it is not quite so near, and so on.

Weitz: You can't do a good free run because you can't starve animals for very long. This is the problem. You get two or three days, but rats and mice don't tolerate more than a couple of days of complete food deprivation.

Van Gelder: We have some relevant data. First, the *math5*$^{-/-}$ mice that lack retinal ganglion cells serve as a similar model to the early postnatal enucleation. We find slightly different results. We do find a profound effect on *tau* free-running period in the *math5*$^{-/-}$ mice. They normally free run at 24.5 as opposed to 23.6 h, but we don't see the variability that you see. In fact, they are as tight as the wild-type in terms of the standard deviation of *tau*. The second thing we find in the *math5*$^{-/-}$ mice is that they are very resistant to non-photic entrainment as well. We tried to get their pacemakers to entrain to probe the phase response curve with restricted wheel access. I think this operates in many ways like restricted food access. We were unable to entrain most of the animals to a 24 h T cycle with wheel availability for 2 h, whereas all our wild-type animals entrained to that regimen. The third thing we found with these mice that was completely unexpected was that when we looked at their behaviour during wheel lock, these animals are free running at 24.5 h. There is a 2 h wheel availability every day. They routinely anticipate the wheel availability by 2 h, and their maximal activity occurs in the two hours before the wheel availability, regardless of the phase of the free-running rhythm. This poses the question as to whether this is truly dissociated from the SCN or is it something that is entrained by the SCN when available. Our data would suggest it really is an independent oscillator, because we have the two rhythms free running right through each other regardless of phase.

Menaker: Ralph Mistlberger has seen the same thing in SCN intact animals, which are free running in constant darkness and which are at the same time entrained to food anticipation. You can entrain them to 24 h and they will free run at 24.5 h, and these cross as if the oscillator is completely independent of the SCN.

Van Gelder: What we haven't done yet is to probe what happens when we give back the wheel and then take it away again, analogous to the food-entrainable situation, to see whether we would regain that same phase relationship. I think this would be a way to test this. You can deprive them of a wheel for long periods of time.

Menaker: I think that the anticipation of the wheel is the same kind of thing physiologically as the anticipation of the food. And it is all motivational stuff: this is why I think it is also probably related to the methamphetamine-induced oscillation. Keeping things as simple as possible, then there is only one other extra SCN oscillator controlling behaviour.

Sassone-Corsi: If this is true, we should somehow be able to identify this in the brain.

Menaker: I think it is really important to identify it in the brain. It has been rather brushed aside.

Schibler: One experiment I never understood but which has been done successfully is to put rodents on heavy water, and this lengthens the period incredibly. Has this been done for the food-anticipatory behaviour?

Menaker: I don't think so.

Schibler: Other oscillators are not affected by the heavy water, such as ultradian oscillators.

Menaker: You'd have to do this after the *ad libitum* feeding.

References

Abe M, Herzog ED, Yamazaki S et al 2002 Circadian rhythms in isolated brain regions. J Neurosci 2002 22:350–356

Renner M 1960 The contribution of the honeybee to the study of time-sense and astronomical orientation. In: Chovnick A (ed) Clocks. Cold Spring Harbor Symp Quant Biol XXV. Cold Spring Harbor Press, New York, p 361–367

On the communication pathways between the central pacemaker and peripheral oscillators

Nicolas Cermakian†, Matthew P. Pando*†, Masao Doi†, Luca Cardone†, Irene Yujnovsky†, David Morse†‡ and Paolo Sassone-Corsi†[1]

*Douglas Hospital Research Center, McGill University, 6875 LaSalle Boulevard, Montréal (QC) H4H 1R3, Canada, *ExonHit Therapeutics, 217 Perry Parkway, Building 5, Gaithersburg, Maryland 20877, USA, †Institut de Génétique et de Biologie Moléculaire et Cellulaire, 1 rue Laurent Fries, 67404 Illkirch, Strasbourg, France and ‡Département des Sciences Biologiques, Université de Montréal, Montréal (QC) H3C 3J7, Canada*

Abstract. Circadian rhythms are regulated by clocks located in specific structures of the CNS, such as the suprachiasmatic nucleus (SCN) in mammals, and by peripheral oscillators present in various other tissues. The expression of essential clock genes oscillates both in the SCN and in peripheral pacemakers. Peripheral tissues in the fly and in the fish are directly photoreceptive. In particular, we have established the Z3 embryonic zebrafish cell line that recapitulates the dynamic light-dependent regulation of the vertebrate clock *in vitro*. In mammals the synchronization to daily light cycles involves neural connections from a subset of light-sensitive receptor-containing retinal ganglion cells. Humoral and/or hormonal signals originating from the SCN are thought to provide timing cues to peripheral clocks. However, alternative routes exist, as some peripheral clocks in mammals can be specifically entrained in a SCN-independent manner by restricted feeding regimes. Thus, not all peripheral tissues are equal in circadian rhythmicity. Testis, for example, displays no intrinsic circadian rhythmicity and the molecular mechanisms of clock gene activation in male germ cells appear to differ from other tissues. The study of the connecting routes that link the SCN to peripheral tissues is likely to reveal signalling pathways of fundamental physiological significance.

2003 Molecular clocks and light signalling. Wiley, Chichester (Novartis Foundation Symposium 253) p 126–139

Neurons of the mammalian suprachiasmatic nucleus (SCN) contain cell-autonomous, self-sustained oscillators, which are able to maintain circadian periodicity even when isolated *in vitro* or when the animal is placed under

---

[1]This chapter was presented at the symposium by Paolo Sassone-Corsi, to whom correspondence should be addressed.

constant conditions (Welsh et al 1995). The mammalian circadian system was thought to be based on this unique centralized clock structure. The discovery that both vertebrates and invertebrates have a widely dispersed circadian timing system challenged this view (Balsalobre et al 1998, Giebultowicz et al 2000, Plautz et al 1997, Tosini & Menaker 1996, Whitmore et al 1998, Yamazaki et al 2000). Indeed, many tissues contain oscillators, and it is likely that every cell in a given tissue contains an intrinsic autonomous clock (Schibler & Sassone-Corsi 2002).

Drosophila (Giebultowicz et al 2000, Plautz et al 1997), zebrafish (Whitmore et al 1998) and mammals (Yamazaki et al 2000) have all been shown to possess circadian oscillators in various tissues, including non-neuronal tissues. For example, cultures of *Drosophila* wings and antennae (Plautz et al 1997), or of zebrafish hearts and kidneys (Whitmore et al 1998), display circadian oscillations of clock genes in constant conditions. Strikingly, these peripheral clocks display independence from the central clock. For example, clock gene oscillations exhibit distinct patterns of expression from tissue to tissue in the zebrafish (Cermakian et al 2000), whereas in *Drosophila* excretory tubules taken from one fly maintain their phase of oscillations even when grafted onto another fly that is entrained on a reversed light–dark (LD) cycle (Giebultowicz et al 2000).

Z3: a light-entrainable vertebrate cell line

Peripheral clocks in *Drosophila* and zebrafish display a striking feature: they are directly light responsive. No need for an eye or other specialized structures, circadian expression of clock genes in cultured *Drosophila* tissues (Plautz et al 1997), and zebrafish organs (Whitmore et al 2000), can be directly reset by LD cycles. An intriguing possibility is that circadian photoreception could employ distinct photopigments in the retina and in peripheral tissues. In the case of zebrafish, light-responsiveness has even been demonstrated for cultured cells (Pando et al 2001). Indeed, we have established a zebrafish embryonic cell line, named Z3, that has the unique feature to recapitulate the light response characteristics of a vertebrate clock. In Z3 cells, oscillations of clock gene expression can be entrained to new LD cycles and *Per2* gene transcription responds acutely to short exposure to light (Pando et al 2001). To establish the signalling pathways that light utilizes to elicit induction of clock gene expression, we have analysed the action spectrum of *Per2* transcriptional induction in Z3 cells. The window of wavelength derived from the action spectrum is compatible with blue-light photoreceptors, implicating cryptochromes (CRYs) as likely photoreceptors in this system. Indeed, our study has revealed that light-induced expression of clock genes in Z3 cells involves a subset of the six known zebrafish CRYs. Using a pharmacological

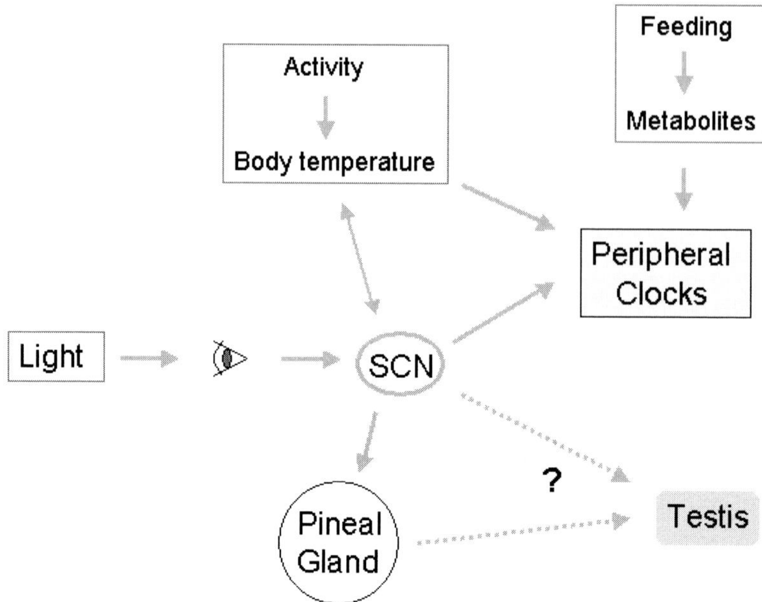

FIG. 1.   Communication pathways between the central and peripheral circadian clocks in nocturnal rodents. SCN neurons receive light information directly from retinal cells via the retinohypothalamic tract (RHT). This photic entrainment corrects the phase of the SCN oscillator every day to ensure synchronization of circadian rhythm with geophysical time. The SCN synchronizes peripheral clocks in organs such as liver, heart, and kidney via direct and indirect routes. Indirect synchronization is accomplished by controlling daily activity-rest cycles and, as a consequence, feeding time. Feeding (or starving) cycles are dominant *zeitgebers* for many, if not most peripheral clocks. Food metabolites, such as glucose, and hormones related to feeding and starvation, are probably the feeding-dependent entrainment cues. Activity cycles also influence body temperature rhythms, which in turn can participate in the phase entrainment of peripheral clocks. Direct entrainment may employ cyclically secreted hormones and perhaps neuronal signals conveyed to peripheral clocks via the peripheral nervous system. Body temperature rhythms, which in part are controlled by the SCN, may also contribute to the synchronization of peripheral clocks. As peripheral tissue, the testis represents a unique case as it was shown to display no rhythmicity (Morse et al 2003). As the hypothalamic–pituitary–gonadal axis governs testicular function in a seasonal/photoperiodic manner, it is possible that alternative routes — different from other peripheral tissues — control the expression of clock genes in male germ cells. In particular, the contribution of the pineal gland in the control of gonadal function and as a relay of clock information into endocrine signalling is of interest.

approach it was established that light-induced *Per2* transcriptional induction in Z3 cells requires the intracellular MAPK (mitogen activated protein kinase) signalling pathway (Cermakian et al 2002). Thus, light signalling has been directly coupled to activation of the MAPK transduction route, leading to stimulated transcription. An exciting challenge for future studies will be to identify light-responsive promoter sequences and subsequently the nuclear factors that bind directly to them.

Peripheral oscillators are (almost) everywhere in mammals

Mammals display no photoreception in peripheral tissues. The effect that light has on peripheral oscillators in mammals is indirect: the SCN integrates photic cues from the retina and the retinohypothalamic tract (RHT), and then synchronizes peripheral oscillators through output pathways (Abe et al 2002, Brown & Schibler 1999, Yamazaki et al 2000). In the absence of SCN signals, oscillations (in clock gene transcripts or in expression of a reporter in *Per1*–luciferase transgenic animals) rapidly dampens in peripheral oscillators (Abe et al 2002, Balsalobre et al 1998, Yamazaki et al 2000). The signals from the central clock must thus entrain these dampened oscillators. These signals could follow neuronal pathways, either to various areas of the brain (Abe et al 2002, LeSauter & Silver 1998) or to tissues via the autonomic nervous system (Ueyama et al 1999). The SCN was also proposed to reset peripheral clocks through humoral signals (Fig. 1). This is supported by the observation that a serum shock can induce oscillations in cultured fibroblasts (Balsalobre et al 1998), and that forskolin, an adenylate cyclase activator, can restart oscillations in dampened tissues *in vitro* (Abe et al 2002, LeSauter & Silver 1998, Yamazaki et al 2000). Moreover, co-culturing of SCN neurons with NIH 3T3 cells induces oscillations in the fibroblasts via a yet unidentified signalling molecule that can pass through a semi-permeable membrane (Allen et al 2001).

What could be the nature of this diffusible signal? Some substances have been proposed as candidates. Glucocorticoids seem to play a role, since dexamethasone was shown to induce the same circadian gene expression in fibroblasts as does serum shock, and can provoke transient changes in the phase of clock gene oscillations in peripheral tissues when injected into mice (Balsalobre et al 2000). Another possible synchronizer is retinoic acid, which can delay the *Per2* rhythm in vascular smooth muscle cells both in culture and *in vivo*, possibly due to an interaction of retinoic acid receptor with CLOCK or its homologue MOP4 (McNamara et al 2001). Thus, different signalling routes may be able to mediate distinct responses in peripheral clocks (Schibler & Sassone-Corsi 2002).

Uncoupling the SCN from peripheral clocks

Food may also result in diffusible signals affecting the phase of peripheral clocks. For example, when feeding of nocturnal animals like mice is restricted to daytime

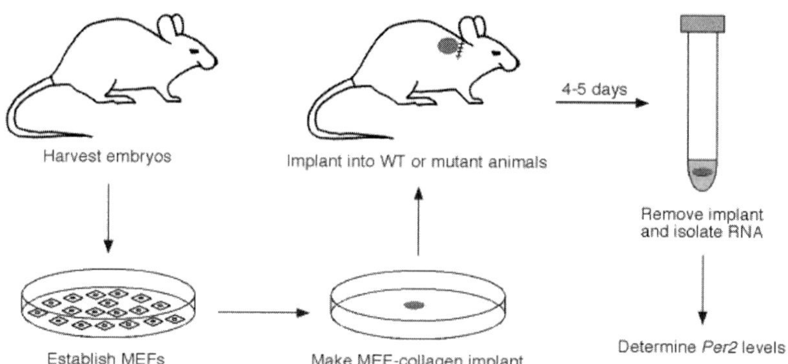

FIG. 2.   A procedure to transplant a peripheral clock. Following the production of mouse embryo fibroblasts (MEFs), the cells are resuspended in a medium supplemented with collagen. After approximately 12 h, a durable collagen matrix forms around the MEFs that confines them in a disk, or implant, with a diameter of approximately 1.5 cm. This collagen disk is then implanted subcutaneously on the back of a host mouse. After several days, implants and mouse tissues are collected for analysis. This approach allows a large number of potential combinations where the host mouse and/or the MEFs may carry genetically targeted mutations. The signalling pathways governing the communications between the SCN and peripheral tissues can be analysed.

instead of being available *ad libitum*, the phase of peripheral oscillators (for example in the liver and in the kidney) is observed to be different from that of the SCN (Damiola et al 2000, Stokkan et al 2001). The mechanism underlying this entrainment is still unknown, but one intriguing observation has been made recently: on shifting from food *ad libitum* to restricted food accessibility, peripheral clocks take some time to entrain to their new phase, and this slow phase resetting is due to glucocorticoids (GCs) (Le Minh et 2001). Mice that do not make GCs have peripheral clocks that entrain much faster, as do organs lacking GC receptor. Thus GCs apparently have opposite roles, first in causing a phase shift when injected in mice, and second, in opposing the phase shift caused by food-induced entrainment. These experiments, in addition to giving insight into how environment impinges on peripheral clock function, provide another possible way by which the central clock could synchronize peripheral oscillators: the SCN indirectly controls the time of feeding by regulating activity rhythms. In nocturnal animals, most of the food is absorbed during the dark period (Damiola et al 2000). In this view, food restriction to daytime would bypass the normal communication between the SCN and the peripheral oscillators. A consequence is that this pathway must be dominant over neuronal and humoral pathways discussed before, as during daytime feeding, the SCN and peripheral tissue rhythms appear completely uncoupled (Fig. 1). Differential food intake directly influences the balance of energy metabolism. Importantly, glucose was shown to suppress *Per1* and *Per2* expression and consequently induce circadian clock function in cultured cells that have been serum-starved (Hirota et al 2002). Day–night oscillation in glucose uptake and utilization could therefore contribute to food-dependent synchronization of peripheral clocks, possibly via a redox mechanism (Hirota et al 2002).

An experimental approach to study SCN–periphery communication

Targeted mutagenesis of the *Per1* gene in the mouse by homologous recombination results in mild phenotypic changes in circadian function (Bae et al 2001, Cermakian et al 2001, Zheng et al 2001). *Per1* mutant mice are able to entrain to standard LD cycles, display a modest shortening of the free-running circadian period, a slight decrease in oscillator precision and no significant alterations of clock gene expression in the SCN. Interestingly, rhythmic expression of clock genes in peripheral tissues appeared slightly delayed in the *Per1* mutant mice. These studies suggested that *Per1* plays a specialized function in the molecular mechanism of peripheral clocks (Cermakian et al 2001). Indeed, in support of this supposition, it was shown that, in drastic contrast to the phenotype of the *Per1* null mice, *Per1*-deficient peripheral oscillators placed in culture display an intrinsic period of only 20 h. This observation was confirmed by isolating mouse embryo fibroblasts (MEFs) from wild-type and *Per1* null mice which were placed in culture and then serum-shocked to induce circadian gene expression. The different function of PER1 in peripheral tissues was again demonstrated by a 20 h period of the *Per1* null MEFs (Pando et al 2002).

This difference was exploited to investigate the functional dependence of the peripheral clocks on the central pacemaker. MEFs originating from different mice were encapsulated in a collagen disc and implanted surgically into host animals of various genotypes. In this setting the MEFs represent a sort of transposable peripheral tissue (Fig. 2). This approach allows *in vivo* analysis of the physiological relationships between the host SCN and the implanted MEFs. It was shown that, under normal physiological conditions, the SCN is able to rescue or compensate for genetic defects affecting the period of peripheral clocks. For example, the *Per1*-deficient MEFs that in culture display a 20 h period, implanted in wild-type mice modify their period to a classical 24 h. Thus, the SCN is able to bypass the intrinsic genetic defect of a peripheral clock. This is provided that rhythmicity is still present, as the complete lack of clock function in MEFs from *Clock c/c* mice cannot be rescued (Pando et al 2002). Overall these results show that peripheral clocks are subordinated to the dominance exerted by the central clock and exhibit the characteristics of the host SCN.

The approach of MEF implants is somewhat complementary to the well-established one of SCN transplantation. The extreme versatility of this technology is based on the possibility of using host mice and/or MEF implants carrying targeted mutations of specific genes encoding any type of signalling molecules. It is evident that this approach will allow the *in vivo* identification and characterization of the signalling pathways regulating the physiological communications between SCN and periphery.

A very special peripheral tissue: the testis

Spermatogenesis is a complex sequence of events that results in the formation of haploid spermatozoa from precursor stem cells. The process starts with the proliferation and differentiation of diploid spermatogonial stem cells to give rise to diploid spermatocytes, which undergo meiosis to produce round haploid spermatids. These cells are sculpted into their final elongate mature shape in the process of spermiogenesis, which involves extensive biochemical and morphological remodelling. The entire process requires about 35 days in mice. A remarkable number of highly specific transcriptional events regulate the differentiation of male germ cells (Sassone-Corsi 2002). The development of germ cells within the seminiferous tubule is tightly regulated by a finely tuned hormonal program orchestrated by the hypothalamic–pituitary–gonadal axis. In many rodents the seasonal variations in gonadotropin synthesis are directly linked to the photoperiod and result in drastic changes in the production of mature germ cells. In regimes of short photoperiod spermatogenesis is arrested, mimicking the natural situation of many wild rodents. This scenario is unique for the testis tissue, placing it in a unique position within the organism.

The study of clock gene expression in the testis is of interest as it may provide clues on whether the circadian clock plays a role in timing some of the developmental events which take place during spermatogenesis. Strikingly, it was found that expression of clock components, such as *Per1*, does not oscillate in the testis. Importantly, the CLOCK/BMAL1 heterodimer (Darlington et al 1998), the usual transactivator of *Per1* expression, appears not to be involved in *Per1* regulation in male germ cells. This conclusion is based on the observations that *Clock* and *Per1* transcripts peak in different cell types within the seminiferous tubules, and that *Per1* expression is not decreased in *Clock c/c* mice (Morse et al 2003). This latter observation suggests that another transcription factor, perhaps one restricted to the first stages of spermiogenesis, might be involved in *Per1* expression instead of CLOCK. One such transcription factor specific to spermiogenesis is CREM, whose protein levels peak in spermatids in a more concerted manner than those of *Per1*. CREM was previously shown to operate as a master-switch for the transcription of a number of post-meiotic genes

(Sassone-Corsi 2002). In addition the *Per1* promoter contains a CREM binding site (Travnickova-Bendova et al 2002).

What is special about the testis that requires the absence of the circadian clock so pervasive in other tissues? The testis has a number of characteristics, which make it quite different from other tissues. The testis contains spermatogenic cells that perform a constant and complex cell differentiation program where reductive cell divisions occur. It may be that the complex pattern of gene expression engendered by the circadian clock leads to unfavourable interactions with the developmental process of paramount importance to the testis. Alternatively, the normal oscillation of clock gene expression may be distorted by other transcriptional regulators or co-activators which are only present in seminiferous tubules (Sassone-Corsi 2002).

It is of course important to note that the circadian timing system does play an important role in determining the reproductive capacity of seasonally breeding mammals. For example, the syrian hamster experiences testicular atrophy during long night photoperiods (LD 12:12) but is normal during short night photoperiods (LD 14:10). The involvement of the circadian clock can be seen from the photoperiodic response of the *tau* mutant hamster, which has a short free-running period of 20 h. Interestingly, these mutant hamsters experience testicular atrophy with a 10 h 'short night' dark period (Stirland et al 1996). It is thought that the circadian clock measures melatonin levels, as pinealectomized mutant hamsters receiving melatonin infusions for 10 h every 20, experienced gonadal atrophy (Stirland et al 1996). It is unlikely however, that the testis is reacting directly to these melatonin signals. It is known that the levels of luteinizing hormone (LH) and follicle-stimulating hormone (FSH), which have profound effects on testis development, are affected by photoperiod (Lincoln et al 1977), suggesting that a circadian clock in the testis is not required to mediate the effects of the photoperiod. In any event, these results indicate that the testis is devoid of the basic molecular circadian clockwork observed both in the suprachiasmatic nucleus and in numerous non-neuronal tissues. Although we cannot rule out the possibility that a circadian clock involving different molecular components may be operating in the testis, our results indicate that the testis is unlikely to employ a circadian clock in timing developmental processes. Looking for the mechanism and the consequences of this unexpected situation may allow us to uncover novel unexpected roles of so-called clock genes as well as to provide crucial data on the timing of male germ cell differentiation.

Conclusions

The wealth of information accumulated over recent years illustrates how complex the nature of the links between the environment and the clock is. It will be important to understand the exact contribution of different input pathways to the entrainment of clocks, in particular when different synchronizers can act. For example, the SCN can be entrained by photic signals, but also by physiological non-photic cues (Hastings et al 1997, Mrosovsky 1996). Peripheral clocks can receive signals from the central clock, but also other signals from the environment, for example when food is restricted to daytime (Fig. 1). How do these cues impact on the oscillators, and how is their combined action processed by the local clock mechanism? How peripheral clocks can be reset and how food restriction can specifically entrain peripheral clocks are additional questions of doubtless physiological relevance. It is likely that we will all gather again at the Novartis Foundation in a few years and be amazed by the progress made and also challenged by new questions to be addressed.

Acknowledgements

Work in P.S.C.'s laboratory is supported by grants from CNRS, INSERM, CHUR, Human Frontier Science Program, Organon Akzo/Nobel, Fondation pour la Recherche Médicale and Association pour la Recherche sur le Cancer.

References

Abe M, Herzog ED, Yamazaki S et al 2002 Circadian rhythms in isolated brain regions. J Neurosci 22:350–356

Allen G, Rappe J, Earnest DJ, Cassone VM 2001 Oscillating on borrowed time: diffusible signals from immortalized suprachiasmatic nucleus cells regulate circadian rhythmicity in cultured fibroblasts. J Neurosci 21:7937–7943

Bae K, Jin X, Maywood ES, Hastings MH, Reppert SM, Weaver DR 2001 Differential functions of mPer1, mPer2, and mPer3 in the SCN circadian clock. Neuron 30:525–536

Balsalobre A, Damiola F, Schibler U 1998 A serum shock induces circadian gene expression in mammalian tissue culture cells. Cell 93:929–937

Balsalobre A, Brown SA, Marcacci L et al 2000 Resetting of circadian time in peripheral tissues by glucocorticoid signaling. Science 289:2344–2347

Brown SA, Schibler U 1999 The ins and outs of circadian timekeeping. Curr Opin Genet Dev 9:588–594

Cermakian N, Whitmore D, Foulkes NS, Sassone-Corsi P 2000 Asynchronous oscillations of two zebrafish CLOCK partners reveal differential clock control and function. Proc Natl Acad Sci USA 97:4339–4344

Cermakian N, Monaco L, Pando MP, Dierich A, Sassone-Corsi P 2001 Altered behavioral rhythms and clock gene expression in mice with a targeted mutation in the Period1 gene. EMBO J 20:3967–3974

Cermakian N, Pando MP, Thompson CL et al 2002 Light induction of a vertebrate clock gene involves signaling through blue-light receptors and MAP kinases. Curr Biol 12:844–848

Damiola F, Le Minh N, Preitner N, Kornmann B, Fleury-Olela F, Schibler U 2000 Restricted feeding uncouples circadian oscillators in peripheral tissues from the central pacemaker in the suprachiasmatic nucleus. Genes Dev 14:2950–2961

Darlington TK, Wager-Smith K, Ceriani MF et al 1998 Closing the circadian loop: CLOCK-induced transcription of its own inhibitors per and tim. Science 280:1599–1603

Giebultowicz JM, Stanewsky R, Hall JC, Hege DM 2000 Transplanted *Drosophila* excretory tubules maintain circadian clock cycling out of phase with the host. Curr Biol 10: 107–110

Hastings MH, Duffield GE, Ebling FJ, Kidd A, Maywood ES, Schurov I 1997 Non-photic signalling in the suprachiasmatic nucleus. Biol Cell 89:495–503

Hirota T, Okano T, Kokame K, Shirotani-Ikejima H, Miyata T, Fukada Y 2002 Glucose down-regulates Per1 and Per2 mRNA levels and induces circadian gene expression in cultured Rat-1 fibroblasts. J Biol Chem 277:44244–44251

Le Minh N, Damiola F, Tronche F, Schutz G, Schibler U 2001 Glucocorticoid hormones inhibit food-induced phase-shifting of peripheral circadian oscillators. EMBO J 20:7128–7136

LeSauter J, Silver R 1998 Output signals of the SCN. Chronobiol Int 15:535–550

Lincoln GA, Peet MJ, Cunningham RA 1977 Seasonal and circadian changes in the episodic release of follicle-stimulating hormone, luteinizing hormone and testosterone in rams exposed to artificial photoperiods. J Endocrinol 72:337–349

McNamara P, Seo S, Rudic RD, Sehgal A, Chakravarti D, FitzGerald GA 2001 Regulation of CLOCK and MOP4 by nuclear hormone receptors in the vasculature. a humoral mechanism to reset a peripheral clock. Cell 105:877–889

Morse D, Cermakian N, Brancorsini S, Parvinen M, Sassone-Corsi P 2003 No circadian rhythms in testis: *Period1* expression is *Clock*-independent and developmentally regulated in the mouse. Mol Endocrinol 17:141–151

Mrosovsky N 1996 Locomotor activity and non-photic influences on circadian clocks. Biol Rev Camb Philos Soc 71:343–372

Pando MP, Pinchak AB, Cermakian N, Sassone-Corsi P 2001 A cell based system that recapitulates the dynamic light-dependent regulation of the vertebrate clock. Proc Natl Acad Sci USA 98:10178–10183

Pando MP, Morse D, Cermakian N, Sassone-Corsi P 2002 Phenotypic rescue of a peripheral clock genetic defect via SCN hierarchical dominance. Cell 110:107–117

Plautz JD, Kaneko M, Hall JC, Kay SA 1997 Independent photoreceptive circadian clocks throughout *Drosophila*. Science 278:1632–1635

Sassone-Corsi P 2002 Unique chromatin remodeling and transcriptional regulation in spermatogenesis. Science 296:2176–2178

Schibler U, Sassone-Corsi P 2002 A web of circadian pacemakers. Cell 111:919–922

Stirland JA, Hastings MH, Loudon AS, Maywood ES 1996 The tau (τ) mutation in the Syrian hamster alters the photoperiodic responsiveness of the gonadal axis to melatonin signal frequency. Endocrinology 137:2183–2186

Stokkan KA, Yamazaki S, Tei H, Sakaki Y, Menaker M 2001 Entrainment of the circadian clock in the liver by feeding. Science 291:490–493

Tosini G, Menaker M 1996 Circadian rhythms in cultured mammalian retina. Science 272: 419–421

Travnickova-Bendova Z, Cermakian N, Reppert SM, Sassone-Corsi P 2002 Bimodal regulation of mPeriod promoters by CREB-dependent signaling and CLOCK/BMAL1 activity. Proc Natl Acad Sci USA 99:7728–7733

Ueyama T, Krout KE, Nguyen XV et al 1999 Suprachiasmatic nucleus: a central autonomic clock. Nat Neurosci 2:1051–1053

Welsh DK, Logothetis DE, Meister M, Reppert SM 1995 Individual neurons dissociated from rat suprachiasmatic nucleus express independently phased circadian firing rhythms. Neuron 14:697–706

Whitmore D, Foulkes NS, Strahle U, Sassone-Corsi P 1998 Zebrafish Clock rhythmic expression reveals independent peripheral circadian oscillators. Nat Neurosci 1:701–707

Whitmore D, Foulkes NS, Sassone-Corsi P 2000 Light acts directly on organs and cells in culture to set the vertebrate circadian clock. Nature 404:87–91

Yamazaki S, Numano R, Abe M et al 2000 Resetting central and peripheral circadian oscillators in transgenic rats. Science 288:682–685

Zheng B, Albrecht U, Kaasik K et al 2001 Nonredundant roles of the mPer1 and mPer2 genes in the mammalian circadian clock. Cell 105:683–694

DISCUSSION

Green: You discuss your data from the transplant experiments as proof of hierarchical dominance of the SCN. Have you done SCN lesions in the host animals to show that the effects on those tissues are actually from the SCN?

Sassone-Corsi: That's a good question, and we haven't done this. I would expect the effects to be from the SCN, but we need to do this experiment. I would love to see a SCN transplant in a *Clock$^{c/c}$* mutant mouse. But if what Bert van der Horst said in his paper (Bonnefont et al 2003, this volume) is true, that peripheral clocks are not working in *Cry* double knockouts where a normal SCN is introduced, this tells me that all peripheral oscillators are not crucial for motor rhythmic activity. Could the SCN be the only thing responsible for all the rhythmic activity? I am not sure how much peripheral tissues are working in those *Cry* double knockouts.

van der Horst: What we cannot say in this respect is to what extent the peripheral oscillator might still function. Analogous to light potentially driving an hour-glass timekeeper in the *Cry*-deficient SCN, we need to determine whether this transplanted intact wild-type SCN can give signals to *Cry*-deficient peripheral oscillators, that therefore may also start to work as an hour-glass type of timekeeper.

Okamura: I would like to add a comment about the transplantation study presented by Bert van der Horst, because I also collaborate with Dr Inouye on this project (Sujino et al 2003). We have only looked at one or two animals so far, so we can't come to any firm conclusions, but the results so far are interesting. When we examined it using the SCN transplant, inside the transplant we found a rhythm similar to the wild-type. Sometimes the transplants are not near the SCN, but these also restored rhythm. In this case, the innervation from the transplant to host brain regions may not be intact.

Sassone-Corsi: Do you know what happens in peripheral tissues?

Okamura: We didn't measure the clock gene expression in peripheral tissues. But there are indications that some signals were transmitted from the SCN transplant to the cortex.

Sehgal: I was curious about the short period in your MEFs. What are the *in vivo* implications of this?

Sassone-Corsi: That is a very good question. We know that in the animal this is somehow 'corrected' by the SCN. I think that once the cells are in culture we are unmasking an effect that was previously hidden. In the fibroblasts within the animal I can imagine that perhaps at specific times when the SCN is not working there could still be a rhythm.

Sehgal: Have you tried any food entrainment?

Sassone-Corsi: We have done food entrainment, but not on the *Per1* knockouts. The implant shows food entrainment, and thereby functions as a *bona fide* peripheral oscillator.

Rosbash: Doesn't this simply suggest that the individual components of the system are more fragile or vulnerable than the system?

Sassone-Corsi: That is obvious. The question was about the *in vivo* implication of the *Per1* mutation for the physiology of the peripheral tissue.

Sehgal: My question was, if you see a short period in culture, what difference does this make to the peripheral clock *in vivo*?

Menaker: Let me comment on that. You would expect that in the intact animal, all the rhythms are going to have the same period as long as things are functioning normally. But if the free-running period of a peripheral oscillator such as this one is 20 h, this doesn't mean that it is fragile, and it doesn't mean that it will be 20 h in the intact animal. But it does mean that if it is entrained by an oscillator of the SCN that the phase of the rhythm is going to depend on its period, other things being equal. Other things being equal, the period will determine the phase relationship. It is possible that it is an adaptive response in order to control the phase.

Weitz: This was a mutant. It is not the native period of the peripheral tissue. Something is different about the structure of the oscillator.

Sassone-Corsi: Something is different but the period is now 4 h shorter if you remove it from the animal. I think this tells us how plastic the system is.

Sehgal: It can't be good for the animal to have these clocks that are running with a different period having a 24 h rhythm imposed on them.

Menaker: Why not?

Sassone-Corsi: Those animals are fine.

Rosbash: When the SCN is removed in culture from those animals, what is the period? This might provide some insight.

Sassone-Corsi: We haven't done that.

Schibler: It will be important to dissociate the SCN from this.

Rosbash: Why don't you like the idea that the period in the animal is an integrated systems output phenomenon? When you start taking the pieces apart, especially under mutant conditions, you get less robust or less well contained periods and amplitudes.

Sassone-Corsi: I think that what these experiments are telling us is that it depends on what damage we do to the clock as to whether this can be fixed. In the case of *Per1* mutations the damage can be fixed, but with *Clock* mutants it cannot be fixed.

Rosbash: The second data point is so extreme. If you put in a stick of dynamite it will not be fixable.

Menaker: I would suggest using the term 'over-ridden' rather than 'fixed'.

Sassone-Corsi: The experiment in the clock heterozygote animals is revealing. The fact that those 20 h period cells can go all the way to 27.5 is telling us that in terms of the transcriptional organization of the promoter, it is possible to modulate things at a remarkable level. This is not surprising to me from the point of view of gene expression.

Van Gelder: Michael Menaker, I wanted to ask your opinion as to the rapidity of synchronization of these transplants compared to the rapidity of phase shifting that you see in your peripheral tissues. Presumably these MEFs are indeterminate in phase when they are transplanted yet they maintain perfect entrainment within 4 d to the peripheral tissue. They entrain within one cycle, which you never see in your phase-shifting experiments. Do you think this is symptomatic of a very easily entrainable oscillator in the *Per1* mutant, or is that a function of the experimental design?

Menaker: There are several possibilities. I suspect it is the experimental design. Essentially you have a phaseless oscillator which can be very rapidly phased.

Rosbash: Wouldn't you guess that it is not a question of phaseless or rapid phasing, but the fact that the liver really has a robust system of its own?

Menaker: That is the other side of the same thing.

Van Gelder: The *Per1* is not a normal oscillator. What we see in the cryptochromes is that *Cry2* has an exaggerated phase response curve probably because the oscillator is not as robust. It may be the case that *Per1* is also a weak oscillator with an accordingly exaggerated phase response curve.

Rosbash: If you take normal MEFs what happens?

Cermakian: Normal MEFs entrain well and as quickly as the *Per1* MEFs.

Rosbash: That argues that it is the robustness of the system and not the fragility of the *Per1* mutant.

Schibler: A liver cell is about 20 times bigger than a splenocyte. This means it has to be about 20 times more active transcriptionally. If you look experimentally, this is what you find. In other words, you can measure cell size by making a ratio of RNA:DNA and this correlates almost perfectly with incorporation of uridine into DNA. Then the question is, is the period maintained against a gradient of transcriptional activity? You could test this by taking spleen and liver, and looking at the free-running period in culture. Have you done that?

Menaker: We have looked at spleen. It is not different.

Green: Michael Menaker, I want to ask you a question related to an issue Paolo Sassone-Corsi raised. Your interpretation is that introduction of these various tissues into culture does not affect the phase. Why not? All the data seem to show that medium changes or similar types of stimulation should set phase.

Menaker: I agree, it is odd. We have not examined this systematically by doing different phases. We know that at the phase at which we take the tissue, culturing doesn't seem to have an effect. If we kill the animals at some other phase this may not be the case. By analogy to light input to the system I would say we are in the dead zone for whatever the effects of culturing are. I wouldn't be surprised if we were to find a phase effect at some other time.

Takahashi: Carla Green, in SCN slice experiments those controls were done. Explants were made at different time phases and there were phases when the tissues would reset. At the end of the day, when they normally make the SCN cultures, it is the dead zone, at least for the SCN.

References

Bonnefont X, Albus H, Meijer JH, van der Horst GTJ 2003 Light signalling in *Cryptochrome*-deficient mice. In: Molecular clocks and light signalling, Wiley, Chichester (Novartis Found Symp 253) p 56–72

Sujino M, Masumoto K, Yamaguchi S, van der Horst GTJ, Okamura H, Inouye S-IT 2003 Suprachiasmatic nucleus grafts restore circadian behavioral rhythms of genetically arrhythmic mice. Curr Biol 13:664–668

Central and peripheral circadian oscillators in *Drosophila*

Paul E. Hardin, Balaji Krishnan, Jerry H. Houl, Hao Zheng, Fanny S. Ng, Stuart E. Dryer and Nick R. J. Glossop

Department of Biology and Biochemistry, University of Houston, Houston, TX 771204-5001, USA

Abstract. Drosophila circadian oscillators comprise interlocked *period* (*per*)/*timeless* (*tim*) and *Clock* (*Clk*) transcriptional/translational feedback loops. Within these feedback loops, CLOCK (CLK) and CYCLE (CYC) bind E-box elements to activate *per* and *tim* transcription, and we now show that at the same time CLK–CYC repress *Clk* by activating the transcriptional repressor *vrille* (*vri*), thus accounting for the opposite cycling phases of these transcripts and identifying *vri* as the negative component of the *Clk*-feedback-loop. The core oscillator mechanism is assumed to be the same for oscillators in different tissues. However, we have shown that CRYPTOCHROME (CRY) has a light-independent function in the oscillator that controls olfaction rhythms, suggesting that CRY may function within the oscillator mechanism itself as it does in mammals. These olfaction rhythms require the function of 'peripheral' oscillators which are distinct from the 'central' lateral neuron (LN) oscillators that mediate locomotor activity rhythms. Preliminary results show that antennal oscillator cells are sufficient and LNs are not necessary for olfaction rhythms, indicating that unlike the situation in mammals, the central oscillator has little impact on the olfaction rhythm oscillator under these conditions.

2003 Molecular clocks and light signalling. Wiley, Chichester (Novartis Foundation Symposium 253) p 140–160

Circadian clocks are molecular time-keeping mechanisms found in a broad range of cell types from a variety of organisms. The primary roles of these clocks are to maintain their own ∼24 hour molecular rhythm and to drive the rhythmic expression of genes that control output processes in physiology, metabolism and behaviour. Core features of the clock are its ability to synchronize to daily environmental *zeitgebers* (e.g. light–dark or temperature cycles), and then maintain rhythmic function when placed in constant conditions.

The circadian timekeeping mechanism in *Drosophila* consists of interlocked *per*/*tim* and *Clk* feedback loops in gene expression (reviewed in Allada et al 2001, Meyer-Bernstein & Sehgal 2001, Young & Kay 2001, Glossop & Hardin 2002). Although CLK–CYC heterodimers bind E-boxes to activate *per* and *tim* transcription, little is known about how CLK–CYC repress *Clk* transcription and, consequently, drive *Clk* mRNA cycling in the opposite phase as *per* and *tim*. This interlocked feedback loop mechanism is thought to drive rhythmic outputs in a variety of cell types. In *Drosophila*, rhythmic outputs in locomotor activity and olfaction are controlled by different oscillator cells. Locomotor activity rhythms are controlled by central oscillator cells in the brain called small ventral lateral neurons (sLN$_v$s) and olfaction rhythms are controlled by peripheral oscillator cells that have not yet been identified (Helfrich-Forster 1996, Krishnan et al 1999). The aim of our work is to define the circadian timekeeping mechanism and determine how it controls rhythmic outputs. In particular, we will discuss our recent work on CLK–CYC-dependent repression of *Clk* and the identification of cells necessary and sufficient to control olfaction rhythms.

Within the interlocked feedback loop mechanism (reviewed in Allada et al 2001, Meyer-Bernstein & Sehgal 2001, Young & Kay 2001, Glossop & Hardin 2002), the *per/tim* feedback loop is initiated when CLK–CYC activate *per* and *tim* transcription during the early/mid day (∼ZT 04) (Fig. 1). PER and TIM proteins then accumulate and enter the nucleus after a substantial (∼6 h) phosphorylation-dependent delay, bind to CLK–CYC, and repress *per* and *tim* transcription during the mid-evening (∼ZT 16). Once PER and TIM are degraded during the early morning (∼ZT 02), the next cycle of *per* and

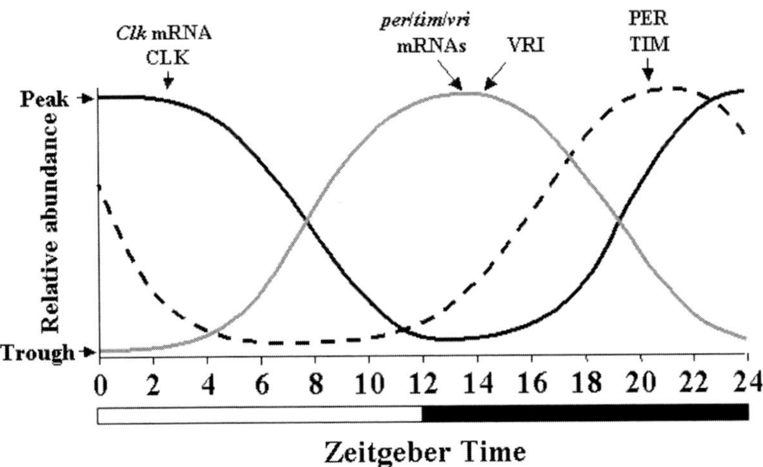

FIG. 1.   Phases of clock gene product cycling. The abundance of clock gene products are shown relative to their peak and trough values. *per*, *tim* and *vri* mRNA and VRI protein levels, grey line; *Clk* mRNA and CLK protein levels, black line; PER and TIM protein levels, dashed line. *Zeitgeber* Time, time in hours during a light-dark cycle where 0 h is lights-on and 12 h is lights-off. Open bar, lights on; closed bar, lights off.

tim transcription can begin. In the *Clk* loop, *Clk* transcription is repressed by CLK–CYC during the early/mid day (~ZT 04), and derepressed by rising levels of PER–TIM during the mid-evening (~ZT 16), which bind CLK–CYC to promote *Clk* transcription. A critical issue is how CLK–CYC repress *Clk*. The lack of canonical E-box binding sites for CLK–CYC in and around *Clk* suggests that repression occurs indirectly, probably through the activation of a transcriptional repressor. A prime candidate for such a repressor is the basic-zipper (bZIP) transcription factor VRILLE (VRI) because (1) *vri* is activated by CLK–CYC, (2) over-expression of VRI reduces or eliminates expression of two CLK–CYC dependent transcripts (*per* and *tim*), and (3) VRI acts as a repressor genetically since over-expression leads to long period rhythms and reducing *vri* gene dosage results in a short period rhythm (Blau & Young 1999).

If VRI directly represses *Clk* expression, this would predict that VRI must cycle in antiphase to *Clk* mRNA, that VRI binding sites will be present in the circadian regulatory region of *Clk*, and that VRI over-expression will repress *Clk* mRNA levels *in vivo*. These predictions were tested in the following ways. First, an antibody generated against VRI shows that it cycles in phase with *vri* mRNA and antiphase to *Clk* mRNA (Fig. 1). Second, the VRI DNA binding domain is almost identical to that of the mammalian transcription factor E4BP4 (George & Terracol 1997), suggesting that it binds the same target sequence. Several perfect or near-perfect E4BP4 target sequences are found within an 8.0 kb *Clk* genomic fragment that mediates *Clk* mRNA cycling, and VRI binds strongly to several of these sites. Third, when VRI over-expression is induced in a cyc^{01} mutant background, in which *Clk* is constitutively expressed at peak levels (Glossop et al 1999), *Clk* mRNA levels fall to about half of their peak value. These results fulfil the predictions above, thus identifying VRI as an integral component of the interlocked feedback loop whose role is to repress *Clk* transcription (Fig. 2).

The mammalian circadian oscillator is also comprised of two interacting feedback loops: a *Per/Cry* loop and a *Bmal1* loop (reviewed in Reppert & Weaver 2001, Glossop & Hardin 2002). The *Per/Cry* loop is analogous to the *per/tim* loop in *Drosophila*. Initially, mammalian CLOCK forms heterodimers with BMAL1 (the mammalian CYC homologue) and drives rhythmic transcription of three *per* homologues (*Per1*, *Per2* and *Per3*) and two *cry* homologues (*Cry1* and *Cry2*). PER proteins (at least PER1 and PER2) then form complexes with CRY proteins, move into the nucleus and repress CLOCK–BMAL1-dependent expression. The *Bmal1* loop is analogous to the *Clk* loop in *Drosophila*. *Bmal1* is first activated in a PER2-dependent manner, and later repressed in a CLOCK–BMAL1-dependent manner. As in *Drosophila*, this repression occurs indirectly, but in this case CLOCK–BMAL1 activates *Rev-Erbα*, which encodes an orphan nuclear receptor that represses *Bmal1* transcription (Preitner et al 2002). Rising levels of PER–CRY in the nucleus then repress *Rev-Erbα*, thus relieving the repression of *Bmal1* and

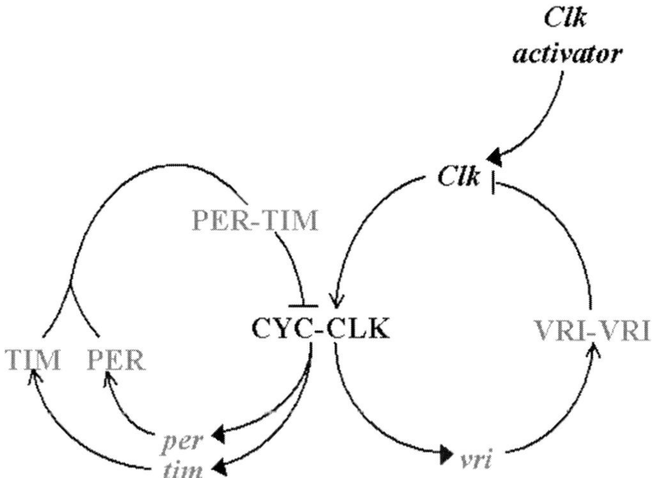

FIG. 2.   Model of the interlocked feedback loop mechanism in *Drosophila*. The *per/tim* loop (left) and *dClk* loop (right) are shown. Transcriptional activator genes, black italics; transcriptional activator proteins, black capitals; transcriptional repressor genes, grey italics; transcriptional repressor proteins, grey capitals. Filled arrows, transcriptional activation; open arrows, translation; bars, repression.

initiating the next cycle (Preitner et al 2002). Although the interlocked feedback loops of flies and mammals are mechanistically similar, they differ in three ways. First, mammalian CRY appears to have taken the place of TIM in that it binds to PER and promotes PER nuclear localization (Reppert & Weaver 2001, Glossop & Hardin 2002). Second, the regulation of *Bmal1* and *Clock* in mammals is switched compared to their fly homologues: *Bmal1* and *Drosophila Clk* are rhythmically expressed whereas *cyc* and mammalian *Clock* are constitutively expressed (Reppert & Weaver 2001, Glossop & Hardin 2002). Third, the bZIP transcription factor VRI represses *Clk* in *Drosophila*, but the orphan nuclear receptor REV-ERBα represses *Bmal1* in mammals (Preitner et al 2002). Each of these differences concerns the identity of factors that carry out conserved regulatory steps within the feedback mechanism, indicating that each of these steps is important for circadian oscillator function regardless of the factor that carries out that step.

The same interlocked feedback loop mechanism is thought to operate in circadian oscillator cells throughout the *Drosophila* circadian system. However, studies of the blue light photoreceptor CRYPTOCHROME (CRY) suggest that central and peripheral oscillator mechanisms in *Drosophila* are not the same. *Drosophila* CRY was initially identified as a photoreceptor that mediates light

a

EAG Response (mV)
Wildtype
cry^b
27°C
18°C
Zeitgeber Time (ZT)

b

EAG Response (mV)
Wildtype
cry^b
Time (hr) at constant low (18°C) temperature

c

EAG Response (mV)
Wildtype
cry^b
Time (hr) at constant high (27°C) temperature

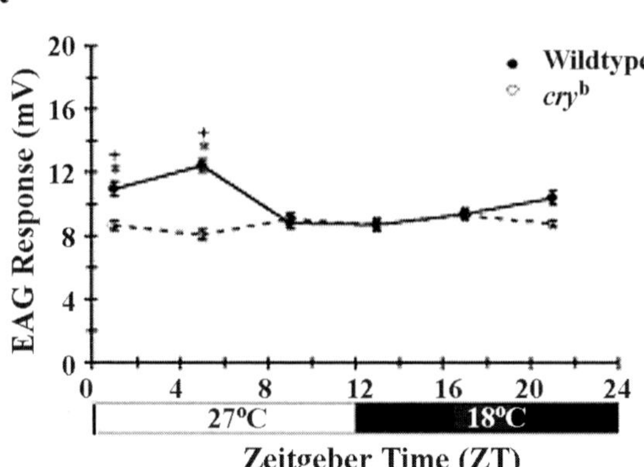

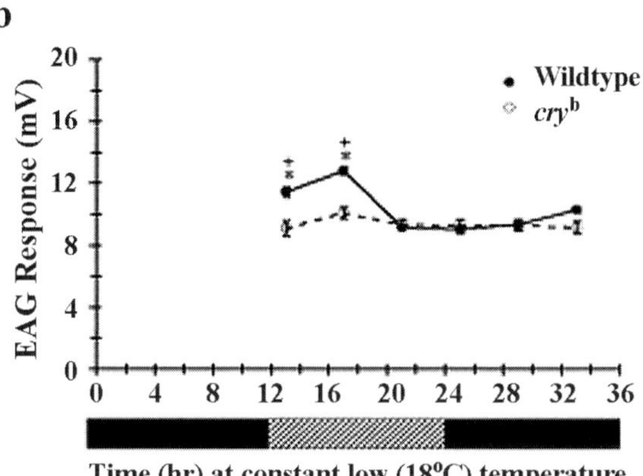

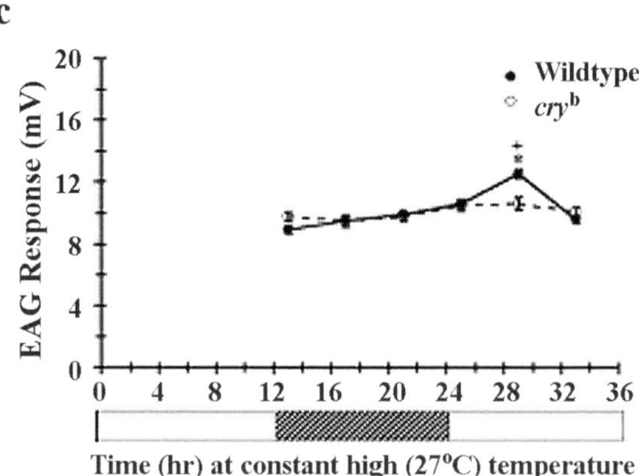

input to central and peripheral oscillators (Emery et al 1998, Egan et al 1999, Ishikawa et al 1999, Emery et al 2000a, Emery et al 2000b). A *Drosophila cry* mutant, *cry*[b], retains central oscillator function, but renders peripheral oscillators largely arrhythmic (Stanewsky et al 1998). Although this arrhythmicity could be due to a loss of light entrainment, it is also consistent with a role for CRY in the oscillator. A peripheral oscillator controls rhythms in olfactory responses in *Drosophila* antennae (Krishnan et al 1999) (see below). As expected, olfactory responses are rendered arrhythmic in the *cry*[b] mutant (Krishnan et al 2001). To determine whether this arrhythmicity is due to an inability to entrain to light, we entrained flies to temperature cycles. Although wild-type flies had robust olfaction rhythms both during and after temperature entrainment, *cry*[b] flies were arrhythmic under these conditions (Fig. 3). Importantly, loss of olfaction rhythms in temperature-entrained *cry*[b] flies results from a non functional oscillator because rhythmic *per* and *tim* transcription in antennae is severely crippled compared with that in temperature entrained wild-type flies (Krishnan et al 2001). These results demonstrate a photoreceptor-independent role for CRY in the peripheral oscillator controlling olfaction rhythms, and imply that the central and peripheral oscillator mechanisms are different. Such mechanistic differences are supported by results showing that PER and TIM oscillations are abolished in the renal Malpighian tubules of *cry*[b] flies, but not in the central sLN$_v$s (Ivanchenko et al 2001). These results support a role for CRY as a component of the timekeeping mechanism in peripheral tissues, in line with the situation in mammals where CRY1 and CRY2 are integral components of the timekeeping mechanism.

Although circadian oscillators are found in many tissues, only two rhythmic outputs have been identified in *Drosophila* adults. The most extensively studied rhythmic output is locomotor activity. A group of 4–5 sLN$_v$s in each hemisphere of the brain is both necessary and sufficient to drive robust activity rhythms (Frisch et al 1994, Renn et al 1999). The other rhythmic output is in olfactory responses, which are measured using an assay of odour-induced

---

FIG. 3.   EAG responses of wild-type and *cry*[b] flies during and after temperature entrainment. (a) Responses during temperature entrainment. White bar, high (27 °C) temperature; black bar low (18 °C) temperature. $n = 24$ flies per point. Overall effects of time of day, genotype, and their interaction are significant ($P < 0.0000001$). Asterisk indicates significant ($P < 0.0005$) difference in wild-type flies at ZT17 versus ZT1 and ZT5. Crosses indicate significant ($P < 0.0001$) difference between wild-type and *cry*[b] flies at the indicated times. (b) Responses in constant low temperature. Black bar, subjective low temperature; hatched bar, subjective high temperature. $n = 48$ flies per point. Overall effects of time of day, genotype and their interaction are significant ($P < 0.0000001$). (c) Responses in constant high temperature. White bar, subjective high temperature; hatched bar, subjective low temperature. $n = 48$ flies per point. Overall effects of time of day, genotype, and their interaction are significant ($P\ 0.0005$). (Reprinted from Krishnan et al 2001, with the permission of Macmillan Publishers Ltd.)

electrophysiological responses in the antennae called an electroantennagram (EAG) (Krishnan et al 1999). EAG responses to ethyl acetate were rhythmic in wild-type flies, but not in *per*[01] or *tim*[01] mutants, under light–dark (LD) cycling conditions and in constant darkness (DD) (Fig. 4). To determine whether these antennal rhythms were controlled by the central oscillator, we tested a transgenic strain of *Drosophila* (i.e. 7.2:2) which rescues oscillator function exclusively in the sLN$_v$s for rhythms in EAG responses (Frisch et al 1994). Although 7.2:2 flies were rhythmic for locomotor activity, their EAG responses were arrhythmic (Fig. 5). This result shows that the sLN$_v$s are not sufficient for EAG rhythms, and that peripheral oscillators are required for EAG rhythms.

To understand how the clock controls olfaction rhythms, cells that are sufficient and necessary for these rhythms must first be identified. Using the bipartite GAL4/ UAS expression system (Brand & Perrimon 1993), we have generated flies that lack sLN$_v$s to determine whether the central oscillator is necessary for olfaction rhythms. Preliminary results from these flies show no gross deficit in EAG responses at the peak and trough timepoints, thus indicating that the sLN$_v$s are not necessary for EAG responses. Since EAG responses are a measure of antennal function and antennae contain peripheral circadian oscillator cells, it is likely that olfaction rhythms are controlled locally by antennal oscillator cells. We have used the GAL4/UAS system to express *per* in antennal cells of *per*[01] flies, thus rescuing circadian oscillator function only in antennae. Using several independent strains to drive *per* expression in antennal cells, our preliminary results show that EAG responses are rhythmic, though the magnitude of the response is lower. This difference in magnitude could be due to several factors such as *per* expression encompassing only a subset of antennal oscillator cells or the contribution of other peripheral oscillators to olfaction rhythms. Nevertheless, this result suggests that antennal oscillator cells are sufficient to drive olfaction rhythms.

The relationship between central and peripheral oscillators is different in flies and mammals. In mammals, these oscillators form a hierarchy in which the central oscillator, which resides in the suprachiasmatic nucleus (SCN), functions as a master clock that is entrained by photic signals from the eye, and in turn drives subservient peripheral oscillators via humoral signals (Moore et al 1995, Yamazaki et al 2000, Kramer et al 2001, Cheng et al 2002). In contrast, both central and peripheral oscillators operate autonomously and are directly entrainable by light in *Drosophila* (Plautz et al 1997), thus obviating the need for a hierarchical system. Our results support the concept of independent oscillators in flies since central (sLN$_v$) oscillators are not necessary for olfaction rhythms and local oscillators in antennae appear to be sufficient.

In summary, we have shown that VRI is an integral component of the interlocked feedback loop mechanism in *Drosophila* whose role is to repress *Clk*

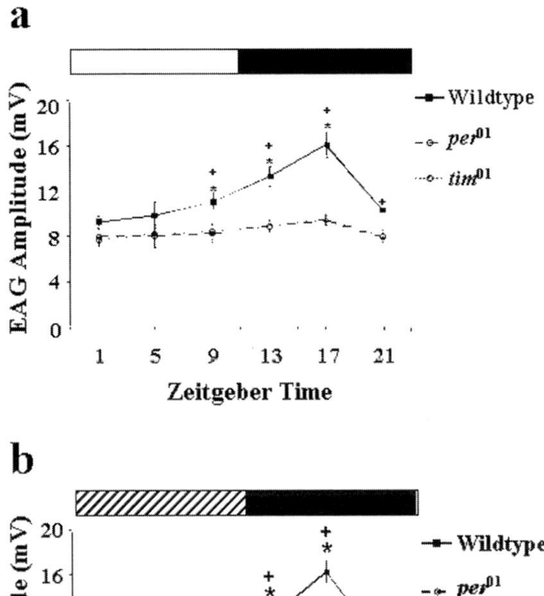

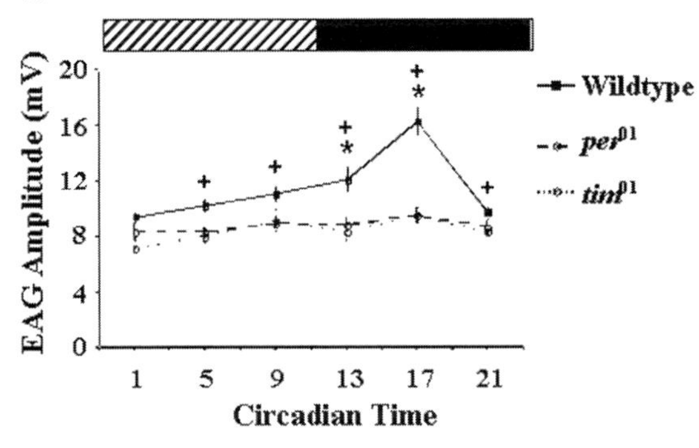

FIG. 4. Olfactory responses at different times of day in three different strains of *Drosophila melanogaster*. Each point represents mean odorant-evoked responses in the specified strain. Error bars denote SEM. (a) Diurnal changes in mean EAG responses to ethyl acetate ($1:10^4$ dilution) during LD 12:12 cycles. Each point represents the mean response of 24 flies. The dark and light bars represent when lights were on or off, respectively. *Zeitgeber* Time (ZT) denotes time during the LD cycle. Overall effects of time of day ($F_{T5,414} = 48.67$), genotype ($F_{G2,414} = 215.13$) and their interaction ($F_{TG10,414} = 14.01$) are statistically significant ($P < 0.0001$) by two-way ANOVA. Asterisks denote significant ($P < 0.05$) increase in EAG responses wild-type flies at ZT13 and ZT17 compared to responses at all other times of day in wild-type flies. Crosses indicate significant increase in EAG responses in wild-type flies compared to per^{01} and tim^{01} flies at the same times of day. No significant differences as a function of time of day were observed in per^{01} or tim^{01} flies by *post hoc* analysis. (b) A similar pattern is observed in flies free-running during day 2 of DD, indicating circadian control. Each point represents the mean response of 24 flies. The hatched and black bars represent subjective lights-on and off, respectively. Circadian time denotes time during constant darkness. Overall effects of time of day ($F_{T5,414} = 69.17$), genotype ($F_{G2,414} = 239.81$) and their interaction ($F_{TG10,414} = 19.09$) are statistically significant ($P < 0.0001$) by two-way ANOVA. (Reprinted from Krishnan et al 1999, with permission from Macmillan Publishers Ltd.)

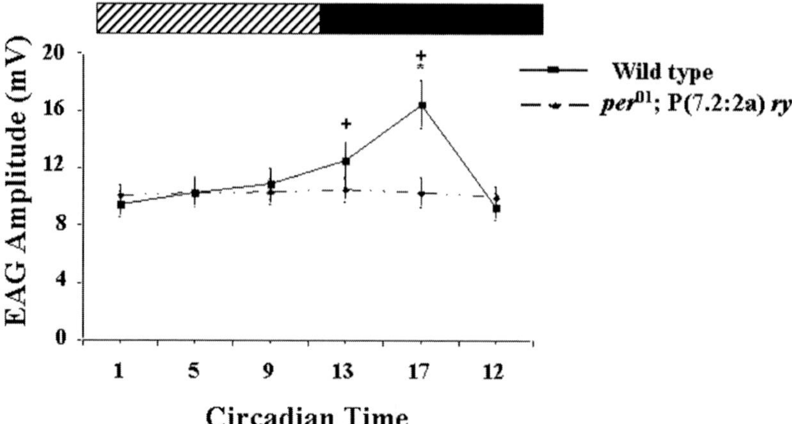

FIG. 5.   Mean olfactory responses in wild type (squares) and *per* 7.2:2 transgenic *Drosophila* (diamonds) during day 2 of DD. Each point represents the mean response to ethyl acetate ($1:10^4$ dilution) in 24 flies. Overall effects of time of day ($F_{T5,414} = 46.52$), genotype ($F_{G5,277} = 54.61$), and their interaction ($F_{TG5,277} = 40.59$) are statistically significant ($P < 0.0001$) by two-way ANOVA. *Post hoc* analyses indicate no significant differences as a function of the time of day in the *per* 7.2:2 transgenic line. Asterisks indicate significant ($P < 0.05$) increase in EAG responses at CT13 and CT17 compared with responses at all other times of day in wild-type flies. Cross indicates significant ($P < 0.001$) increase in EAG responses in wild-type flies compared to *per* 7.2:2 transgenic flies at CT17. (Reprinted from Krishnan et al 1999, with permission from Macmillan Publishers Ltd.)

transcription. This repression, which occurs at the same time that CLK-CYC activates *per*, *tim* and *vri* transcription, ensures that *Clk* mRNA cycles in the opposite phase as *per*, *tim* and *vri* mRNAs. Similar feedback loop oscillators operate in sLN$_v$s and peripheral tissues, but only oscillators in peripheral tissues require CRY function. The only clock output known to be dependent on peripheral oscillators is olfaction rhythm. Our studies suggest that the central oscillator is not necessary for olfaction rhythms and that antennal oscillators are sufficient for olfaction rhythms. These results imply that olfaction rhythms are controlled locally in the antenna rather than by the central oscillator, in contrast to the situation in mammals where peripheral oscillators (and consequently their rhythmic outputs) are controlled by the central oscillator.

References

Allada R, Emery P, Takahashi JS, Rosbash M 2001 Stopping time: the genetics of fly and mouse circadian clocks. Annu Rev Neurosci 24:1091–1119

Blau J, Young MW 1999 Cycling vrille expression is required for a functional *Drosophila* clock. Cell 99:661–671

Brand AH, Perrimon N 1993 Targeted gene expression as a means of altering cell fates and generating dominant phenotypes. Development 118:401–415

Cheng MY, Bullock CM, Li C et al 2002 Prokineticin 2 transmits the behavioural circadian rhythm of the suprachiasmatic nucleus. Nature 417:405–410

Egan ES, Franklin TM, Hilderbrand-Chae MJ et al 1999 An extraretinally expressed insect cryptochrome with similarity to the blue light photoreceptors of mammals and plants. J Neurosci 19:3665–3673

Emery P, So WV, Kaneko M, Hall JC, Rosbash M 1998 CRY, a *Drosophila* clock and light-regulated cryptochrome, is a major contributor to circadian rhythm resetting and photosensitivity. Cell 95:669–679

Emery P, Stanewsky R, Hall JC, Rosbash M 2000a A unique circadian-rhythm photoreceptor. Nature 404:456–457

Emery P, Stanewsky R, Helfrich-Forster C, Emery-Le M, Hall JC, Rosbash M 2000b *Drosophila* CRY is a deep brain circadian photoreceptor. Neuron 26:493–504

Frisch B, Hardin PE, Hamblen-Coyle MJ, Rosbash MR, Hall JC 1994 A promoterless *period* gene mediates behavioral rhythmicity and cyclical *per* expression in a restricted subset of the *Drosophila* nervous system. Neuron 12:555–570

George H, Terracol R 1997 The *vrille* gene of *Drosophila* is a maternal enhancer of *decapentaplegic* and encodes a new member of the bZIP family of transcription factors. Genetics 146: 1345–1363

Glossop NRJ, Hardin PE 2002 Central and peripheral circadian oscillator mechanisms in flies and mammals. J Cell Sci 115:3369–3377

Glossop NRJ, Lyons LC, Hardin PE 1999 Interlocked feedback loops within the *Drosophila* circadian oscillator. Science 286:766–768

Helfrich-Forster C 1996 *Drosophila* rhythms: from brain to behavior. Semin Cell Dev Biol 7: 791–802

Ishikawa T, Matsumoto A, Kato T Jr et al 1999 dCRY is a *Drosophila* photoreceptor protein implicated in light entrainment of circadian rhythm. Genes Cells 4:57–65

Ivanchenko M, Stanewsky R, Giebultowicz JM 2001 Circadian photoreception in *Drosophila*: functions of cryptochrome in peripheral and central clocks. J Biol Rhythms 16: 205–215

Kramer A, Yang FC, Snodgrass P et al 2001 Regulation of daily locomotor activity and sleep by hypothalamic EGF receptor signaling. Science 294:2511–2515

Krishnan B, Dryer SE, Hardin PE 1999 Circadian rhythms in olfactory responses of *Drosophila melanogaster*. Nature 400:375–378

Krishnan B, Levine JD, Lynch MK et al 2001 A new role for cryptochrome in a *Drosophila* circadian oscillator. Nature 411:313–317

Meyer-Bernstein EL, Sehgal A 2001 Molecular regulation of circadian rhythms in *Drosophila* and mammals. Neuroscientist 7:496–505

Moore RY, Speh JC, Card JP 1995 The retinohypothalamic tract originates from a distinct subset of retinal ganglion cells. J Comp Neurol 352:351–366

Plautz JD, Kaneko M, Hall JC, Kay SA 1997 Independent photoreceptive circadian clocks throughout *Drosophila*. Science 278:1632–1635

Preitner N, Damiola F, Lopez-Molina L et al 2002 The orphan nuclear receptor REV-ERB alpha controls circadian transcription within the positive limb of the mammalian circadian oscillator. Cell 110:251–260

Renn SC, Park JH, Rosbash M, Hall JC, Taghert PH 1999 A pdf neuropeptide gene mutation and ablation of PDF neurons each cause severe abnormalities of behavioral circadian rhythms in *Drosophila*. Cell 99:791–802

Reppert SM, Weaver DR 2001 Molecular analysis of mammalian circadian rhythms. Annu Rev Physiol 63:647–676

Stanewsky R, Kaneko M, Emery P et al 1998 The *cry*[b] mutation identifies cryptochrome as a circadian photoreceptor in *Drosophila*. Cell 95:681–692

Yamazaki S, Numano R, Abe M et al 2000 Resetting central and peripheral circadian oscillators in transgenic rats. Science 288:682–685

Young MW, Kay SA 2001 Time zones: a comparative genetics of circadian clocks. Nat Rev Genet 2:702–715

DISCUSSION

Cermakian: I have a general question. In the *Drosophila* system can we really talk about peripheral and central oscillators, or should we consider the lateral neurons as one clock among many other clocks that are fairly independent?

Hardin: I think the lateral neurons are one clock, the antennae are another clock, the legs are yet another, and so on. At least from the evidence here, they appear to be independent.

Sehgal: I disagree. We found that there is a clock in the prothoracic gland that controls eclosion and the lateral neurons are required for the cycling of the prothoracic gland. There are some clocks in *Drosophila* that are not autonomous.

Kay: Was the VRI that you used in the gel shift assays bacterially expressed?

Hardin: No, in the gel shifts it was an *in vitro* translated VRI and the second one was produced by baculovirus.

Kay: Presumably VRI is binding there as a homodimer?

Hardin: Yes.

Kay: And is that what E4BP4 does?

Hardin: Yes.

Kay: So there is no straightforward heterodimeric partner for this?

Hardin: It is capable of binding as a homodimer, and E4BP4 does bind as a homodimer to effect its function.

Sassone-Corsi: What is upstream from VRI? What regulates its stability or function?

Hardin: VRI is activated by Clock and Cycle.

Sassone-Corsi: Is the protein phosphorylated, and does this change in the cycle?

Hardin: It is phosphorylated, but it is like Clock in that we see it phosphorylated the entire time. We don't see its phosphorylation state changing with the cycle. As soon as you see it, it is phosphorylated and this doesn't change a lot.

Kyriacou: I find it curious that you see a peak in the EAG rhythm in the middle of the night. Why is the peak in the night? What does it represent? Is it turnover of receptors?

Hardin: There are two ways to explain that. First, in the middle of the night it is dark and you can't use your major sensory system, and if you need to detect something such as a predator, this would be one way of achieving this — through a heightened sense in the olfactory system. Another possibility is that during the day, when it is warmer outside, there are many more odorants in the environment. It is not so much that it is ramping up its olfactory system, but it is ramping it down during the day so it can separate what is important from what is not important.

Van Gelder: Does *per* RNA oscillate when you drive it under the olfactory-specific GAL4-UAS?

Hardin: The RNA does not.

Van Gelder: So there is a little disconnection between the first part of your paper, where RNA oscillation is central to the entire clock mechanism, and the second part where you rescue rhythmicity in this other clock with a static RNA level.

Hardin: We can rescue locomotor activity rhythms with a static *per* RNA level. What is important here are the protein levels. The other answer we need to find for this story is looking in antennae with anti-PER antibodies to see whether the protein is cycling. The other experiment is to use a luciferase reporter to see whether there are rhythms in PER expression.

Van Gelder: Do you think the RNA cycling is necessary for any of the components? You could test this in the system by putting each of them under control in the appropriate mutant background.

Hardin: It may not be important for the raw function of the oscillator. There may be enough post-transcriptional regulatory events there which can maintain function even without RNA cycling. For constant expression of all the core components Amita Sehgal has come the closest in that she constantly expressed *tim* and *per*, and the oscillator works: you get some rhythmic animals.

Takahashi: What is your feeling about the importance of the transcriptional side, if you can get a rather full rescue with constitutive *tim* and *per*? Indeed, is the rescue done in the constitutive *tim* and *per* experiments by Amita Sehgal really complete? Is it wild-type with high-amplitude robust rhythms? If it is completely normal, what is your whole thinking of the whole model of transcription versus post-transcription?

Sehgal: One of the things we did find in the flies is that the PRC is altered. The resetting is not the same as the wild-type fly would have. The other thing, and this is more speculative, is that the RNA cycling contributes to the temporal precision. This would be yet another function. It could also be a redundant mechanism to ensure protein cycling. I should say that I have talked to a couple of modelling people: one person emailed me out of the blue. He hadn't even seen the paper where we reported rescue by constitutive expression of *per* and *tim* RNA (Yang & Sehgal 2001), but indicated that it wasn't possible to build an oscillator based on a transcription–translation feedback loop and that it had to be at the level of protein phosphorylation and stability. Apparently, he modelled this.

Kyriacou: I got that email too!

Hardin: You said the phase response curves (PRCs) are altered. Is that independent of which component you constitutively express?

Sehgal: No, the two together are constitutive.

Hardin: So if you express *per* plus *tim* constitutively you get a PRC change. What happens if you do one at a time?

Sehgal: We haven't looked at one at a time.

Rosbash: It is important to add here that my reading of these rhythms is that you clearly achieve rescue, but only 40% of the flies are rhythmic. Moreover, we would do a slightly different analysis on those flies. I would bet anything that these are low amplitude rhythms. This would be the simplest explanation for the low penetrance. It clearly works, but whether you decide it is 50% or 10% of flies that achieve wild-type potency is dependent on what kind of analysis you do. By conservative criteria they are far from wild-type flies.

Sehgal: We saw rescue in 40–50% of the flies. I agree that the numbers are not that high. I was talking about period, but the amplitude might also be somewhat altered. However, it was within the wild-type range for these 40–50% or else they would have been characterized as weakly rhythmic, if not arrhythmic. We divided all the flies we tested into three categories: rhythmic, weakly rhythmic and arrhythmic.

Rosbash: There is a further point which ties in with the previous discussion. This is very far from the fibroblasts in culture and the individual cells. I would not be surprised if the animal benefits from the systems assay. In other words, we think about this as the subcellular molecular oscillator, but of course what is being measured is period, phase and behavioural amplitude. This may benefit from the fact that it is an integrated system.

Sehgal: I'm not sure what you mean.

Rosbash: It is fly behaviour. If we did Paolo Sassone-Corsi's experiment and took out the *Drosophila* embryonic fibroblasts, we could study individual cell function.

Schibler: Why not just use the luciferase assay with peripheral tissue in culture?

Rosbash: Even more accurately, we could dissociate the individual tissues. Then we could ask how the individual cells function. My guess would be that those cells would be dramatically impaired in comparison with the wild-type. This is speculation, though.

Menaker: It depends on what level you are talking about the system. You are talking about it at the tissue level, where there is likely to be interaction. But if I understand Paul Hardin's point, the fly itself doesn't look like a system. He doesn't see interaction among the oscillators.

Hardin: In the antenna, the oscillator is sufficient for driving that rhythm. It doesn't seem to be affected by the lack of lateral neurons.

Rosbash: But of course, it gets more. Amita Sehgal's experiment was the double rescue. People were saying that the reason that PER works is because TIM mRNA is still cycling.

Weitz: What is the nature of the PRC problem? It has been argued from an evolutionary point of view that maintaining the correct phase angle for whatever is being regulated in an animal might be the most important adaptive feature of clocks.

Sehgal: I don't remember exactly. There were no significant delays.

Van Gelder: What was the phase angle of locomotion?

Sehgal: We didn't look carefully at this.

Weitz: That alone could be a reason why you would need this other feature, even if the free-running oscillation looked OK.

Sehgal: There are a lot of things that are probably wrong with that kind of oscillation.

Schibler: As I remember from Woody Hastings' work in *Gonyaulax*, where the clock genes are not known, everything seems to be regulated on the level of translation and protein stability.

Rosbash: He has talked to me about this, and they have just completed an Affymetrix gene array experiment with *Gonyaulax*, and there is lots of circadian transcriptional regulation. I presume that they concentrated on translation all these years because that is where their original work led them. Paul Hardin, is there enough known for you to use a driver in a neuron which deals with another odorant to show that the rhythms are relevant for olfaction?

Hardin: There are other drivers available for other subsets of sensillae, and we are testing those to see whether they are insensitive to that particular odorant. The other thing is to use other odorants with these flies and do the converse.

Rosbash: I presume that there are some odorants that won't work with those cells. Would those EAGs be insensitive?

Hardin: Basoconic sensillae are sensitive to a huge array of odours, but other sensillae have more limited sensitivity.

Weitz: Are these other sensillae part of the pheromone system, or are they olfactory too?

Hardin: No one knows much about the pheromone system in flies.

Young: Joe Takahashi's earlier question reminds us that we now generally assume that the primary output of the clock is through a set of transcriptional controls. These are thought to control a wide variety of specific, timed gene responses downstream of the clock. Could a primitive clock have operated without transcriptional feedback from these factors? After all several studies have shown that constitutive regulation of genes such as *per* still supports cycling protein and substantial circadian rhythmicity. Perhaps an ancient system for producing protein cycles post-transcriptionally was made more robust by adding transcriptional feedback as a secondary adaptation. This would be taking the benefits of the activity of these cycling outputs to add special features to the core oscillatory mechanism, which nevertheless may not be completely dependent upon them.

Takahashi: That is a good point. The transcriptional feedback loop has been selected because of all the downstream outputs that it controls. It would be hard to do this with a post-transcriptional mechanism.

Sehgal: Why can't you have the clock proteins controlled post-transcriptionally? Under all these conditions, they are capable of effecting transcription of downstream genes, just not their own because their promoters don't exist.

Menaker: That is one evolutionary scenario. You could gain certain properties by allowing these to become targets of the system as well as origins. But how can we ever figure this out?

Hardin: Whether they will cycle or not depends on whether all of the proteins within the system are regulated at the post-transcriptional level. For example, if VRI is a major regulator of outputs in that phase of the cycle, if you remove VRI cycling and if the protein is dependent on the RNA, does that mean that all of the outputs subservient to VRI will then be arrhythmic? This would require that VRI would also be under some sort of circadian control to preserve its downstream targets.

Young: VRI is a special case. Constitutive expression of VRI will give you primarily arrhythmicity in adult behaviour. While we can fiddle with *per* and *tim*, VRI stands as a counterpoint to add to the uncertainty.

Van Gelder: There is a presumption here that rhythmic gene expression output is the major output of the clock, and yet for locomotor rhythms *pdf* is clearly the funnel point, and *pdf* is not rhythmically regulated.

Rosbash: That's a temporary conclusion. I doubt that is true for *pdf*. We think of *pdf* as being an anti-damping function, converting a non-damped system into liver in culture. That is the way that the behaviour looks.

Stanewsky: I want to comment on *pdf*. It is published that PDF protein accumulates rhythmically, so there is a rhythmic component of PDF expression controlled by the clock.

Van Gelder: PDF RNA did not show up on any of the five published gene chip experiments The rhythmic RNA output is not the output of this molecule that seems to be critical for rhythmic locomotor activity.

Rosbash: It could be that there is a RNA oscillation of some other gene — a kinase or secretory model, for example, which is required for PDF terminal accumulation and release. We have no idea what the molecular gate is for PDF terminal accumulation.

Sehgal: You gave two possible explanations for why the effect of VRI over-expression on CLOCK was not as strong as you had expected. Both of these make sense. But if I remember correctly, the effect of VRI on *per* showed by Justin Blau and Mike Young was much greater. If the effects on *per* are mediated by CLOCK, wouldn't you have expected to see a similar effect on CLOCK?

Hardin: It is a different driver and a different stage of development, so I don't know how comparable they are. They didn't see *per* present in those neurons by immunohistochemistry or *in situ* hybridization.

Rosbash: What was the effect of VRI over-expression on *per* and *tim* cycling?

Young: It wipes it out and suppresses PER and TIM accumulation.

Sehgal: Is it possible that the effects on *per* and *tim* are not entirely mediated by CLOCK?

Rosbash: Would the idea then be that there is only a little CLOCK, and this would then lead to low PER and TIM?

Young: Sufficiently reduced CLOCK that would kind of cascade down. Going back to an earlier issue, we sort of assume that control of downstream transcription is the only output mechanism of the clock; that this oscillator exists to make transcription factors oscillate to give function. But there are other things to look at. When we see DOUBLE-TIME being rhythmically moved between cytoplasm and the nucleus by its PER associations, if there is anything other than PER in the cytoplasm (or nucleus) that requires that kinase's activity you might expect these rhythmic changes in kinase location would produce rhythmic information flow through those protein targets as well, depending on their localization. One of the challenges is to imagine the different pathways that might be in here. Transcription is the easiest thing to follow but there could be other mechanisms for rhythmic output as well.

Dunlap: It also lends itself inherently well to regulated output. Circadian regulation requires rapid turnover or else you don't see regulation. RNAs are made, travel and then disappear.

Rosbash: Most other things that people study turn over even more rapidly. I would argue the reverse. The RNA is well suited to these very long time-courses.

Stanewsky: I would like to return to the question of peripheral versus central oscillators. There is a difference between the lateral neuron clock and the peripheral oscillators. The only rhythm known to persist in flies is the behavioural rhythm, which is driven by the lateral neurons. Whatever other tissue you look at, *per* and *tim* expression dampens fast in constant darkness. Do you know anything about the antennal rhythms?

Hardin: We can't do EAGs on a single animal for more than one time-point. It is a population-based rhythm. All those experiments I showed were on the second day of DD. We have measured out to the fourth day, where amplitude is a little lower, but we haven't measured beyond that point.

Stanewsky: Have you studied gene expression in the antennae?

Hardin: We can see a clear expression rhythm about a week out into DD.

Weitz: It seems that what we have heard about is damped and sustained oscillations, but we haven't yet heard anything telling us whether there are damped or sustained oscillators. So far, very few of these assays have employed single-cell studies. It is conceivable that the main difference between the SCN and other tissues, or between lateral neurons and other tissues, is that those are coupled together by synaptic coupling.

Schibler: It is true that everyone is waiting for the single-cell assays. But at least in serum shocked cells, it is clear that the cells before the shock have no oscillator working.

Weitz: You are talking about the input of your experiment when you first put them in the dish prior to serum shock. When you start there is no functioning clock in the tissue.

Kay: It is like flies. Their lateral neuron rhythms will damp 99% of the signal in the peripheral tissues. If you take those flies from the topcount luminescence assay and put them in locomotor and DD, you will see that their locomotor rhythms are still fine. It is not just a culture effect that you see damping of these rhythms.

Weitz: Is it or is it not desynchronization of the single-cell molecular oscillator? We just don't know.

Kay: The only time we have done that is with Malpighian tubules in culture, where we can almost get single-cell distinction in culture. I wouldn't trust our conclusions on that, which is that we saw *tau* polymorphism in individual Malpighian tubules in culture.

Sassone-Corsi: There are experiments on the *per*-luciferase transgenics showing that each single neuron in the SCN has an independent single-cell clock. These appear to be self-sustaining for a long time. In fact, there was a difference from one cell to another.

Menaker: The question Chuck Weitz is raising is whether the arrhythmicity seen after a while in culturing peripheral tissue is a consequence of all the oscillators damping out, or whether the cells are simply coming apart. I think Ueli Schibler has an answer to this as far as fibroblasts go, but this doesn't tell us what happens in liver.

Rosbash: It isn't even an answer for fibroblasts, because you want the end of the experiment, not the beginning.

Sassone-Corsi: I don't understand that.

Rosbash: You start out with the fibroblasts and you are making an interpretation. They are already damped and then they have months to get even more weird. You are making an argument about the fact that those cells are not synchronous at the beginning. But they are so far along from having been rhythmic that they might be a poor reflection of the original state.

Schibler: Now you are repeating what Michael Menaker said. This is true only for fibroblasts.

Rosbash: It is true for a cell line which has been running for years since it last had normal rhythms.

Weitz: Part of this argument is whether or not one needs to postulate a different kind of clock mechanism for the SCN.

Kay: In the case of the SCN neuron, the possibility is that a major output of that particular cell — membrane activity — feeds back. This is how you get that single cell robustness.

Rosbash: If you trawl the literature, how many studies have taken a tissue, dissociated it into single cells, prevented the cells from having any kind of contact with each other (including signalling molecules), and then seen long-term self-sustained oscillations? Very few, I suspect.

Takahashi: No-one has done that.

Dunlap: If the cells were not winding down but were simply asynchronous and there is a normal PRC with a dead zone, then the re-entraining cue would have to split the population into two peaks, not one.

Menaker: What specific re-entraining cue are you referring to?

Dunlap: Anything that is going to entrain. In the case of fibroblasts, this could be serum.

Van Gelder: It has to be something with a biphasic phase response curve, which may not be the case for some of these agents.

Rosbash: When you did the RNA cycling originally, my recollection was that in the damping in DD over several days, the trough comes up but the peak doesn't go down beneath the trough. It is an argument that the system isn't becoming asynchronous, but that something molecularly is working less well yet continues to oscillate with reduced amplitude.

Menaker: That argument is very indirect.

Schibler: I have one more comment about the fibroblasts. If one looks at the livers of voles every 2.5 h, the same thing is seen. Some molecules are low and some are high. So it doesn't look like it is a mix of all phases.

Young: In flies, with behaviour lasting as long as it does, I remember rumours of someone looking at oscillations in lateral neurons specifically. We are making assumptions about how lateral neurons work, but does anyone have data on lateral neurons after 10 days in DD, for example, to show that they really do cycle?

Rosbash: Yes. The current dogma is even more shaky than that. We have done 8 day DD *in situs* very carefully. It is not just the lateral neurons: the entire brain is totally robust with no damping at all.

Young: How do you account for the original head data?

Rosbash: The eyes. If you grind up the heads, which includes eyes and fat body, then you will find damping. But all the neural centres in the brain are totally non-damping.

Sehgal: Not with the large lateral neurons.

Rosbash: You did one or two day experiments. We repeated this work and found that one day you get a big damping effect, but if you go out to 3, 6 or 8 days it comes back completely. It comes back to full robustness by the 4th day, and the phase is perfect. The interpretation is that the large cells are very light sensitive, and when the lights go off they freak out. Then, because the whole brain is an integrated system, it adapts and gets set in DD. No one bothered to look after one or two days before.

Van Gelder: In the fly, would it be worth going back to the old chimera experiments, with the new twist of making chimeras between a *per*-short and a wild-type, both with luciferase reporter. Then you could ask where the systems level integration is. You should be able to see both rhythms if the peripheral oscillators were staying synchronized to each other, which would be the systems interpretation, or dissociating within the time frame of a few days, given the difference in *tau* between *per*-short and wild-type.

Rosbash: We are lacking a richer source of neural drivers to be able to do this. But it would be an interesting experiment.

Young: You could do gynandromorphs.

Rosbash: Each fly would be a one-off. You'd need a real time assay looking at just one fly.

Kyriacou: Coming back to Joe Takahashi's comment about constant RNA and the effect that this has on the clock, it clearly has some effect on the behavioural output. But one thing that Michael Rosbash showed with *doubletime* mutants and that we have also seen in the housefly is that it is possible to move the RNA cycle and the PER protein cycle absolutely on top of each other without any delay. In the case of the housefly we get a behavioural rhythm that is an hour shorter, but it is a perfectly good robust cycle. So, if we are talking about RNA and protein cycles, they can be shifted so that there is no 4 h delay between them and we still get very robust rhythms.

Sehgal: How about nuclear localization?

Kyriacou: We haven't looked at that yet.

Takahashi: That is important for the modelling of the system. It suggests that transcription and protein are overlapping.

Stanewsky: We have to be careful with the interpretation of this experiment. That work on RNA and protein together was only tested in LD.

Kyriacou: I think, but am not sure, that we have also shown it in DD.

Takahashi: To sort of summarize, what I heard was that the rhythms in Amita Sehgal's constitutive *tim* and *per* flies are not completely wild-type. Only 40% of flies are rhythmic and the amplitude wasn't studied, so they might have low amplitude. They certainly have low penetrance.

Sehgal: We did do FFTs on them but I don't recall the numbers.

Rosbash: In our experience there is almost universal correspondence between low penetrance and low power. This is the usual explanation for low penetrance.

Takahashi: Then we had Mike Young's nice idea that we want transcription to be driven on a more global basis for lots of outputs.

Menaker: Paul Hardin, do you think that the extent to which the *Drosophila* system is composed of separate oscillators is related to the lifespan of the animal? In other words, if you were to do the rough translation of the phase-shifting trajectories that we see in mammals and apply those to flies, if those oscillators were all connected the fly would never get back in synchrony if it flew from London to New York! The question is, will a long-lived insect be organized in a mammalian way, with connected oscillators, or not? Is this an insect characteristic or a lifespan characteristic?

Hardin: I don't think it is necessarily an insect characteristic. In zebrafish there are the same independent light-dependent oscillators around the body, and their lifespan is considerably longer than that of flies.

Weitz: The transparency of the animal might have more to do with this independence than the lifespan.

Rosbash: Wouldn't it be the other way round? Don't you think that the circadian system has co-opted the complex neuroendocrine systems such as the HP axis that exist in mammals? Presumably much of the integration in mammals — the non-autonomy of the systems — makes use of the complex neuroendocrine relationships between organs and systems. Then the question is, if those systems existed first, might the circadian system have taken advantage of them?

Menaker: Insects do have complex neuroendocrine systems, and animals had oscillators long before we had a hypothalamic–pituitary axis.

Rosbash: These were not necessarily integrated oscillators in the way that we think about them in mammals. I would guess that insect neuroendocrine communication is much more primitive.

Hardin: I think Chuck Weitz is right in that a lot probably has to do with the transparency of the animal and its ability to entrain peripheral oscillators directly.

Sehgal: What about the less transparent non-mammalian organisms?

Green: Xenopus doesn't seem to show evidence for light sensitivity in the peripheral tissues.

Rosbash: Have people looked at melanocytes?

Green: We haven't looked at melanocytes but we have looked in other organs.

Schibler: Have you looked in young transparent larvae or just adult frogs?

Green: We have looked at both. The very young transparent larvae don't seem to be very rhythmic at the whole animal level. This is in contrast to the eyes, where we can measure light-sensitive rhythms early on in development.

Van Gelder: We shouldn't over-simplify. We are considering *Drosophila* as a collection of independently entrainable clocks. The HB-islet still contributes to at least some behavioural organization in circadian entrainment. In the *cry* mutants there is still this non-cell autonomous effect on the lateral neurons through a visual system effect.

Ishida: With respect to lifespan, work by Paul Shaw on *cyc*[0] mutant flies showed that they became very sensitive to desiccation and sleep stress (Shaw et al 2002). *cyc*[0] flies died early after sleep deprivation. So the molecular mechanism of lifespan might be related to that of the circadian system. But another possibility is that BMAL1 and CLOCK regulate many other output factors, such as stress-sensitive proteins.

References

Shaw PJ, Tononi G, Greenspan RJ, Robinson DF 2002 Stress response genes protect against lethal effects of sleep deprivation in *Drosophila*. Nature 417:287–291

Yang Z, Sehgal A 2001 Role of molecular oscillations in generating behavioral rhythms in *Drosophila*. Neuron 29:453–467

Integration of molecular rhythms in the mammalian circadian system

Hitoshi Okamura

Division of Molecular Brain Science, Department of Brain Sciences, Kobe University Graduate School of Medicine, Chuo-ku, Kobe 650-0017, Japan

Abstract. The discovery of clock genes and the general principles of their oscillation have made research on biological clocks a highly interesting field in the life sciences. As in other species, the mammalian circadian core oscillator is thought to be composed of an autoregulatory transcription–(post)translation-based feedback loop involving a set of clock genes. The production, phosphorylation, ubiquitination and proteasome-dependent degradation of clock proteins has a key role in generating the clock oscillation. The generation of internal clock time occurs in the hypothalamic suprachiasmatic nucleus (SCN), where clock gene oscillation in each neuron is coupled and amplified. These well synchronized oscillatory signals are spread into the whole brain and to peripheral organs which contain peripheral clocks. The important feature of the circadian system is that the rhythm of gene transcription of clock genes in the SCN reflects the behavioural rhythm almost perfectly. Investigations on biological clocks present the fascinating prospect of analysing the integrational mechanism of 'time' from genes to the living organism.

2003 Molecular clocks and light signalling. Wiley, Chichester (Novartis Foundation Symposium 253) p 161–170

Most organisms have an internal clock and thus circadian rhythm represents a basic feature of life. The discovery of clock genes and the general principle of their oscillation has stimulated research on biological clocks and this research has provided a major impact on the field of life sciences. In many organisms, the oscillation is driven by a transcription–translation-based core feedback loop of a set of clock genes (Young & Kay 2001). One feature of the circadian system is the prevalence of the oscillation at the levels of genes which reflects the cell, tissue and system level oscillations. In mammals, the central oscillator resides in the small paired oval shaped suprachiasmatic nucleus (SCN) which is located in the base of the anterior hypothalamus (Okamura et al 2002). This contains thousands of clock oscillating cells which generate standard internal time, and spread the time signal to the whole of the body. These circadian changes then result in altered behaviour and hormone secretion.

Core feedback loop and accessory feedback loops of clock genes

Clock oscillation occurs first at a cellular level. The clock genes so far identified in mammals are structurally similar to those in *Drosophila* (Young & Kay 2001). This suggests that mammals and *Drosophila* utilize similar components to generate circadian ($\sim$24 h) rhythms. Mammalian clock research is now showing whether the core feedback loop of clock genes speculated to be present in *Drosophila* (Hardin et al 1990) is also conserved in mammals.

The core feedback loops generating circadian oscillation in mammals can be summarized as follows (Fig. 1). *mPer1* and *mPer2* are two main oscillators, since (1) the targeted disruption of both *mPer1* and *mPer2* results in the complete loss of circadian rhythms (Bae et al 2001, Zheng et al 2001) and (2) introduction of *mPer1* or *mPer2* gene into arrythmic *per^{01}* mutants of *Drosophila* that are otherwise arrhythmic due to a lack of endogenous PER protein, restored rhythm (Shigeyoshi et al 2002). The heterodimer formed by the bHLH-PAS proteins (CLOCK and BMAL1) binds to the E-box of *mPer1* and *mPer2* promoters and initiates transcription (Gekakis et al 1998). Activated transcription results in the formation of *mPer1* and *mPer2* mRNAs, which are translated in the cytoplasm to mPER1 and mPER2 proteins. These proteins translocate into the nucleus, and form negative complexes that comprise mCRY1, mCRY2, mPER1, mPER2, mPER3 and mTIM, and that suppress the transcription of the *mPer* genes by binding to the positive factors (CLOCK/BMAL1). Since *mCry1/mCry2* double knockout mice (van der Horst et al 1999, Okamura et al 1999) and *Bmal1* (*Mop3*) knockout mice (Bunger et al 2000) show the immediate loss of circadian rhythm in constant darkness, *mCry1/mCry2*, and *Bmal1* play a key role in making up the core loop.

The core feedback loop is very stable and accurate for counting 24 h intervals. The stability and high amplitude of this rhythm will be ascertained at multiple stages including intracellular, intercellular and systemic levels. The complementary molecular loops assisting the core feedback loop will contribute the stability of the rhythm at the level of gene transcription. One such loop consists of a positive factor BMAL1 which shows a pattern of daily expression inverse to that of *mPer1/2* through the transcription suppressor REV-ERBα (Preitner et al 2002, Ueda et al 2002). Another type of regulation is the antagonistic regulation of PAR proteins (HLF, TEF and DBP) and E4BP4 (Mitsui et al 2001, Yamaguchi et al 2000a) through CLOCK/BMAL1: PAR proteins and E4BP4 competitively bind to a specific sequence ATTACGTAAC which is located in just the upstream region of the second transcription initiation site of the *mPer1* gene (Yamaguchi et al 2000a, Mitsui et al 2001). As for the *dbp* gene, CLOCK/BMAL1 binds to the E-box of the second intron, and increases the transcription (Yamaguchi et al 2000b). Therefore, PAR proteins, such as DBP, expressed highly at the beginning of the

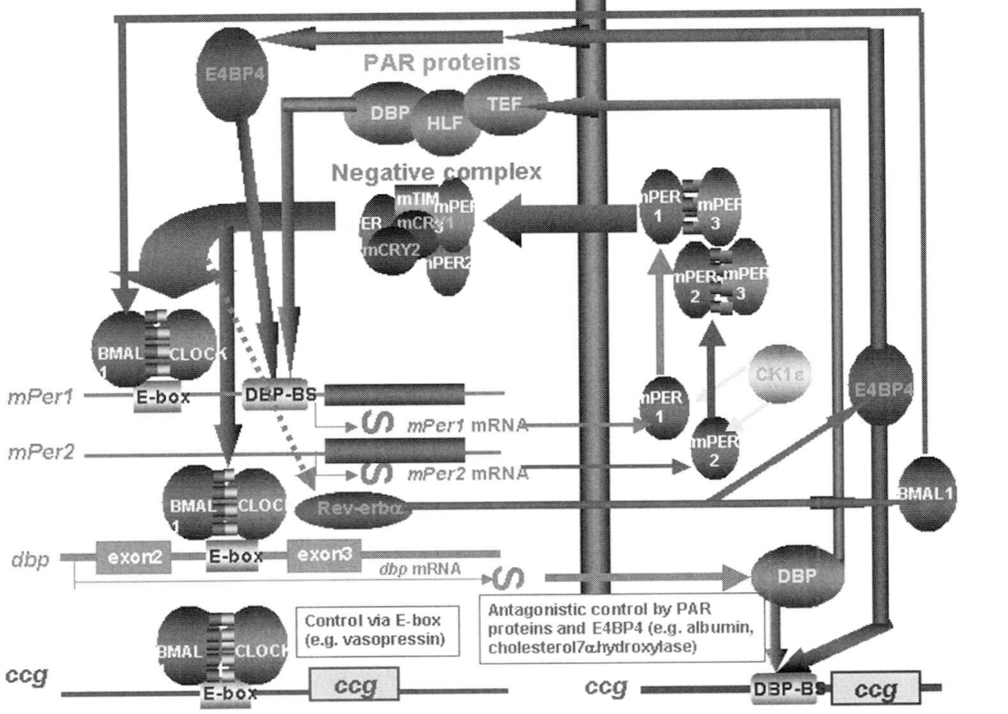

FIG. 1.   The core, accessory and output molecular mechanisms of the mammalian circadian clock. BMAL1/CLOCK heterodimer binds to E-box in clock oscillating genes and clock control genes, accelerating their transcription. The core negative autoregulatory feedback loop provided by *mPer1* and *mPer2* is regulated at protein level by negative complex consisting of mPER1, mPER2, mPER3, mCRY1 and mCRY2. This core loop is supplemented by accessory loops; (1) BMAL1 loop mediated by REV-ERBα suppressing the BMAL1 transcription during daytime, (2) PAR protein loop mediated by PAR nuclear proteins accelerating the daytime *mPer1* transcription. Time information of core loops flows out to clock-controlled genes (*ccg*) via CLOCK-BMAL1 binding to E-box (Jin et al 1999) or the antagonistic regulation of PAR proteins and E4BP4 (Mitsui et al 2001).

daytime assist the enhancement, but E4BP4 expressed abundantly at early night assists the suppression of *mPer1* transcription.

Regulation of clock proteins

There is growing evidence that clock proteins are regulated dynamically in both temporal (production and degradation) and spatial (nuclear and cytoplasmic) dimensions. The phosphorylation of mPER1 and mPER2 by casein kinase Iε (CK1ε) is known as an important step for the accumulation of negatively active clock proteins (Lowrey et al 2000) as in *Drosophila* (Kloss et al 1998).

Recently, it has been shown that ubiquitination and proteasome-dependent degradation of mPER proteins occurs in mammalian cells (Yagita et al 2002). It is also evident that the ubiquitination of mPER proteins is inhibited in the presence of mCRY proteins and the mPER proteins appear to be more fragile if they do not dimerize with mCRY proteins (Fig. 2). Moreover, mCRY protein,

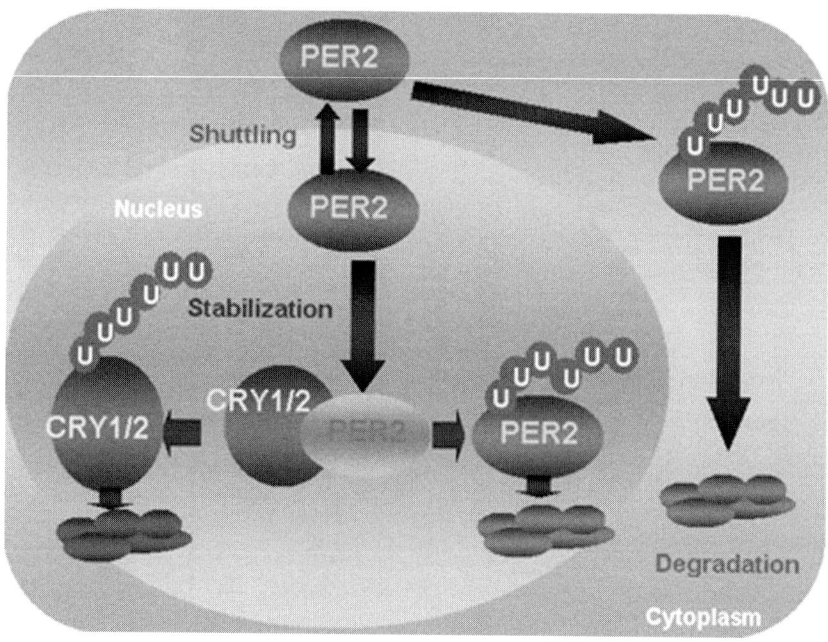

FIG. 2.   A model for mCRY-mediated nuclear accumulation of mPER2. mPER2 protein is produced in the cytoplasm and translocated into the nucleus. The protein keeps on shuttling between nucleus and cytoplasm via the CRM1/Exportin 1 nuclear export system until (i) mPER2 is ubiquitinated and subsequently degraded by the proteasome system or (ii) the stabilization of nuclear mPER2 by the binding of mCRY1 or mCRY2. mPER2, in turn, stabilizes the nuclear mCRY protein by preventing ubiquitination and the following degradation.

which is the strongest suppressor of *mPer1* transcription, can be ubiquitinated when mPER proteins are absent (Yagita et al 2002). Thus, the negative complex is able to suppress *mPer1* and *mPer2* transcription. The nuclear export machinery may contribute by taking out mPER proteins released from the negative complex. The decrease of mPER concentration in the nucleus facilitates the breakdown of mCRY, and it will lead to the start of the gene transcription of *mPer1* and *mPer2*.

Output to clock-controlled genes

It is now clear that many mammalian genes are controlled by the circadian clock (Panda et al 2002). At present, two routes are speculated for the signal transduction from the core clock loop. The first is directly controlled by the central loop by CLOCK/BMAL1 heterodimers through an E-box in the promoter region such as vasopressin (Jin et al 1999). The second route is an indirect pathway, consisting of antagonistic regulation of PAR proteins and E4BP4 (Mitsui et al 2001), which is also used as the accessory feedback loop of clock genes. Aromatic L-amino acid decarboxylase in the SCN may be regulated by the latter mechanism (Ishida et al 2002), in which the positive PAR proteins and the negative E4BP4 will switch back and forth between the on–off conditions of the target genes.

Clock oscillation in the suprachiasmatic nucleus

Studies of mammals subjected to SCN destruction and transplantation have revealed that the hypothalamic SCN contains a master circadian oscillator which is involved in a number of behaviours and hormonal secretions. The circadian oscillatory activity of SCN neurons is directly demonstrated by the measurement of [^{14}C]glucose metabolic activity and field potentials assessed by electro-physiological devices. The clock oscillatory genes *mPer1* and *mPer2* are expressed rhythmically in most neurons in the SCN. Thus, thousands of clock cells in the SCN might generate the rhythm.

To examine the *mPer1* gene expression occurring at each SCN cell, we generated transgenic mice carrying a luciferase reporter gene under the control of the two oscillating promoters of the *mPer1* gene (Yamaguchi et al 2000b). Investigations of brain slice cultures of *mPer1-luc* mice with a two-dimensional photon counting camera showed that the luciferase-mediated bioluminescence follows a robust circadian fluctuation in the SCN (Asai et al 2001). At each cell level, most cell clocks in the SCN undergo an orchestrated expression pattern. Since circadian fluctuation of bioluminescence was detected through the inserted optical fibre just above the SCN of a *mPer1-luc* transgenic mouse (Yamaguchi et al 2001), it is clear that there is a harmonized ticking of clock genes in the SCN of living mammals (Fig. 3).

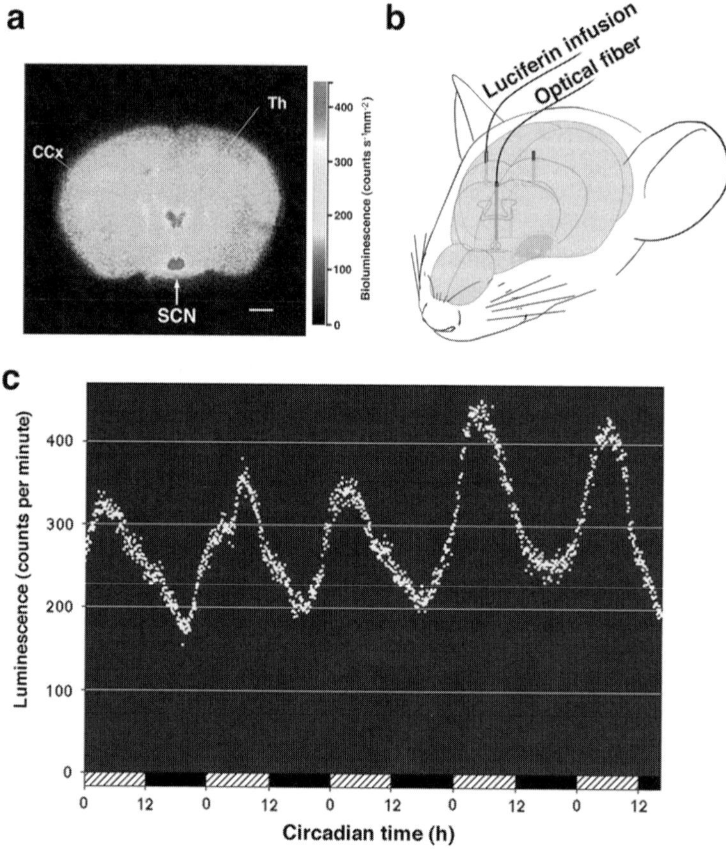

FIG. 3.   *In vivo* recording of luciferase luminescence by inserting an optical fibre into the brain of *mPer1-luc* transgenic mouse. (a) Bioluminescence of coronal brain slice transversing the SCN. Images displayed with scale representing accumulated photocounts. Th, thalamic nuclei; CCx, cerebral cortex. Scale bar = 1 mm. (b) Schematic presentation of the insertion of optical fibre onto the mouse SCN. (c) Circadian fluctuation of luminescence in the SCN. A transgenic mouse, previously housed under a 12 h light/12 h dark cycle, was continuously infused with luciferin (10 μM) at a rate of 15 μl/h. Luminescence was recorded under constant dark conditions via an optical fibre positioned above the SCN. One dot represents an average of the values of 5 minutes. Hatched and closed bars at the bottom of the figure represent subjective day and subjective night, respectively. Adapted from Yamaguchi et al (2000b, 2001).

Although the molecular analyses draw the common molecular mechanism for the mammalian SCN oscillation, there are topographic differences among SCN cells. *mPer1* and *mPer2* in dorsomedial cells showed a strong autonomous expression with no light response, while those in ventrolateral cells showed a weak autonomous expression with strong induction by resetting the light signals conveyed by the direct and indirect retinohypothalamic tracts (Shigeyoshi et al 1997, Yan et al 1999, Yan & Okamura 2002). The coupling processes among SCN cell groups will contribute the phase-resetting and circadian generation in the SCN.

Integration of clock oscillation in the living organism

The molecular clocks previously thought to exist only in the SCN are now also found in cells of many peripheral organs. In fibroblast cell lines, external stimuli such as high concentration of serum and endothelin can induce the circadian expression of the clock genes in a few cycles (Balsalobre et al 1998, Yagita et al 2000), by the similar transcriptional and translational mechanisms of core feedback loops found in the SCN (Yagita et al 2001). In the vascular smooth muscle cells, angiotensin II is a possible inducer of rhythm (Nonaka et al 2001). In the liver, the peripheral clock is entrained by a restriction of feeding (Damiola et al 2000), which is independent from the SCN (Hara et al 2001). Pineal *mPer* gene

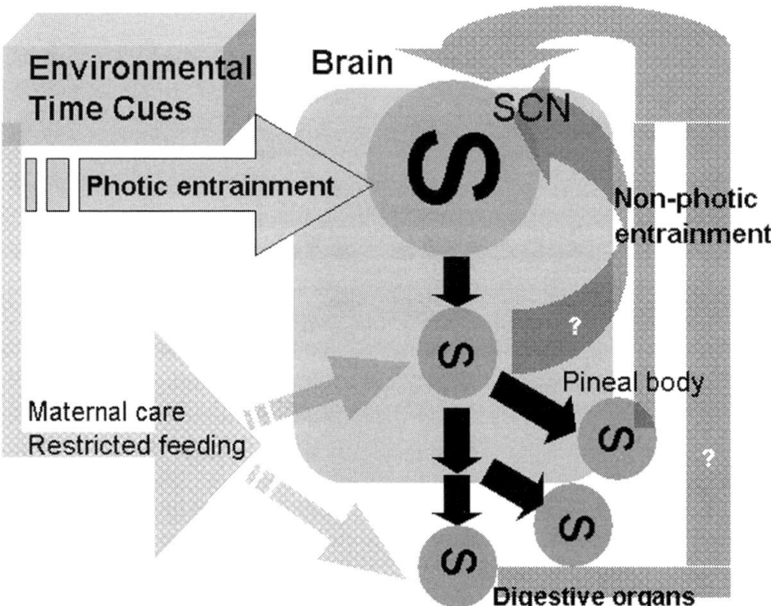

FIG. 4. Diagram of a circadian system in mammals. The master clock in the SCN entrains all non-SCN clocks in the brain and peripheral organs. Light information enters into the SCN, and non-photic information such as restriction feeding and maternal care enter into the peripheral clocks. Additionally, this scheme proposes the existence of non-photic entrainment from information from non-SCN clocks.

expression was regulated by the adrenergic inputs (Takekida et al 2000). Nevertheless, the rhythms induced in these peripheral cells dampen after a few cycles (Yamazaki et al 2000), and thus it is now speculated that clock systems in mammals display a complex hierarchical structure headed by the SCN (Fig. 4). The master clock in the SCN entrains all non-SCN clocks in the brain (Yamamoto et al 2001) and peripheral organs (Yamazaki et al 2000). Light information enters the SCN, and non-photic information such as restriction feeding and maternal care enter the peripheral clocks.

The astonishing discovery of circadian biology is that the core transcription–translation oscillatory loop consisting of a small number of clock genes reflects

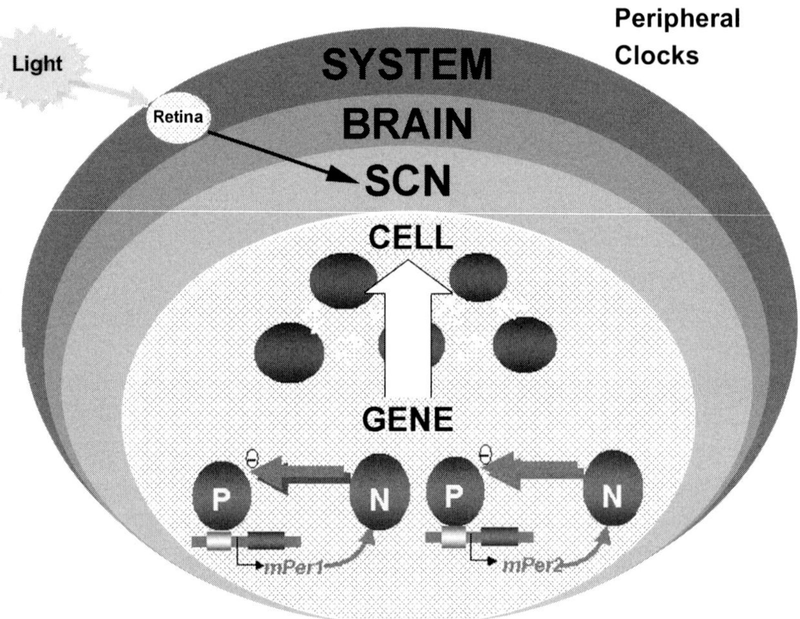

FIG. 5. Integration of circadian rhythm from gene to cell, to nerve nuclei, to brain, and to system. 'Gene' depicts rhythmic transcription of *mPer1* and *mPer2*. 'Cell' represents neuronal electrical activities of single SCN neurons. 'SCN' indicates the sum of the local neuronal and glial circuits. 'Brain' symbolizes functions produced by neuronal circuits in the brain such as sleep and recognition. 'System' symbolizes behaviour, peripheral neuronal activities, and hormonal secretion. *P* and *N* at gene level represent positive and negative elements, respectively. Positive factors stimulate the transcription of clock genes, and their translational products negatively regulate the transcription of their own gene. At SCN, cell clocks interact with each other, and harmonize to make a strong rhythm in the SCN as a whole. At system level, many of the peripheral organs have their own 'peripheral clock'. The master clock in the SCN receives light information via retina, the presumed peripheral clocks in the digestive system such as in the liver receive feeding information.

the behavioural rhythm almost perfectly. This means that the clock gene and protein oscillation generated by the core loop in each SCN neuron is coupled and amplified at the level of the SCN (Fig. 5). From the SCN, harmonized strongly oscillating activities are spread out to the whole brain and to all the peripheral organs that contain clocks. Even though each neuron in the SCN generates circadian oscillation, the system of amplification and transmission needs to be well organized to transmit the temporal information to the peripheral organs effectively. By pursuing how molecular rhythms are integrated into the behavioural circadian rhythm, clock research presents the fascinating prospect of analysing the integrational mechanisms of 'time' from genes to the living organism.

Acknowledgements

The research from my laboratory reported in this review was supported by grants from the Special Coordination Funds and the Grant-in-Aid for the Scientific Research on Priority Areas of the Ministry of Education, Culture, Sports, Science and Technology of Japan.

References

Asai M, Yamaguchi S, Isejima H et al 2001 Visualization of mPer1 transcription *in vitro*: NMDA induces a rapid phase-shift of mPer1 gene in cultured SCN. Curr Biol 11:1524–1527

Bae K, Jin XW, Maywood ES, Hastings MH, Reppert SM, Weaver DR 2001 Differential functions of *mPer1*, *mPer2*, and *mPer3* in the SCN circadian clock. Neuron 30:525–536

Balsalobre A, Damiola F, Schibler U 1998 A serum shock induces circadian gene expression in mammalian tissue culture cells. Cell 93:929–937

Bunger MK, Wilsbacher LD, Moran SM et al 2000 Mop3 is an essential component of the master circadian pacemaker in mammals. Cell 103:1009–1017

Damiola F, Le Minh N, Preitner N, Kornmann B, Fleury-Olela F, Schibler U 2000 Restricted feeding uncouples circadian oscillators in peripheral tissues from the central pacemaker in the suprachiasmatic nucleus. Genes Dev 14: 2950–2961

Gekakis N, Staknis D, Nguyen HB et al 1998 Role of the CLOCK protein in the mammalian circadian mechanism. Science 280:1564–1569

Hara R, Wan K, Wakamatsu H et al 2001 Restricted feeding entrains liver without participation of suprachiasmatic nucleus. Genes Cells 6:269–278

Hardin PE, Hall JC, Rosbash M 1990 Feedback of the *Drosophila period* gene product on circadian cycling of its messenger RNA levels. Nature 343:536–540

Ishida Y, Yokoyama C, Inatomi T et al 2002 Circadian rhythm of aromatic L-amino acid decarboxylase in the rat suprachiasmatic nucleus: gene expression and decarboxylating activity in clock oscillating cells. Genes Cells 7:447–459

Jin X, Shearman LP, Weaver DR, Zylka MJ, de Vries GJ, Reppert SM 1999 A molecular mechanism regulating rhythmic output from the suprachiasmatic circadian clock. Cell 96:57–68

Kloss B, Price JL, Saez L et al 1998 The Drosophila clock gene double-time encodes a protein closely related to human casein kinase Iε. Cell 94:97–107

Lowrey PL, Shimomura K, Antoch MP et al 2000 Positional syntenic cloning and functional characterization of the mammalian circadian mutation *tau*. Science 288:483–492

Mitsui S, Yamaguchi S, Matsuo T, Ishida Y, Okamura H 2001 Antagonistic role of E4BP4 and PAR proteins in the circadian oscillatory mechanism. Genes Dev 15:995–1006

Nonaka H, Emoto N, Ikeda K et al 2001 Angiotensin II induces circadian gene expression of clock genes in cultured vascular smooth muscle cells. Circulation 104:1746–1748

Okamura H, Miyake S, Sumi Y et al 1999 Photic induction of *mPer1* and *mPer2* in *Cry*-deficient mice lacking a biological clock. Science 286:2531–2534

Okamura H, Yamaguchi S, Yagita K 2002 Molecular machinery of the circadian clock in mammals. Cell Tissue Res 309:47–56

Panda S, Antoch MP, Miller BH et al 2002 Coordinated transcription of key pathways in the mouse by the circadian clock. Cell 109:307–320

Preitner N, Damiola F, Lopez-Molina L et al 2002 The orphan nuclear receptor REV-ERBα controls circadian transcription within the positive limb of the mammalian circadian oscillator. Cell 110:251–260

Shigeyoshi Y, Taguchi K, Yamamoto S et al 1997 Light-induced resetting of a mammalian circadian clock is associated with rapid induction of the *mPer1* transcript. Cell 91:1043–1053

Shigeyoshi Y, Meyer-Bernstein E, Yagita K et al 2002 Restoration of circadian behavioral rhythms in a *period* null *Drosophila* mutant (*per^{01}*) by mammalian period homologues *mPer1* and *mPer2*. Genes Cells 7:163–171

Takekida S, Yan L, Maywood ES, Hastings MH, Okamura H 2000 Differential adrenergic regulation of the circadian expression of the clock genes Period1 and Period2 in the rat pineal gland. Eur J Neurosci 12:4557–4561

Ueda HR, Chen W, Adachi A et al 2002 A transcription factor response element for gene expression during circadian night. Nature 418:534–539

van der Horst GT, Muijtjens M, Kobayashi K et al 1999 Mammalian *Cry1* and *Cry2* are essential for maintenance of circadian rhythms. Nature 398:627–630

Yagita K, Yamaguchi S, Tamanini F et al 2000 Dimerization and nuclear entry of mPER proteins in mammalian cells. Genes Dev 14:1353–1363

Yagita K, Tamanini F, van der Horst G, Okamura H 2001 Molecular mechanisms of the biological clock in cultured fibroblasts. Science 292:278–281

Yagita K, Tamanini F, Yasuda M, Hoeijmakers JHJ, van der Horst GTJ, Okamura H 2002 Nucleocytoplasmic shuttling and mCRY-dependent inhibition of ubiquitylation of the mPER2 clock protein. EMBO J 21:1301–1314

Yamaguchi S, Mitsui S, Yan L, Yagita K, Miyake S, Okamura H 2000a Role of DBP in the circadian oscillatory mechanism. Mol Cell Biol 20:4773–4781

Yamaguchi S, Mitsui S, Miyake S et al 2000b The 5′ upstream region of *mPer1* gene contains two promoters and is responsible for circadian oscillation. Curr Biol 10:873–876

Yamaguchi S, Kobayashi M, Mitsui S et al 2001 View of a mouse clock gene ticking. Nature 409:684

Yamamoto S, Shigeyoshi Y, Ishida Y et al 2001 Expression of the *Per1* gene in the hamster: brain atlas and circadian characteristics in the suprachiasmatic nucleus. J Comp Neurol 430:518–532

Yamazaki S, Numano R, Abe M et al 2000 Resetting central and peripheral circadian oscillators in transgenic rats. Science 288:682–685

Yan L, Okamura H 2002 Gradients in the circadian expression of *Per1* and *Per2* genes in the rat suprachiasmatic nucleus. Eur J Neurosci 15:1153–1162

Yan L, Takekida S, Shigeyoshi Y, Okamura H 1999 *Per1* and *Per2* gene expression in the rat suprachiasmatic nucleus: circadian profile and the compartment-specific response to light. Neuroscience 94:141–150

Young MW, Kay SA 2001 Time zones: a comparative genetics of circadian clocks. Nat Rev Genet 2:702–715

Zheng B, Albrecht U, Kaasik K et al 2001 Nonredundant roles of the mPer1 and mPer2 genes in the mammalian circadian clock. Cell 105:683–694

Circadian transcriptional output in the SCN and liver of the mouse

John B. Hogenesch, Satchidananda Panda, Steve Kay* and Joseph S. Takahashi†[1]

*The Genomics Institute of the Novartis Research Foundation, San Diego, CA, 9212, *Department of Cell Biology, The Scripps Research Institute, La Jolla, CA, 92037 and †Howard Hughes Medical Institute, Department of Neurobiology and Physiology, Northwestern University, Evanston, IL 60208, USA*

Abstract. Circadian oscillators orchestrate daily rhythms in behaviour and physiology to adapt to the predictable daily appearance of light. Identifying the complement of circadian-regulated transcripts in major organs is critical in the understanding of both the biochemical targets of clock regulation and the mechanism of such control. Recent analysis of temporal gene expression patterns in peripheral and central oscillators have revealed hundreds of circadian-regulated transcripts, most of which are tissue-specific. Mapping of these transcripts to physiological processes and pathways has revealed that major functions of those organs tested are under circadian regulation, and importantly, key and rate-limiting steps in these processes are often the targets of circadian control. Overall, nearly 10% of the mammalian genome may be regulated by the clock, demonstrating the pervasive control of the circadian oscillator in temporal coordination of transcription throughout the organism. This wealth of circadian outputs offers exciting challenges to deciphering systems-level transcriptional regulatory mechanisms that underlie spatiotemporal gene expression.

2003 Molecular clocks and light signalling. Wiley, Chichester (Novartis Foundation Symposium 253) p 171–183

In mammals, the circadian clock regulates many physiological and behavioural processes in synchrony with the environment, including locomotor activity and metabolism (reviewed in Panda et al 2002a). In the absence of environmental cues, a master clock resident in the suprachiasmatic nuclei (SCN) of the hypothalamus coordinates oscillators in peripheral organs like the liver, directing rhythmic changes in gene expression and the resultant physiology. Biochemical approaches, as well as reverse and forward genetics in mammals, have shown that the clock comprises a transcriptional–translational feedback loop similar to

---

[1]This paper was presented at the symposium by Joseph Takahashi to whom all correspondence should be addressed.

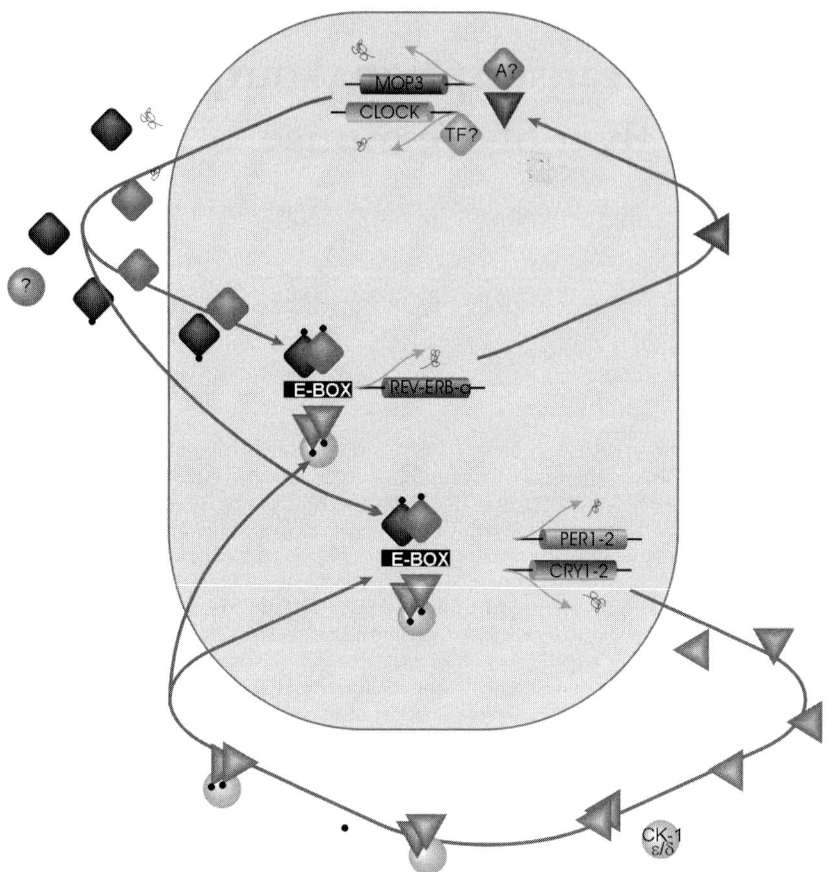

FIG. 1.   Mechanism of circadian transcriptional generation. Two bHLH-PAS transcriptional activators, CLOCK (mid grey squares) and MOP3/BMAL (dark grey squares) heterodimerize and bind E-box enhancers present in the structural genes of several repressor proteins, cryptochromes (light grey triangles), *Per* genes (mid grey triangles), REV-ERBα (dark grey triangles). These repressor proteins eventually feed back on the activity of the CLOCK/MOP3 complex causing a waveform of activation of the complex with a period length of 24 h. Several ancillary factors also participate in this feedback loop, such as casein kinase (CK)1ε (or TAU) (light grey circles). Evidence is emerging that modifications such as phosphorylation (•) participate in aspects of clock function such as protein localization and interaction.

that of fruit flies (Fig. 1) (King & Takahashi 2000, Young & Kay 2001). The transcriptional activators CLOCK and MOP3 heterodimerize and bind an E-box element present in the structural genes of the repressors, *Cry1* and *Cry2*. The CRY proteins are translated, translocate to the nucleus, and eventually repress the CLOCK/MOP3 complex resulting in a transcriptional oscillation with a period length of approximately 24 h. Several other factors including the PER proteins, PER1, PER2 and PER3, CK(casein kinase)1ε (*Tau* or *Double-time*), REV-ERBα, NPAS2/MOP4 and MOP9/BMAL2 have been shown to interact with these factors in the SCN or periphery, fine-tuning the core oscillator into an accurate time-keeping mechanism (Fig. 1) (reviewed in Reppert & Weaver 2002).

Despite our growing mechanistic understanding of the generation, maintenance and modulation of the core circadian clock, deciphering the molecular output genes that encode clock-controlled physiology and behaviour has proven difficult. Recently, several proteins have been described that contribute to modulation of locomotory activity. In an elegant screen for secreted factors that modulate locomotory activity, Kramer et al (2001) identified transforming growth factor (TGF)α and epidermal growth factor (EGF) receptors in mediating the light suppression of activity observed in many nocturnal rodents. Meanwhile, Cheng et al (2002) described prokineticin 2 as rhythmically expressed in the SCN and suggested that it plays a central role in inhibition of locomotor activity during the subjective day. Although these studies have begun to characterize the mechanism of circadian gating of locomotory activity by the clock, they also underscore the need for a comprehensive understanding of transcriptional output of the clock and its relation to physiology. To begin to comprehensively describe circadian transcriptional output, we (and others) have applied oligonucleotide arrays and computational methods. Here we report those findings and summarize the current state of the field.

Identification of circadian transcriptional output genes

To comprehensively describe circadian transcriptional output in the SCN and liver of mice, we used an experimental design similar to that of an earlier study performed in *Arabidopsis* (Harmer et al 2000). Briefly, animals were entrained to a 12:12 h light–dark (LD) cycle for two weeks, then placed in constant darkness for one full day and subsequently harvested for tissue dissection and RNA extraction every 4 h from circadian time (CT)18 to CT72. To balance experimental design issues with cost, we harvested tissue from ten animals per time point, pooled the RNA samples, then labelled and hybridized the samples to high-density oligonucleotide arrays in duplicate. Primary image analysis and probe-set condensation were performed using standard methods, and the data were analysed using a modified version of CORCOS (Harmer et al 2000). This process resulted in a goodness-of-fit measure to a cosine wave accommodating multiple measures, and ultimately generated a multiple-measures-corrected minus β (MMCβ) value. To set our cut-off value for this MMCβ value, we examined the expression patterns of known cycling genes in our dataset and the corresponding MMCβ values generated by the algorithm (Kita et al 2002, Kornmann et al 2001).

This analysis revealed that a MMCβ value of 0.1 provided the best balance between type 1 (false negative) and type 2 (false positive) error rates, and at this level roughly 300 genes from each tissue were considered to have circadian expression patterns. To validate the circadian expression of the genes fitting the MMCβ threshold, *in situ* hybridizations and real-time PCR were performed. Finally, a subsequent study aimed at describing the circadian expression pattern of genes in the SCN and liver confirmed the circadian rhythmicity of more than 1/3 of the cycling genes at the conservative MMCβ value of 0.1, and the majority at more permissive MMCβ values (Ueda et al 2002a). Taken in sum, these various analyses confirmed the cycling of most (but not all) of the circadian transcripts and supported the validity of the methodologies employed.

It had been previously observed that circadian clock components tend to be rhythmically expressed in multiple tissues, persisting even in *ex vivo* tissue culture experiments (Balsalobre et al 2000). This observation prompted us to investigate our dataset for genes cycling in common between the SCN and liver — reasoning that these transcripts could potentially encode for clock components. We found 28 genes whose mRNAs cycled in both the SCN and liver, including the known clock components PER2 and BMAL1/MOP3. Most of these were delayed in their expression level in the liver with respect to the SCN, as previously reported. Surprisingly, the sets of cycling transcripts in the SCN and in liver are largely non-overlapping (Fig. 2) (Panda et al 2002b, Ueda et al 2002a), while a similar observation comparing circadian gene expression in the heart and liver has been reported elsewhere (Storch et al 2002). What could explain the tissue-specific nature of circadian gene regulation? The simplest explanation arises from the observation that cycling genes in the SCN tend to have higher expression levels in that tissue than liver (Panda et al 2002a). Stated another way, approximately 25% of SCN or liver circadian output genes are not expressed in the reciprocal tissue. Another 25% of genes show reduced, albeit detectable, levels of expression in the alternate tissue. The remaining 50% of genes demonstrated tissue-specific circadian regulation, where the median expression level of the genes are well above the limit of detection in both tissues, but cyclic expression is restricted to only one tissue type. An excellent example of this is *Ccr4/Nocturnin*, whose circadian mRNA expression occurs only in liver, but not in the SCN — despite its expression in SCN (Panda et al 2002b, Wang et al 2001). Therefore, the circadian clock conscripts the tissue-specific transcriptional machinery to effect specific expression patterns for the vast majority of cycling genes.

The observation that a relatively small number of cycling genes were common to SCN and liver (or liver and heart) suggested that direct clock regulation of circadian output may be a somewhat rare event. To address this point, we profiled the livers of *Clock/Clock* mutant mice at a time point where the CLOCK/MOP3 complex is transcriptionally active. We reasoned that direct targets of the

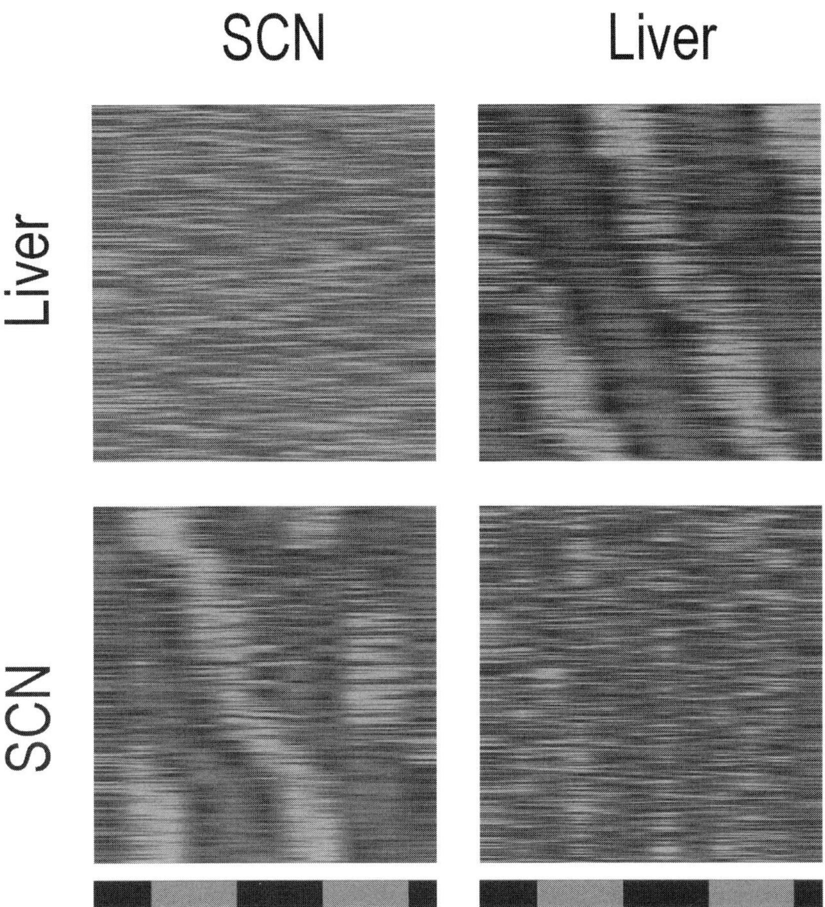

FIG. 2.   Tissue-specific circadian gene regulation. Probe set identifiers for cycling genes (MMCβ < 0.1) in the SCN and liver were retrieved from a relational database, and used to query for expression patterns of both tissues. The results were clustered as previously described using Cluster and visualized using Treeview (Eisen et al 1998, Su et al 2002) such that the expression patterns for liver cycling genes are shown in both liver and SCN and vice versa. Expressed genes are indicated in dark grey, while light grey denotes lower than median levels of expression. At the bottom is a bar denoting circadian time where subjective day is indicated in grey, while subjective night is indicated in black.

transcriptional activator CLOCK should be down-regulated in *Clock/Clock* mutant mice. Our analysis revealed 56 genes that were misregulated in the *Clock/Clock* mutant mice; however, the expression patterns of 45 of these were not found cycling in wild-type mice. Another way to address the direct transcriptional regulation of the CLOCK/MOP3 complex is to determine whether their cognate binding site, the CACGTG E-box, is present in the structural genes of circadian-regulated transcripts. To compensate for the frequency of occurrence of this element in random genomic sequence, we took advantage of the recently available mouse genome sequence and considered elements that were conserved between mouse and human. These results indicate that the *Clock/Mop3* consensus sequence is present in only about 10% of circadian-regulated transcripts. Ueda and colleagues took a similar approach and investigated the number of cycling genes with consensus cAMP response elements (CREs) and REV-ERBα/ROR elements proximal to the transcriptional start sites of human orthologues of mouse cycling genes. The authors found consensus CREs in the proximal regions of seven genes whose expression peaked during subjective night or dawn and ten genes with REV-ERBα/ROR elements whose expression peaked during the subjective day (Ueda et al 2002a). Taken in sum, these data support a model by which the core clock complex initiates rhythmic transcription of tissue-specific transcription factors; these factors in turn activate their specific target genes to control tissue and phase specific expression.

Relating chronobiology to transcriptional output

Next we wanted to categorize the functions of circadian output genes in relation to the physiologies mediated by the SCN and liver. The SCN serves as the master circadian oscillator in mammals, exerting its control via neuronal signalling involving both synaptic transmission and the release of diffusible factors (Silver et al 1996). For example, a circadian rhythm in the action potential of individual SCN neurons has been described (Welsh et al 1995). In addition, the rhythmic release of several neuropeptides such as somatostatin and vasopressin has been characterized in the SCN (reviewed in Inouye 1996). Characterization of the circadian output genes in the SCN revealed that its principal role in neuropeptide release is recapitulated by the transcriptional output there. For example, one of the largest functional groups of cycling transcripts in the SCN contains genes implicated in peptide synthesis, processing and release (Panda et al 2002b). Transcripts encoding several neuropeptides including pro-opiomelanocortin, pituitary adenylate cycle activating polypeptide 1, cholecystokinin, platelet-derived growth factor (PDGF), arginine vasopressin, somatostatin, enkephalin, galanin and calcitonin gene-related peptide are circadianly regulated in the SCN. In addition, genes involved in synthesis of non-peptide neurotransmitters glutamate (glutamic acid decarboxylase 1) and histamine (histidine decarboxylase) were also rhythmically expressed in the SCN. Rhythms in a few genes participating in neuropeptide processing may accentuate the rhythms in active neuropeptide production in the SCN. Examples of these include secretory granule neuroendocrine protein 1, which activates prohormone convertase 2 involved in neuropeptide processing, and secretogranin III, another secretory granule protein, as well as components of the vesicle fusion (synaptosomal-associated protein, 25 kDa; syntaxin binding protein 1), recycling (epidermal growth factor receptor pathway substrate 15) and transport (vacuolar protein sorting 4B). Finally, a potassium large conductance calcium-activated channel, subfamily M, α member 1, plays a critical role in neurosecretion, and cycles with a peak level of transcription in the late subjective night; notably, the *Drosophila* homologue, *Slowpoke*, is also regulated by the circadian clock in flies. Thus, in the SCN the circadian clock regulates neurosecretory components and thus the signalling output of the SCN.

The principal role of the liver is in nutrient, endobiotic and xenobiotic metabolism, all of which are regulated by the circadian clock. Mice gate most of their feeding to the night and subjective night under constant darkness. Rhythmic expression of the transcripts for several proteins involved in nutrient metabolism was observed with a peak expression in the early subjective night. The protein products of these transcripts included glucose transporters, the glucagon receptor, and components of the hexose sugar metabolism pathways. Such regulation may promote the use of hexose sugars during the post-adsorptive period. Evidence for the transcriptional regulation of intermediate metabolism by the clock was found in the characterization of the cholesterol biosynthetic pathway. The activity of the rate-limiting enzyme in the pathway, HMG-CoA reductase, was found to be under the transcriptional regulation of the clock (Shapiro & Rodwell 1969). Likewise, we found its steady state mRNA expression displayed a circadian pattern of expression as well as several other enzymes within this pathway, including HMG-CoA lyase, isopentenyl-diphosphate delta isomerase, farnesyl-diphosphate farnesyl-transferase, and lanthosterol oxidase. Coordinated regulation of multiple steps in a pathway may help to ensure circadian control of its ultimate product (in this case cholesterol), or may reflect circadian regulation of parallel pathways that utilize shared components. Xenobiotic metabolism has likewise long been known to be under circadian control (Lake et al 1976). We found circadian oscillations in the transcription of several genes involved in xenobiotic metabolism including four methyltransferases, betaine-homocysteine methyltransferase, nicotinamide N-ethyltransferase, thioether S-methyltransferase, and thiopurine methyl-transferase, as well as S-adenosylhomocysteine hydrolase—a regulator of most methyltransferase activity. These examples highlight the pervasive circadian regulation of metabolism in the mammalian liver.

Interestingly, clock regulation of these physiologies seems to have evolved to exert transcriptional control of key rate-limiting steps. In the SCN for example, transcriptional regulation of metallothionein 1 activator, a component of all three subunits of RNA polymerase I, may ensure coordinated regulation of rRNA and ribosomal protein transcription, ultimately leading to a rhythm in the size and morphology of the nucleoli. Neurotransmitter synthesis and neuronal signalling are also regulated at key steps by the circadian clock. In the liver, examples of circadian regulation of rate-limiting steps are abundant as HMG-CoA reductase, delta-aminolevulinate synthase and cytochrome P450 7a1 are rate-limiting steps in cholesterol, haeme and bile acid biosynthesis, respectively, while glycerol kinase is rate-limiting in the regulation of the use and uptake of glycerol as an energy source. Furthermore, circadian transcriptional regulation of rate-limiting steps is conserved across species, as both HMG-CoA reductase and delta-aminolevulinate synthase are circadianly regulated in flies as well. Indeed, in plants HMG-CoA lyase is circadianly regulated, the key rate-limiting step in HMG-CoA metabolism. Thus, a hallmark of the circadian clock in several species is that it has evolved the efficient process of targeting key rate-limiting steps in biological pathways.

Conclusion

In the past two years, transcriptional profiling has been extensively applied to the study of circadian systems (Akhtar et al 2002, Ceriani et al 2002, Claridge-Chang et al 2001, Duffield et al 2002, Grundschober et al 2001, Harmer et al 2000, Kita et al 2002, McDonald & Rosbash 2001, Panda et al 2002a, Storch et al 2002, Ueda et al 2002a,b). Collectively, these works are beginning to suggest general rules and underlying themes in clock-mediated transcription. In mammals, results from several groups suggest that as many as 10% of protein-encoding transcripts are regulated by the circadian clock. The availability of the genome sequences for several model organisms has enabled use of sequence analysis tools to begin to analyse systems level controls at the level of DNA elements. These studies suggest that the core clock complex in mammals regulates relatively few transcriptional output genes directly (Panda et al 2002a), while elegant transcriptional networks are being constructed for transcriptional output factors such as CREB and REV-ERBα (Ueda et al 2002a). Furthermore, word-based searching tools are enabling *de novo* response element discovery, and have identified at least one complex involved in regulation of evening phased circadian transcriptional output (Harmer et al 2000). Much of this regulation is tissue-specific, both in flies and in mammals, implying that the clock has enlisted the use of the tissue-specific transcriptional machinery to exert its actions (Ceriani et al 2002, Panda et al 2002a, Storch et al 2002, Ueda et al 2002a). Key rate-limiting steps in biochemical pathways and processes are often sites of circadian control (Ceriani et al 2002, Panda et al 2002a). In mammals, the extension of these works will include the application of transcriptional profiling to the remaining 80% of genes in the genome and the multitude of tissues, organs, and systems that have yet to be investigated. Finally, the further refinement of sequence and expression analysis, as well as experimental tools (chromatin immunoprecipitation in combination with array profiling) should enable the construction of a systems-level description of the output of the clock.

References

Akhtar RA, Reddy AB, Maywood ES et al 2002 Circadian cycling of the mouse liver transcriptome, as revealed by cDNA microarray, is driven by the suprachiasmatic nucleus. Curr Biol 12:540–550

Balsalobre A, Brown SA, Marcacci L et al 2000 Resetting of circadian time in peripheral tissues by glucocorticoid signaling. Science 289:2344–2347

Ceriani MF, Hogenesch JB, Yanovsky M, Panda S, Straume M, Kay SA 2002 Genome-wide expression analysis in *Drosophila* reveals genes controlling circadian behavior. J Neurosci 22:9305–9319

Cheng MY, Bullock CM, Li C et al 2002 Prokineticin 2 transmits the behavioural circadian rhythm of the suprachiasmatic nucleus. Nature 417:405–410

Claridge-Chang A, Wijnen H, Naef F, Boothroyd C, Rajewsky N, Young MW 2001 Circadian regulation of gene expression systems in the *Drosophila* head. Neuron 32:657–671

Duffield GE, Best JD, Meurers BH, Bittner A, Loros JJ, Dunlap JC 2002 Circadian programs of transcriptional activation, signaling, and protein turnover revealed by microarray analysis of mammalian cells. Curr Biol 12:551–557

Eisen MB, Spellman PT, Brown PO, Botstein D 1998 Cluster analysis and display of genome-wide expression patterns. Proc Natl Acad Sci USA 95:14863–14868

Grundschober C, Delaunay F, Puhlhofer A et al 2001 Circadian regulation of diverse gene products revealed by mRNA expression profiling of synchronized fibroblasts. J Biol Chem 276:46751–46758

Harmer SL, Hogenesch JB, Straume M et al 2000 Orchestrated transcription of key pathways in *Arabidopsis* by the circadian clock. Science 290:2110–2113

Inouye ST 1996 Circadian rhythms of neuropeptides in the suprachiasmatic nucleus. Prog Brain Res 111:75–90

King DP, Takahashi JS 2000 Molecular genetics of circadian rhythms in mammals. Annu Rev Neurosci 23:713–742

Kita Y, Shiozawa M, Jin W et al 2002 Implications of circadian gene expression in kidney, liver and the effects of fasting on pharmacogenomic studies. Pharmacogenetics 12: 55–65

Kornmann B, Preitner N, Rifat D, Fleury-Olela F, Schibler U 2001 Analysis of circadian liver gene expression by ADDER, a highly sensitive method for the display of differentially expressed mRNAs. Nucleic Acids Res 29:E51–1

Kramer A, Yang FC, Snodgrass P et al 2001 Regulation of daily locomotor activity and sleep by hypothalamic EGF receptor signaling. Science 294:2511–2515

Lake BG, Tredger JM, Burke MD, Chakraborty J, Bridges JW 1976 The circadian variation of hepatic microsomal drug and steroid metabolism in the golden hamster. Chem Biol Interact 12:81–90

McDonald MJ, Rosbash M 2001 Microarray analysis and organization of circadian gene expression in *Drosophila*. Cell 107:567–578

Panda S, Hogenesch JB, Kay SA 2002a Circadian rhythms from flies to human. Nature 417: 329–335

Panda S, Antoch MP, Miller BH et al 2002b Coordinated transcription of key pathways in the
    mouse by the circadian clock. Cell 109:307–320
Reppert SM, Weaver DR 2002 Coordination of circadian timing in mammals. Nature 418:
    935–941
Shapiro DJ, Rodwell VW 1969 Diurnal variation and cholesterol regulation of hepatic HMG-
    CoA reductase activity. Biochem Biophys Res Commun 37:867–872
Silver R, LeSauter J, Tresco PA, Lehman MN 1996 A diffusible coupling signal from the
    transplanted suprachiasmatic nucleus controlling circadian locomotor rhythms. Nature
    382:810–813
Storch KF, Lipan O, Leykin I et al 2002 Extensive and divergent circadian gene expression in
    liver and heart. Nature 417:78–83
Su AI, Cooke MP, Ching KA et al 2002 Large-scale analysis of the human and mouse
    transcriptomes. Proc Natl Acad Sci USA 99:4465–4470
Ueda HR, Chen W, Adachi A et al 2002a A transcription factor response element for gene
    expression during circadian night. Nature 418:534–539
Ueda HR, Matsumoto A, Kawamura M, Iino M, Tanimura T, Hashimoto S 2002b Genome-
    wide transcriptional orchestration of circadian rhythms in *Drosophila*. J Biol Chem
    277:14048–14052
Wang Y, Osterbur DL, Megaw PL et al 2001 Rhythmic expression of Nocturnin mRNA in
    multiple tissues of the mouse. BMC Dev Biol 1:9
Welsh DK, Logothetis DE, Meister M, Reppert SM 1995 Individual neurons dissociated from
    rat suprachiasmatic nucleus express independently phased circadian firing rhythms. Neuron
    14:697–706
Young MW, Kay SA 2001 Time zones: a comparative genetics of circadian clocks. Nat Rev
    Genet 2:702–715

DISCUSSION

Hastings: We have been involved in some work with Andrew Louden in the *tau*
mutant hamster. Looking at the peripheral gene expression in those mutant
hamsters, surprisingly *Per1* and DBP take up phases which you wouldn't predict
simply on the basis of the period shortening in the mutant. Looking at the phase in
circadian time we should correct for that difference in period, but the gene
expression cycles remain inappropriately phased in the mutant. They are
internally desynchronized.

Rosbash: Is this molecular phase, or behavioural phase?

Hastings: Your reference point is the activity cycle, and then we are measuring
gene expression in peripheral tissue relative to that onset.

Menaker: If period doesn't explain the phasing, which it clearly doesn't, this
means either there are different phase–response curves for the individual tissues
in response to a single signal, or there are a set of different signals.

Rosbash: The paradigm from the fly genetics work has been that a lot of mutants
have been analysed where there is excellent tracking of period and phase in
advances and delays. I can't think of an exception where the evening activity
peak doesn't shift in the proper direction and with the appropriate magnitude in
response to a period mutant. Can one think about these dislocations in terms of

other interpretations, such as subtle tweaking in different tissues which are not really circadian?

Weitz: Joe Takahashi, is your expectation that you would only see this at the tissue level? The idea is presumably that in a cell-autonomous sense, period and phase are still coupled because of the way we imagine the oscillator works, but when you are now talking about how a collection of coupled oscillators integrate this into behavioural outputs, there are now other factors that must come into play. Is this fair to say? Is it only in the chimeras, and not in the heterozygotes, that you see these unusual combinations and dissociations of phase and period?

Takahashi: I'm not saying we throw out the coupling. Obviously in a long period mutant you get later phase under the appropriate light cycle. In *Clock* we can show that the entrained phase can be later, or in *tau* the entrained phase is much earlier. I want to point out that there are many cases where we look at some mechanistic aspect, such as the loci that control these two features, and they are not overlapping. This is interesting.

Weitz: There is no assumption of cell autonomy in that analysis.

Takahashi: That is true.

Weitz: This means that from the chimeric analysis one imagines that what you are beginning to see are genetic contributions to the tissue-level organization of the oscillator.

Dunlap: A further difference might also be in the synchronizing cue. When we think about coupling period with phase it is always in the discreet model for resetting, rather than a parametric model. Certainly, for all of the peripheral pacemakers like liver cells, getting a parametric cue such as a slow-wave change in temperature or a slow-wave change in hormone levels would seem more likely than getting an abrupt change in anything as is demanded for discrete phase shifting.

Kay: There are lots of orphan receptor-type molecules cycling in these different target tissues. You don't even need a cycling signal. One would imagine there is a suite of humoral signals which are more or less relevant in different tissues. We have done temporal profiling in the aorta, and we see a completely different set of signals, many of which are relevant to cyclic control of bloodflow.

Dunlap: Since you have this beautiful data series of long and short periods in peripheral tissues, have you looked at temperature compensation?

Takahashi: No. You should ask Mike Menaker that question.

Menaker: We are now doing this in our laboratory. The preliminary results are interesting: some tissues are compensated, others aren't.

Young: Does some of the separation of phase and period in the chimeras have to do with where the cells giving wild-type contributions are ending up in the SCN?

Takahashi: That was our original hope: that we could see an anatomical basis for this separation. We analysed 12 different regions of the SCN. There is a weak correlation for amplitude and period in one of the regions, but it is very weak. The nature of the chimeras is much more fine-grained than the size of the nucleus, so we couldn't get, for example, a big patch on the front end of the SCN. The biggest differences are right–left differences, for some reason.

Schibler: Do you make your chimeras by aggregation or injection?

Takahashi: We use morula aggregation, drop culture overnight, and then they are implanted.

Schibler: This should be better for your purposes than injection into blastocysts. It should give a coarser distribution.

Takahashi: It should do, but there is something about the way that the cells migrate in the embryo of the mouse that mixes them up. The only way that people have been able to get really patchy chimeras in mouse is to go cross-species. They use a different species of mouse to make a chimera, and then you can get patches. Within a species it turns out to be pretty fine-grain.

Rosbash: A decade ago we did a study in flies with transgenes. We were putting different promoters in and got one set that displayed a highly altered amplitude and had very good periods, and another construct which had good periods and weak amplitudes. These were localized to different parts of the brain. In the end, we were unable to figure out quite what this meant, other than describing the phenomenon.

Menaker: If you make a partial SCN lesion and leave 10% of the SCN behind, you get a pretty normal looking rhythm. This makes it surprising that you have this democratic interaction among all the mutant versus wild-type cells in the SCN. One would think that a subpopulation within that democracy would take over, since you only need 10% of the SCN for normal function.

Takahashi: There's no conflict in your 10% remnant situation.

Menaker: Your situation suggests that there is conflict and interaction among the kinds of cells that are present in the SCN.

Takahashi: As we get more information it is becoming clear that the subdivisions are different. More anterior cells were shorter and had earlier phases; more ventrolateral cells have lower amplitude and later phases. I think there is going to be anatomical specificity. Perhaps we just couldn't see it in those experiments.

Van Gelder: You said that the *Per2*–luciferase knock-in behaviour is due to the presence of a single copy of wild-type *Per2* in this mouse. Does this fusion gene actually behave as a null if you put it over the *Per2⁻*?

Takahashi: Because it is a knock-in we can make a homozygote mouse and there is no wild-type copy there, or course. This mouse is completely normal and has no phenotype that we can find yet. It was a C-terminal fusion so all of *Per2* is there. It just has this thing dangling off the end, which you might think could compromise its function, but it didn't.

Van Gelder: So there really are two copies.

Takahashi: What I showed was heterozygous normals.

Van Gelder: Presumably that would still be the same if the homozygote doesn't show any phenotype. Does *Per* localize normally even with this C-terminal tag?

Takahashi: We have not looked, but I assume it must do since we didn't see any aberrant phenotype. There is no other *Per*. The knock-in is nice because it is so clean compared with the transgene.

Rosbash: Can I ask one clarification question about this period and phase clustering in the chimeras. I take it that all along the gradient from 90% wild-type, 10% mutant, right down to the 10% wild-type, 90% mutant, there was no distinction with regard to period and amplitude. This means that they must self-organize in some way that favours one or the other independent of the proportions of wild-type and mutant tissue.

Takahashi: Yes, this is not what we expected. The *Clock* mutant has a longer period and a low amplitude. We expected it to be like a *Clock* mutant, but this didn't happen.

Loros: Interestingly, long-period FRQ mutants in *Neurospora* are long-period and high-amplitude. When we look at molecular rhythms, feedback is less strong, and so you get a greater phase of transcriptional build-up and then you have more protein. The whole thing is not only long, it is much higher amplitude.

Takahashi: This might happen in a *Per2*-long mutant.

Sehgal: We have always thought in *Drosophila* that the shorter mutants are higher amplitude.

Loros: Short FRQ mutants show a lower amplitude of molecular cycling. Amounts don't have much time to build up before you get repression.

Sehgal: And these are both negative regulators.

Loros: We do have a period length mutant in White Collar (WC) 2, the ER24 allele that is both long-period and not temperature compensated. We haven't looked at ER24 in terms of the amplitude of the molecular rhythms of either FRQ or WC-1.

Hardin: It depends on why they are short. For *Per*-short, it is short because the protein goes away prematurely. Everything accumulates to the proper level; it just starts the next cycle prematurely. In FRQ, it could be that things don't accumulate to the right levels.

Loros: I think FRQ accumulates and continues to do so, but it doesn't have the ability to repress as well in this situation.

The molecular workings of the
Neurospora biological clock

Allan C. Froehlich*, Antonio Pregueiro*, Kwangwon Lee*, Deanna Denault*, Hildur Colot*, Minou Nowrousian*, Jennifer J. Loros*† and Jay C. Dunlap*[1]

Department of Genetics and †Department of Biochemistry, Dartmouth Medical School, Hanover, NH 03755, USA

Abstract. In *Neurospora crassa* the FRQ/WC feedback loop has been shown to be central to the function of the circadian clock. Similar to other eukaryotic systems it is based on a transcription–translation PAS heterodimer type feedback. FRQ levels cycle with a period identical to that of the *Neurospora* circadian cycle and its expression is rapidly induced by light. A complex of White Collar 1 (WC-1) and White Collar 2 (WC-2) (the WCC) is required for the transcriptional activation of *frq*. The oscillation in *frq* message is transcriptionally regulated via a single necessary and sufficient *cis*-acting element in the *frq* promoter, the Clock-Box (CB) bound by WCC. Light-induction of *frq* transcription is mediated by WCC binding to two *cis*-acting elements (LREs) in the *frq* promoter. WC-1, with flavin adenine dinucleotide (FAD) as a cofactor, is the blue-light photoreceptor. The original description of a *frq*-null strain, *frq⁹*, (Loros et al 1986) included a description of oscillations in asexual conidial banding that occasionally appeared following 3 to 7 days of arrhythmic development now referred to as FLO for FRQ-less oscillator. Unlike the intact clock, FLO period is sensitive to media composition. We have identified a circadianly regulated gene whose mutation interferes with FLO even under temperature entrainment conditions. This same mutation affects the circadian clock in a *frq+* background causing a shorter period length as well as temperature response defects. This gene may be an entry point to study the connection between the biological clock and other basic cellular mechanisms.

2003 Molecular clocks and light signalling. Wiley, Chichester (Novartis Foundation Symposium 253) p 184–202

Life on earth has evolved under the continual 24 h fluctuations in light and temperature that constitute a day. Many organisms have evolved the ability to anticipate these external changes in their environment using endogenous 'biological clocks'. In recent years, the molecular components that make up these intracellular clocks have begun to be identified, and similarities among a wide

---

[1]This paper was presented at the symposium by Jay C. Dunlap to whom all correspondence should be addressed.

range of organisms have emerged (Dunlap 1999, Cermakian & Sassone-Corsi 2000, Allada et al 2001, Loros & Dunlap 2001). Studies using the filamentous fungus *Neurospora crassa* have played a central role in building the current clock paradigm through identification of central players, their interactions with one another and their interactions with the external environment.

Components of the *Neurospora* circadian system

The *frequency* (*frq*) gene was the first clock component isolated in *Neurospora* and the second clock component isolated in any organism (the first being the *period* gene from *Drosophila*). Rhythmic expression of both *frq* message and FRQ protein is central to the functioning of the clock (Dunlap 1999, Cermakian & Sassone-Corsi 2000, Allada et al 2001, Loros & Dunlap 2001), and a greater understanding of the mechanism underlying the generation and regulation of *frq* rhythms has led to a greater understanding of the clock itself. Central to the regulation of *frq* are the products of the *white collar 1* (*wc-1*), and *white collar 2* (*wc-2*) genes. WC-1 and WC-2 are predominately nuclear transcription factors containing *trans*-activation domains and zinc-finger (Zn-finger) DNA binding domains. They form a white collar complex (WCC) by heterodimerizing via PAS domains (Linden et al 1999) and act as positive elements in the activation of *frq* (Crosthwaite et al 1997); in a $wc\text{-}1^{KO}$ or a $wc\text{-}2^{KO}$ strain, very limited, unregulated transcription of *frq* occurs (K. Lee, J.J. Loros and J.C. Dunlap, unpublished data; Cheng et al 2001a, Collett et al 2002). The positive action of the WCs is counter-acted by FRQ itself which acts as a negative element, repressing the levels of its own transcript (Aronson et al 1994a). FRQ, WC-1 and WC-2 therefore comprise a negative feedback loop central to clock function.

This negative feedback loop is interconnected with several positive feedback loops resulting from additional interactions, both direct and indirect, of FRQ, WC-1 and WC-2 (Fig. 1). FRQ plays a positive role in the post-transcriptional production of rhythmic WC-1 from a constitutively expressed *wc-1* transcript (Lee et al 2000). The steady-state level of WC-1 protein is also positively regulated by WC-2, but through an apparently different posttranscriptional mechanism than FRQ; neither mechanism is clearly understood at this time (Cheng et al 2002). At least part of WC-2's positive effect on WC-1 levels is through the direct interaction of the two proteins via WC-2's PAS domain. FRQ also plays a positive role in the regulation of WC-2, at least partially achieved by increasing the abundance of *wc-2* transcript (Cheng et al 2001a), a mechanism apparently different from FRQ's post-transcriptional effect on WC-1. WC-2 is constitutively expressed, the most abundant of the three proteins, and physically interacts with both FRQ and WC-1, acting as a bridge between WC-1 and FRQ (Denault et al 2001). FRQ forms a homodimer through its coiled-coil domain with

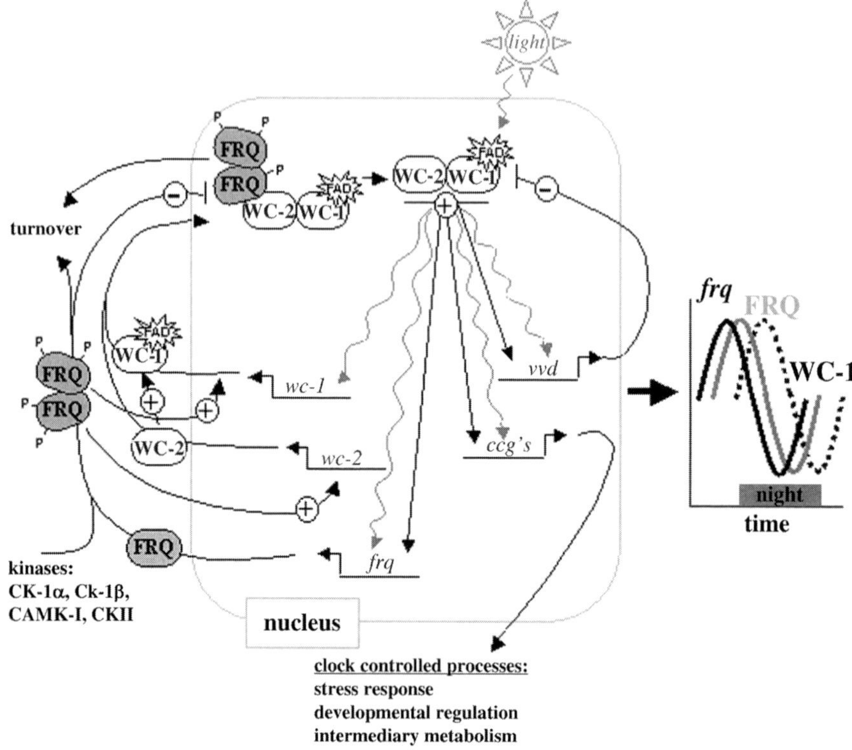

FIG. 1.   Model of the molecular mechanisms involved in the *Neurospora* circadian clock. Multiple interlocked feedback loops are involved in generating rhythms: a negative autoregulatory loop of FRQ on its own gene (through inhibition of the transcriptional activator WC-1/WC-2), a positive effect of FRQ on *wc-2* transcript and WC-1 protein, and a positive effect of WC-2 on WC-1. VVD identifies an autoregulatory negative feedback loop outside of the core oscillator. Multiple kinases phosphorylate FRQ leading to its turnover. WC-1/WC-2 drive transcription by directly binding to DNA: in the dark, rhythmic transcription of *frq* and possibly of *vvd* and other *ccg*s (clock-controlled genes), and in the light, absorbed by WC-1 through a bound FAD molecule, increased transcription of *frq* as well as other light-induced genes.

this self-association enabling the physical interaction between FRQ and the WCC (Cheng et al 2001b). The complex interactions of FRQ, WC-1 and WC-2 result in the rhythmic expression of *frq* message and protein as well as the antiphasic oscillations in WC-1, ultimately contributing to the robustness of the *Neurospora* circadian clock.

VIVID (VVD), a novel member of the PAS protein superfamily, identifies an autoregulatory negative feedback loop that closes outside of the core oscillator but which impacts all aspects of circadian timing (Heintzen et al 2001). Expression of *vvd* is controlled by the clock, but it is not required for circadian rhythmicity. Nonetheless, loss of the VVD protein has far reaching effects on the perception of light and on the entire circadian system ranging from input, as seen in the phase response curve, to oscillator function as measured by period length, to output as manifested in the phasing and expression levels of clock-controlled genes.

Coincident in time with its interaction with the WCC, FRQ is phosphorylated (Garceau et al 1997, Liu et al 2000), an event that appears to govern the time-of-day regulation of FRQ stability, of central importance to the kinetics of the circadian cycle. There are multiple phosphorylation events which may be processive in a manner such that one event elicits the next (Garceau et al 1997, Liu et al 2000). Several kinases responsible for phosphorylation of FRQ have been identified, including: (1) two forms of casein kinase 1 (CK-1a and CK-1b) (Gorl et al 2001), (2) casein kinase II, (CKII) (Yang et al 2002), and (3) a Ca/CaM-dependent kinase (CAMK-1) (Yang et al 2001). CAMK-1 appears to account for nearly half of the FRQ kinase activity *in vitro*, but disruption of CAMK-1 *in vivo* has only slight effects on phase, period, and light-induced phase shifting of circadian conidiation rhythm (Yang et al 2001). In contrast, disruption of CKII results in hypo-phosphorylation and increased levels of FRQ protein leading to abolished molecular and overt circadian rhythms (Yang et al 2002). Attempts to inactivate CK-1a were unsuccessful suggesting that the gene may be essential for cell viability, but CK-1a together with CK-1b are able to phosphorylate *in vitro* FRQ's two PEST sequences (Gorl et al 2001). Deletion of FRQ's PEST-1 sequence results in loss of overt rhythmicity, but interestingly retains rhythmicity of FRQ, but with a longer period (Gorl et al 2001). Mutation of another site of FRQ phosphorylation, Ser513, also leads to a reduction in the rate of FRQ degradation and subsequently to an increased period length (Liu et al 2000). Additional kinases may be involved in the phosphorylation of FRQ and the identification of these clock components along with their sites of kinase activity will give a clearer picture of the precise roles that phosphorylation plays in regulating FRQ and the clock.

Temperature regulation

Recent DNA microarray experiments have found that FRQ, and by extension the clock, is required for temperature-regulated gene expression suggesting an unexpected role of the circadian circuitry in environmental temperature sensing (Nowrousian et al 2003). Of the genes on the array, 1.3% were found to be regulated by a 12 h 22 °C/12 h 27 °C temperature regime (well below that eliciting a heat shock response), but in a *frq^null^* strain all temperature regulation was lost. The temperature-regulated genes were also all clock regulated, as were a total of 2% of the genes on the array.

Interaction between temperature and the clock is also seen with the temperature-regulated production of the two forms of the FRQ protein, a long form of 989 amino acids and a form lacking the first 100 amino acids (Liu et al 1997). The form-specific functions have remained elusive, but both forms are needed for robust rhythmicity.

The interplay of temperature and the *Neurospora* clock continues on a third level with temperature steps resetting the clock through a post-transcriptional mechanism (Liu et al 1998). *frq* transcript oscillations at different temperatures are close to superimposable, but FRQ amounts oscillate around higher levels at higher temperatures—the lowest FRQ level (late night) at 28 °C is higher than the highest level (late day) at 21 °C—so the 'time' associated with a given number of molecules of FRQ is different at different temperatures. Thus a shift in temperature corresponds to a shift in the state of the clock (literally a step to a different time) although initially no synthesis or turnover of components occurs. After the step, relative levels of *frq* and FRQ are assessed in terms of the new temperature and they respond rapidly. Thus, temperature changes reset the circadian cycle instantaneously and from within the circadian loop (Liu et al 1998).

Light regulation and the identity of the circadian photoreceptor

Like temperature, light signals can entrain and reset the *Neurospora* clock. The central means through which light exerts its influence on the clock is by causing a rapid induction of *frq* message (Crosthwaite et al 1995). Induction is mediated by two light response elements (LREs) in the 3 kb *frq* promoter (Froehlich et al 2002). Electrophoretic mobility shift assays (EMSA) using the LREs as probes and *Neurospora* nuclear protein extracts reveal two distinct complexes for each LRE,

---

FIG. 2.   The *in vitro* light-induced shift of the WCC occurs at biologically relevant fluence and wavelengths. (A) Dose–response curve generated for the *in vitro* light shift. Aliquots of dark-grown *Neurospora* nuclear protein extracts were exposed to varying amounts of white light and then used in a series of binding reactions with a LRE probe. Densitometric analysis of WCC/LRE complexes shows shift with increasing amounts of light from the faster migrating (closed squares) to slower migrating complex (open squares) ($n=3$ ±SEM). A representative gel is shown. (B) Equal-intensity action spectrum generated for *in vitro* light shift. Aliquots of dark-grown extracts were exposed to the same fluence of light at wavelengths varying from 410 nm to 540 nm and then used in a series of binding reactions with LRE probe. Densitometric analysis of the slower migrating/light-induced complex (left axis, open squares and dashed line) shows a peak in sensitivity ~455 nm–470 nm and no response to wavelengths above 500 nm ($n=3$ ±SEM). A representative gel is shown. The original *in vivo* action spectrum for inhibition of circadian banding by continuous light, to which the *in vitro* response is quite similar, is replotted (right axis, grey line) from (Sargent & Briggs 1967). Reprinted from Froehlich et al 2002 with permission from the American Association for the Advancement of Science.

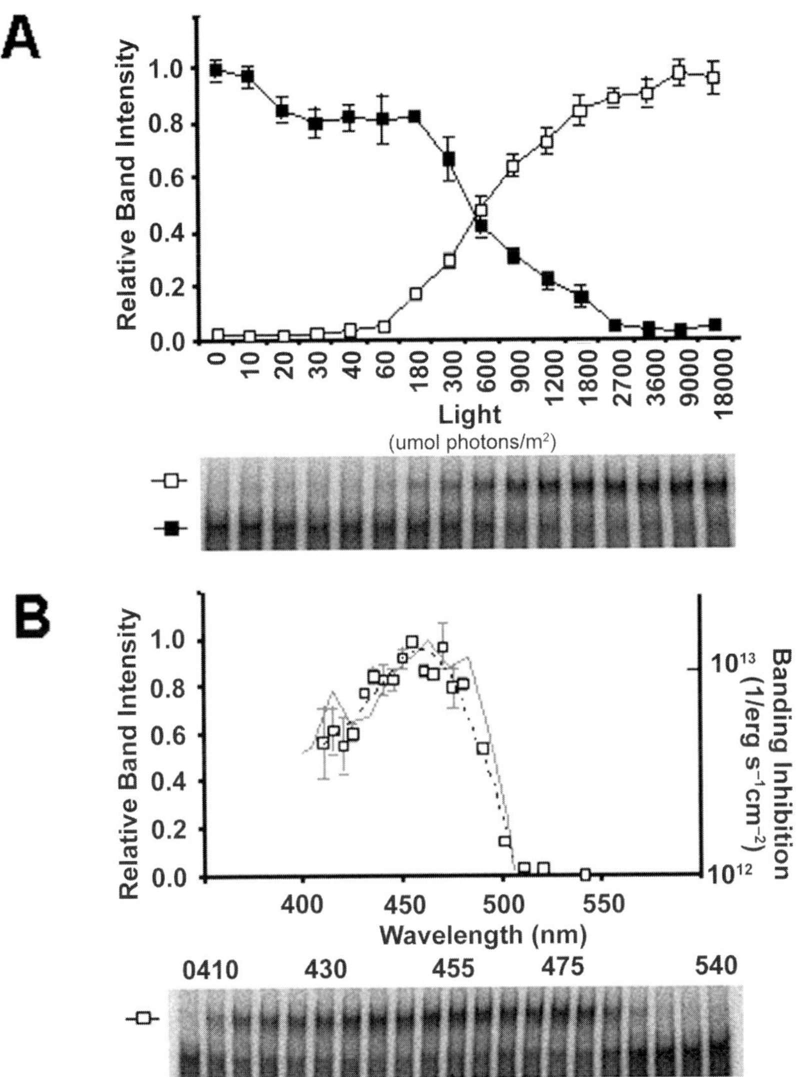
A
Relative Band Intensity
1.0
0.8
0.6
0.4
0.2
0.0
0
10
20
30
40
60
180
300
600
900
1200
1800
2700
3600
9000
18000
Light
(umol photons/m²)
B
Relative Band Intensity
1.0
0.8
0.6
0.4
0.2
0.0
Banding Inhibition
(1/erg s⁻¹cm⁻²)
10¹³
10¹²
400
450
500
550
Wavelength (nm)
0410
430
455
475
540

a faster-migrating complex seen using extracts from cultures grown in the dark and a slower-migrating complex seen using extracts from light-treated cultures (Froehlich et al 2002). WC-1 and WC-2 are found in the LRE-bound complexes, both the dark- and light-induced complexes (Froehlich et al 2002), which is not surprising since WC-1 and WC-2 are necessary for light-induction of *frq* transcript (Crosthwaite et al 1997, Collett et al 2002, Lee et al 2003). WC-1 and WC-2 are localized to the nucleus, and are capable of binding to DNA *in vitro* (Linden et al 1999).

Extracts from dark-grown cultures retain light sensitivity *in vitro*, forming the slower migrating 'light' complex when used in binding reactions even if exposed to white light hours after extraction in the dark (Froehlich et al 2002). The close agreement among the *in vitro* action spectrum (Fig. 2B), the *in vitro* dose–response curve (Fig. 2A) and previously published *in vivo* data for light effects on clock responses suggests that the *in vitro* light shift is a true reflection of the *in vivo* light responsiveness of *Neurospora*, that of a blue-light photoreceptor, potentially flavin-based, with a peak activity at ∼465 nm and no response above 520 nm (Froehlich et al 2002).

Using WC-1 and WC-2 proteins produced *in vitro* using a coupled transcription/translation reticulocyte system, we have shown that WC-1 and WC-2 together are able to bind to the LREs as not just one but two distinct complexes with mobilities similar to those seen using nuclear extracts, suggesting that the light and dark complexes consist exclusively of WC-1 and WC-2 (Fig. 3) (Froehlich et al 2002). WC-1's LOV domain suggests that WC-1 may itself be the photoreceptor. Usually bound to a flavin molecule, LOV domains (a subgroup of the PAS domain family) are associated with environmental sensing including light, oxygen and voltage (Briggs & Huala 1999). Addition of flavin adenine dinucleotide (FAD) to the WC-1 translation reaction, confers light sensitivity to the *in vitro* translated proteins. Additionally, WC-1 with FAD, exposed to light in the absence of WC-2, can initiate the mobility shift when subsequently combined in the dark with WC-2. These data suggest that a WC-1/WC-2 dimeric complex is located directly on the *frq* promoter LREs in the dark where WC-1 is poised to absorb blue light using its bound FAD chromophore, triggering the multimerization of the WCC and subsequently increasing *frq* transcription.

The functions of the LREs can be distinguished. The proximal LRE is necessary for maintaining elevated levels of *frq*/FRQ in prolonged light exposure as well as eliciting the initial rapid light-induced increase in *frq* transcript; its loss thus affects phase. Loss of the distal LRE eliminates rhythmicity (Froehlich et al 2003), and the distal LRE is sufficient to drive rhythmic transcription of a reporter. This *cis*-acting element is thus responsible for generating rhythmic *frq* message, earning the additional title of Clock-Box (C-Box) (Froehlich et al 2003). Recalling that the distal LRE (C-Box) is bound by WC-1/WC-2, it becomes clear that the WCs (which are known to play a dual role in the regulation of *frq*, one in light

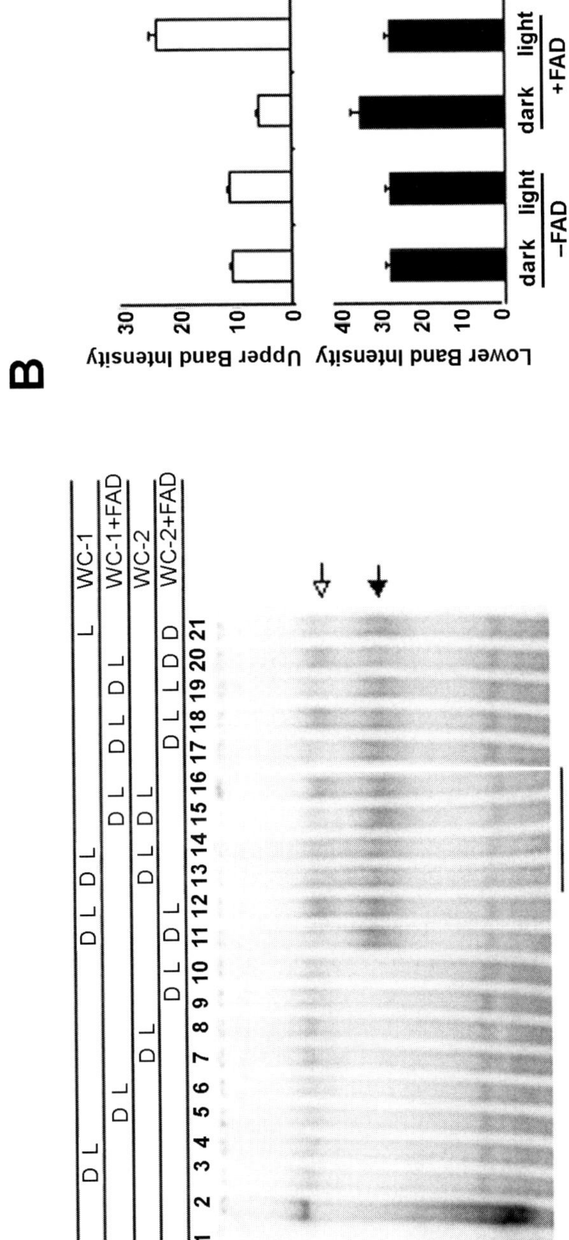

FIG. 3. *In vitro* expressed WC-1 in the presence of FAD is light sensitive. (A) WC-1 and WC-2 were separately produced using an *in vitro* transcription–translation reticulate lysate system with FAD added to some reactions, as indicated, and then used in a series of binding reactions with LRE probe. Addition of WC-1 or WC-2 to a reaction is indicated above each lane; 'D' indicates that protein added was not exposed to white light, and 'L' indicates that protein added was exposed to white light. Arrows highlight WCC bound to the LRE probe. Lane 1, no protein or lysate. Lane 2, unprogrammed reticulate lysate. Lanes 3 to 10, WC-1 or WC-2. Lanes 11 to 18, WC-1 and WC-2 incubated together before indicated light treatment and before addition to binding reactions. Lanes 19 to 21, WC-1 and WC-2 given light treatments individually, as indicated, and then incubated together in the dark before being added to binding reaction. (B) The reactions in the lanes indicated by the line at the bottom of the gel in (A) (lanes 13 to 16) were repeated in triplicate (data not shown) and the WCCs were densitometrically quantified. The white bars are the upper/slower migrating complex, and the black bars are lower/faster migrating complex. Bars are ±SEM. Reprinted from Froehlich et al 2002 with permission from AAAS.

induction and the other in rhythmic dark transcription) carry out both roles through a single element in the *frq* promoter.

The WC-1/WC-2/C-Box complex is present throughout the day (Fig. 4A,B) (Froehlich et al 2003) although the amount varies with peaks in binding occurring around subjective dawn ∼CT22-24 and a trough near dusk ∼CT12. This closely matches the changes in *frq* transcript levels (Aronson et al 1994a, Crosthwaite et al 1995) and is appropriate given the WC's positive role in *frq* transcription. Rhythmic *frq* transcript is therefore presumably the result of oscillations in WCC binding and subsequently activating transcription.

There is no correlation between the WC levels and the amount of WCC bound to the C-Box (Froehlich et al 2003), but FRQ's negative regulation of its own expression and direct physical interaction with WC-1 and WC-2 make it a strong candidate for regulation of WCC binding (Aronson et al 1994a, Cheng et al 2001b, Denault et al 2001). Additionally, oscillations in FRQ levels are phased appropriately and suggest an attractive means of generating oscillations in WCC binding. Increasing FRQ levels, using either a *Neurospora* strain containing an inducible copy of FRQ or *in vitro* generated FRQ, results in a strong dose-dependent reduction in WCC binding to the C-Box (Fig. 4C,D) (Froehlich et al 2003) demonstrating a direct role for FRQ in reducing the ability of the WCC to bind to the C-Box, thereby providing the molecular basis for the negative feedback of FRQ on its own expression.

The FRQ-less oscillator and cloning of *prd-4*

As a natural part of the maturation of molecular chronobiology, research is beginning to examine circadian systems of oscillators rather than only core oscillators. The first decade of work on 'clock molecules' (1984 through 1994) focused on the cloning of putative clock genes from flies and fungi, describing their regulation, and establishing (to an initially sceptical field) that transcription–translation feedback loops (TTFL) lay at the core of circadian clocks. However, even early on there was general appreciation that the circadian system would likely involve an interconnected set of feedback loops (see references in Dunlap 1998); in fact there exists a sound theoretical basis for such models (Pittendrigh & Bruce 1959, Pavlidis 1969, Winfree 1976). The initial report of the *frq⁹* allele (Loros et al 1986) revealed that such strains retained the ability to express a rhythm (albeit one lacking circadian characteristics), later shown to be the *frq*[null] phenotype (Aronson et al 1994b). The rhythm appears in only a fraction of cultures — about 20% — more or less randomly after a few days in constant conditions: it (1) displays a highly variable (SD of 4 h) period length ranging from 12–35 h (depending on temperature, carbon source and concentration, and other factors), (2) cannot be entrained by light cycles and thus

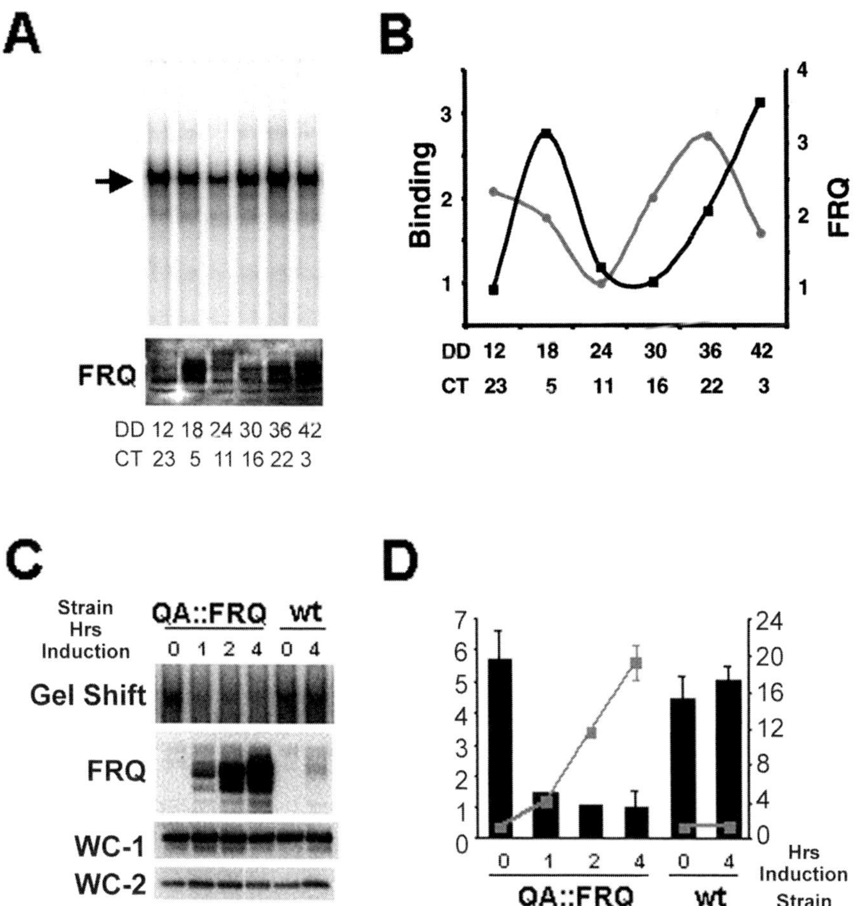

FIG. 4.   The WC-1/WC-2 complex rhythmically binds the Clock-Box and is negatively regulated by FRQ. (A, *top*) EMSA using the C-Box probe and nuclear protein extracts from cultures harvested across a circadian day reveals changes in the amount of WC-1/WC-2 complex bound (arrow) but not in its apparent size. (A, *bottom*) Western blot analysis of FRQ in the extracts used for EMSA confirms appropriate rhythmicity. (B) Densitometric analysis of the C-Box bound complex and FRQ shows that the amount of C-Box bound complex, peaks approximately 6 h before the peak in FRQ. (C) An inducible FRQ construct (QA::FRQ) was used to determine the effects of FRQ concentration on WCC binding to the C-box. Extracts from QA::FRQ and wt strains with and without QA induction were used with the C-Box probe in a series of binding reactions (*top panel*). FRQ, WC-1, and WC-2 levels in the extracts were analysed by Western blot (*bottom 3 panels*). (D) Densitometric analysis of C-Box bound complex and FRQ. Binding indicated by black bars, and FRQ levels indicated by grey line and boxes.

adopts random phases, and (3) lacks the temperature and nutritional compensation characterizing the circadian system. More recently temperature cycles were used to show that this occasional rhythmicity in frq^{null} strains was due to an oscillator that supposedly could be coupled to the FRQ/WCC feedback loop (Merrow et al 1999).

This oscillator(s)-known as the FRQ-less oscillator or FLO (Iwasaki & Dunlap 2000) may contribute to the operation of the *Neurospora* circadian system or to output. One of the earliest references to coupled oscillators in circadian systems (Pittendrigh & Bruce 1959) describes a model in which the driving 'A' oscillator (here FRQ/WCC) entrains a slave 'B' oscillator that regulates major aspects of output. Both original (Loros et al 1986) and recent (Aronson et al 1994b, Merrow et al 1999) studies place (by extrapolation) the intrinsic period length of the FLO at about 12–13 h, but otherwise without the FRQ/WCC loop the rhythm bears few circadian characteristics (lacking compensation, entrainability by light, sustainability, or consistency of period length). Thus few would assert that the FLO can produce circadian rhythms. We (A. Pregueiro, J. Loros & J. Dunlap, unpublished results) have not been able to reproduce an earlier report (Merrow et al 1999) claiming that FLO can be entrained (rather than simply driven) by temperature cycles, and frequency demultiplication to temperature cycles is lost without the FRQ/WCC loop. Without FRQ/WCC the rhythm is not circadian and with FRQ/WCC, FLO is either masked or non-existent.

There is no way to know, absolutely, whether the FLO is exclusively output or whether it can influence the FRQ/WCC oscillator: if the latter — that is, FLO is coupled to the FRQ/WCC oscillator — then some period length mutants might identify FLO components. However, since the FRQ/WCC is the driving oscillator, loss-of-function FLO mutations might be silent. Therefore we reasoned that the best bet for identifying a FLO component might be in dominant (that is, prospective gain-of-function mutations) clock mutants. Further, these should be examined in the temperature entrainment protocol in a frq^{null} background to fully expose the FLO. We constructed double mutants between frq^{10} and all of the existing uncloned *Neurospora* clock mutants, one of which, *prd-4*, displays a semidominant 3–4 h period shortening of the rhythm and a partial loss of temperature compensation (Gardner & Feldman 1981). These frq^{10}, *clock mutant X* strains were examined under temperature entrainment, and one, *prd-4*, typically revealed a loss of FLO over a week's growth even under temperature entrainment (see Fig. 5) although weak expression of FLO sometimes appeared in a time series spanning several weeks. The strain 'flo-1' contains the mutant *prd-4* gene. *prd-4* thus provides a molecular entrée, perhaps the first, into a FLO, a 'B' oscillator.

Detailed genetic mapping allowed identification of the *prd-4* defining mutation as a single base pair change in an ORF containing a forkhead associated (FHA) domain N-terminal to a kinase domain. FHA domains are specific for binding to phospho-Thr and find a role in the regulatable assembly of reversible protein

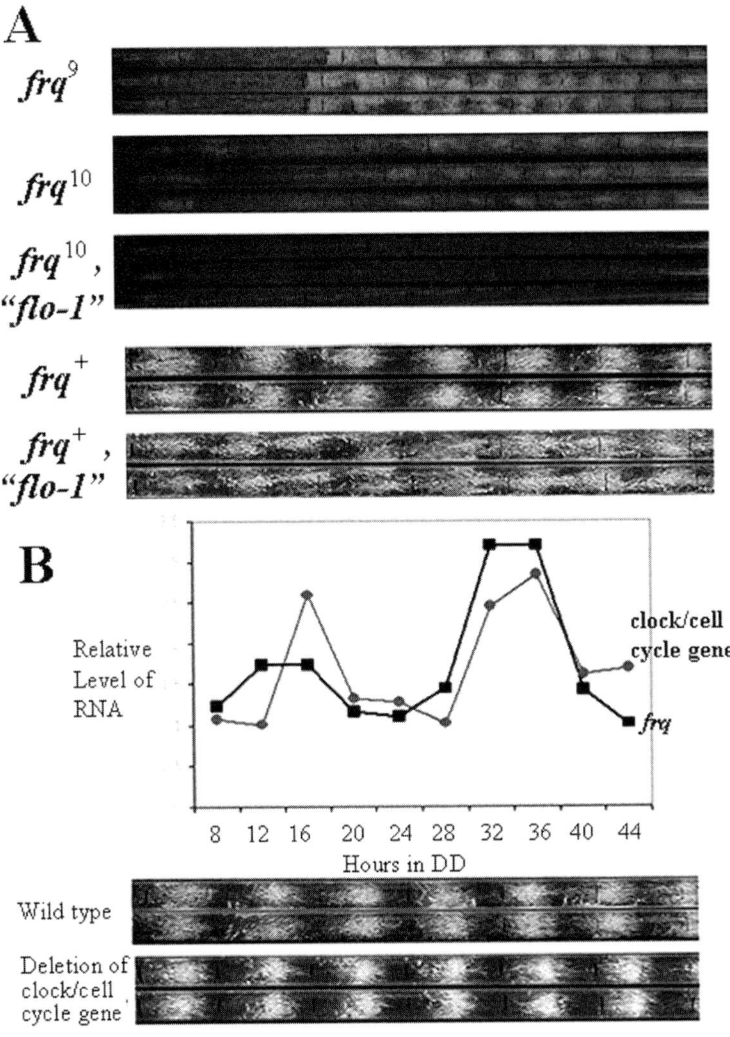

FIG. 5. Mutation of *prd-4* interferes with the FLO. Shown are race tubes (Loros & Dunlap 2001) of strains bearing mutations in *frq* and/or *prd-4*. Growth is from left to right. (*Top*) Tubes were entrained to a full 12 h 27 °C: 12 h 22 °C cycle, conditions shown to dependably visualize FLO (Merrow et al 1999). Expression of FLO is lost in strains bearing the canonical *prd-4* mutant allele. (*Bottom*) Strains of the genotype shown express a circadian rhythm under free-running conditions indicating that PRD-4 is not required for rhythmicity.

complexes associated with transcriptional regulation, mRNA splicing, DNA damage responses and cell cycle checkpoints. Founding members of this family are the yeast checkpoint kinase RAD53p and CHK orthologues in *Schizosaccharomyces pombe* (*Cds1*), *Drosophila melanogaster* (*Dmnk*), mice and humans (*hCds1, Chk2*) (reviewed in Yaffe & Elia 2001). BLAST searches with the PRD-4 protein sequence identify all of these RAD53/Chk2 homologues from different species as its closest homologue (with BLAST values from e^{-43} for Rad53 to e^{-67} for human Chk2). In DNA damage responses, Chk2 is phosphorylated, thereby enhancing its latent kinase activity $\sim$fivefold (e.g. Lee & Chung 2001). Chk2 autophosphorylates in *trans* and phosphorylates Cdc25A and C leading to cell cycle arrest.

prd-4[KO] strains grow normally and still have a clock. Importantly, the *prd-4* transcript is circadianly regulated in synchrony with *frq*. This and the fact that it is not essential for rhythmicity strongly suggests that *prd-4* operates as an output factor, and its latent and (auto-or-*trans*) activatable kinase function and FHA domain suggest myriad possibilities for regulatory loops. PRD-4 could thus be a link between the biological clock and other fundamental cell processes, specifically including the cell cycle. It is well known that in systems where the cell division cycle is longer than a day (that is, a doubling time of, say 36 h), the cell cycle is gated—the so-called GET effect (Ehret & Wille 1970), named for *Gonyaulax*, *Euglena* and *Tetrahymena*, three slow-growing circadianly regulated unicellular protists in which the phenomenon was first described. There can still be gating even under conditions where cells are growing more rapidly (Mori et al 1996) and the phenomenon is seen in humans (Moore-Ede et al 1982) where it may (along with circadian regulation of kidney filtration and liver detoxification) be the basis for circadian changes in the efficacy of chemotherapeutic agents. There has never been a plausible molecular connection between cell cycles and circadian rhythms; the intuitive connection we see with PRD-4, the clock and the cell cycle may provide a way to understand this regulation.

Acknowledgements

This work was supported by grants from the National Institutes of Health R37GM34985 to J.C.D. and MH44651 to J.C.D. and J.J.L., and the National Science Foundation MCB-0084509 to J.J.L., and the Norris Cotton Cancer Center core grant at Dartmouth Medical School.

References

Allada R, Emery P, Takahashi JS, Rosbash M 2001 Stopping time: the genetics of fly and mouse circadian clocks. Annu Rev Neurosci 24:1091–1119
Aronson BD, Johnson KA, Loros JJ, Dunlap JC 1994a Negative feedback defining a circadian clock: autoregulation in the clock gene *frequency*. Science 263:1578–1584

Aronson BD, Johnson KA, Dunlap JC 1994b The circadian clock locus *frequency*: protein encoded by a single open reading frame defines period length and temperature compensation. Proc Natl Acad Sci USA 91:7683–7687

Briggs WR, Huala E 1999 Blue-light photoreceptors in higher plants. Annu Rev Cell Dev Biol 15:33–62

Cermakian N, Sassone-Corsi P 2000 Multilevel regulation of the circadian clock. Nat Rev Mol Cell Biol 1:59–67

Cheng P, Yang Y, Heintzen C, Liu Y 2001a Coiled-coil domain-mediated FRQ-FRQ interaction is essential for circadian clock function in *Neurospora*. EMBO J 20:101–108

Cheng P, Yang Y, Liu Y 2001b Interlocked feedback loops contribute to the robustness of the *Neurospora* circadian clock. Proc Natl Acad Sci USA 98:7408–7413

Cheng P, Yang Y, Gardner KH, Liu Y 2002 PAS domain-mediated WC-1/WC-2 interaction is essential for maintaining the steady-state level of WC-1 and the function of both proteins in circadian clock and light responses of *Neurospora*. Mol Cell Biol 22:517–524

Collett MA, Garceau N, Dunlap JC, Loros JJ 2002 Light and clock expression of the *Neurospora* clock gene frequency is differentially driven by but dependent on WHITE COLLAR-2. Genetics 160:148–158

Crosthwaite SK, Loros JJ, Dunlap JC 1995 Light-induced resetting of a circadian clock is mediated by a rapid increase in *frequency* transcript. Cell 81:1003–1012

Crosthwaite SK, Dunlap JC, Loros JJ 1997 *Neurospora wc-1* and *wc-2*: transcription, photo-responses, and the origins of circadian rhythmicity. Science 276:763–769

Denault DL, Loros JJ, Dunlap JC 2001 WC-2 mediates WC-1-FRQ interaction within the PAS protein-linked circadian feedback loop of *Neurospora*. EMBO J 20:109–117

Dunlap JC 1999 Molecular bases for circadian clocks. Cell 96:271–290

Ehret CF, Wille JJ 1970 The photobiology of circadian rhythms in protozoa. In: Halldal P (ed) Photobiology of microorganisms. New York, Wiley, p 369–416

Froehlich AC, Liu Y, Loros JJ, Dunlap JC 2002 White Collar-1, a circadian blue light photoreceptor, binds to the *frequency* promoter. Science 297:815–819

Froehlich AC, Loros JJ, Dunlap JC 2003 Rhythmic binding of a WHITE COLLAR containing complex to the frequency promoter is inhibited by FREQUENCY. Proc Natl Acad Sci USA 100:5914–5919

Garceau NY, Liu Y, Loros JJ, Dunlap JC 1997 Alternative initiation of translation and time-specific phosphorylation yield multiple forms of the essential clock protein FREQUENCY. Cell 89:469–476

Gardner GF, Feldman JF 1981 Temperature compensation of circadian periodicity in clock mutants of *Neurospora crassa*. Plant Physiol 68:1244–1248

Gorl M, Merrow M, Huttner B, Johnson J, Roenneberg T, Brunner M 2001 A PEST-like element in FREQUENCY determines the length of the circadian period in *Neurospora crassa*. EMBO J 20:7074–7084

Heintzen C, Loros JJ, Dunlap JC 2001 The PAS protein VIVID defines a clock-associated feedback loop that represses light input, modulates gating, and regulates clock resetting. Cell 104:453–464

Iwasaki H, Dunlap JC 2000 Microbial circadian oscillatory systems in *Neurospora* and *Synechococcus*: models for cellular clocks. Curr Opin Microbiol 3:189–196

Lee K, Loros JJ, Dunlap JC 2000 Interconnected feedback loops in the *Neurospora* circadian system. Science 289:107–110

Lee CH, Chung JH 2001 The hCds1 (Chk2)-FHA domain is essential for a chain of phosphorylation events on hCds1 that is induced by ionizing radiation. J Biol Chem 276:30537–30541

Lee K, Dunlap JC, Loros JJ 2003 Roles for WHITE COLLAR-1 in circadian and general photoperception in *Neurospora crassa*. Genetics 163:103–114

Linden H, Ballario P, Arpaia G, Macino G 1999 Seeing the light: news in *Neurospora* blue light signal transduction. Adv Genet 41:35–54

Liu Y, Garceau NY, Loros JJ, Dunlap JC 1997 Thermally regulated translational control of FRQ mediates aspects of temperature responses in the *Neurospora* circadian clock. Cell 89:477–486

Liu Y, Merrow M, Loros JJ, Dunlap JC 1998 How temperature changes reset a circadian oscillator. Science 281:825–829

Liu Y, Loros J, Dunlap JC 2000 Phosphorylation of the *Neurospora* clock protein FREQUENCY determines its degradation rate and strongly influences the period length of the circadian clock. Proc Natl Acad Sci USA 97:234–239

Loros JJ, Dunlap JC 2001 Genetic and molecular analysis of circadian rhythms in *Neurospora*. Annu Rev Physiol 63:757–794

Loros JJ, Richman A, Feldman JF 1986 A recessive circadian clock mutant at the *frq* locus in *Neurospora crassa*. Genetics 114:1095–1110

Merrow M, Brunner M, Roenneberg T 1999 Assignment of circadian function for the *Neurospora* clock gene *frequency*. Nature 399:584–586

Moore-Ede MC, Sulzman FM, Fuller CA 1982 The clocks that time us. Harvard University Press, Cambridge, MA, p 219–233

Mori T, Binder B, Johnson CH 1996 Circadian gating of cell division in cyanobacteria growing with average doubling times of less than 24 hours. Proc Natl Acad Sci USA 93:10183–10188

Nowrousian M, Duffield G, Loros JJ, Dunlap JC 2003 The frequency gene is required for temperature-dependent regulation of many clock-controlled genes in *Neurospora crassa*. Genetics 164: 923–933

Pavlidis T 1969 Populations of interacting oscillators and circadian rhythms. J Theor Biol 22:418–436

Pittendrigh C, Bruce V 1959 Daily rhythms as coupled oscillator systems and their relation to thermoperiodism and photoperiodism. In: Withrow RB (ed) Photoperiodism and related phenomena in plants and animals. AAAS, Washington DC, p 475–505

Sargent ML, Briggs WR 1967 The effect of light on a circadian rhythm of conidiation in *Neurospora*. Plant Physiol 42:1504–1510

Winfree A 1976 On phase resetting in multicellular clockshops. In: Hastings JW, Schweiger H-G The molecular basis of circadian rhythms. Abakon Verlagsgesellschaft, Berlin, Germany, p109–129

Yaffe MB, Elia AE 2001 Phosphoserine/threonine-binding domains. Curr Opin Cell Biol 13:131–138

Yang Y, Cheng P, Zhi G, Liu Y 2001 Identification of a calcium/calmodulin-dependent protein kinase that phosphorylates the *Neurospora* clock protein FREQUENCY. J Biol Chem 276:41064–41072

Yang Y, Cheng P, Liu Y 2002 Regulation of *Neurospora* circadian clock by casein kinase II. Genes Dev 16:994–1006

DISCUSSION

Foster: The WC-1 photoreceptor analysis is lovely. In the binding assay you generated your action spectrum from, I was amazed to see that you got sufficient resolution in it to see the absolutely characteristic double peaks that you'd expect from a flavoprotein-like pigment. If you ever get a chance to replot those data, fit a flavoprotein nomogram to it. I'm sure that it will fit perfectly.

Menaker: Is this the only circadian photoreceptive molecule that is positively identified at the moment?

Dunlap: Although it's true that the circadian photoreceptor story is most completely developed for *Neurospora* and WC-1, I would certainly say in the case of flies that cryptochrome very likely is a photoreceptor.

Weitz: That has not been proved *in vitro*.

Dunlap: It is worth adding that since we published this (Froehlich et al 2002), which is not that long ago, VVD has been identified as another photoreceptor. VVD acts only through VVD/WC-1, so in the absence of WC-1 there is also no VVD response. VVD confers photoadaptation on the basic light response, but it also binds a flavin and senses light *in vitro* (Schwerdtfeger & Linden 2003).

Sehgal: What about the phytochromes and cryptochrome?

Dunlap: We have knocked them out singly and in pairs, and there are phenotypes.

Loros: We think there are some very subtle phenotypes in terms of light-regulated gene expression. We are hoping definitive phenotypes may be easier to find in the WC knockout backgrounds.

Stanewsky: Didn't Merrow and Roenneberg show recently that these double WC-1 deletions are rhythmic under LD conditions, so that they can sense light (Dragovic et al 2002). This would suggest that there is another photoreceptor in addition to WC-1: the story is more complicated.

Dunlap: We have been in touch with Merrow and the good news here is that we all now agree on the facts, although statements they made in their paper (Dragovic et al 2002) to the effect that *frq* can be light-induced in a *frq*[null] strain are clearly in error. After Merrow spoke at The Society for Research on Biological Rhythms Conference in May we sent her true null strains of *wc-1* to use along with the partial-function strains she and Roenneberg had mistakenly represented as nulls. She wrote to us, confirming our data, that there is no light-induction of *frq* in a *wc-1* null. This confirms the error in her talk in May, and confirms that part of what they published after her email was still wrong. In fact, in a strain truly lacking the *wc-1* gene there is no acute light induction of FRQ. So we now agree on this. In the paper they published (Dragovic et al 2002) there is confusion about the alleles: strains that were called '*wc-1* delta' in the paper were instead just partial loss-of-function alleles, not knockouts as their name implied since delta is shorthand for deletion. Thus, inappropriate genetic nomenclature led to misleading conclusions. So now all parties agree that there is no acute-light induction of *frq* in strains lacking WC-1.

Stanewsky: Didn't they just look at conidiation in LD, and this was still observable?

Dunlap: Merrow wrote to us before their paper was published and said that they could confirm that there wasn't any light induction of *frq* in the true knockout strain.

Rosbash: Is there a conidiation rhythm in the WC knockout under LD conditions?

Loros: We sent them our definitive knockout. They told us that they think they can reproduce their conidiation data. This may be a high-light fluence response. There may be specific light responses in *Neurospora* for development or other metabolic functions specifically in the absence of WC-1. This doesn't surprise any of us considering that there are these other putative photoreceptors.

Dunlap: We don't know that this has been looked at yet in the definitive knockout strain. We haven't seen the data, but we assume this is the case and that a full photoperiod LD cycle can still drive photoresponses, but not entrain them since there's no circadian clock. As Jennifer said, this is not a surprise since we have found other photoreceptors that don't require WC-1; they just don't appear to be the circadian photoreceptor.

Van Gelder: Is FLO light-entrainable?

Dunlap: No.

Kay: Have you knocked out the cysteine that binds the flavin? Is this cysteine conserved?

Loros: Yes, it is conserved.

Kay: It might be interesting to see what functions remain if you prevent the binding of flavin in terms of what parts of the biochemistry are really light dependent.

Dunlap: That's a good experiment, and the prediction would be that we would still get circadian regulation, but no light regulation. We haven't done this yet.

Loros: These experiments are in progress.

Sassone-Corsi: At a more general level, how many light-regulatory elements are there in the *Neurospora* genome?

Dunlap: That's a good question. Are you asking what percentage of genes would be light induced? There are about 10 000 genes, and I guess about 2% would be light-regulated, which makes 200.

Sassone-Corsi: Do you think they would all contain the same kind of LREs?

Dunlap: To the extent that a large number of light-induced genes are regulated directly or indirectly through WC-1/WC-2, we'd expect many of them to have the LRE sequence that Allan found in front of *frq* (Froehlich et al 2002). But there are those other genes still light induced without WC-1.

Loros: WC-1 has zinc fingers and is a GATA-type transcription factor. When Allan Froehlich mapped the DNA binding sites we found they were not actually GATA sites: they are degenerate GATN repeat sites. They do line up with other known LRE regions from light-inducible promoters in *Neurospora*.

Schibler: I have a question concerning the mechanism of repression: interfering with the DNA binding activity. Did you do that with FRQ-overexpressing cells?

Dunlap: Yes.

Schibler: How about if you take purified protein?

Dunlap: We haven't done that, but I expect it would work the same way.

Schibler: If you look at the FRQ protein in *Neurospora*, is it a large complex?

Dunlap: It is at least a dimer. The coiled coil domain that is responsible for dimerization is essential for carrying out its function.

Young: What is the status of transformation assays to test for function by complementation?

Dunlap: There is no selection for clock-affecting mutations; it is just a screen that is available to identify variants.

Young: Is part of the *Neurospora* genome project's goal an attempt to make mutants all along each of the chromosomes?

Dunlap: It would be an enormous amount of work to execute a screen for mutants in the FLO. When you think about it, the situation now with respect to the non-circadian rhythms represented by FLO and similar oscillators is very similar to where we all were with circadian clocks in the early 1980s: we had some putative mutants that affected clock expression/function but we had no idea whether they were core circadian clock molecules or simply had pleiotropic effects on the expression of the circadian rhythm. We had no selection, only screens. As we all know it was a lot of work to convincingly show that molecules like FRQ and PER were essential for the circadian clock. I'm not sure it's worth making the investment in positional cloning of an allele that might not inform you at all of a mechanism for FLO. And even if you did, since the FLO rhythm is not a circadian rhythm, where is this going to be published? In our experience, a gene that affects an oscillator like the FLO that is really not circadian and that no one thinks is involved with the circadian mechanism won't interest a student.

Van Gelder: Do you have any idea about the kinetics of the gel shift? How long do you have to keep that extract in the light before you see the shift? What is the relationship of this to the *in vivo* kinetics of phase shifting?

Dunlap: That is a good question. It takes very little light and it happens fast. Once it happens it is very stable. Sue Crosthwaite showed that 2 s of 25 lux, corresponding to around 24 μmoles of photons/m^2, is sufficient *in vivo* to see a photoresponse of the clock and to see light induction of *frq* (Crosthwaite et al 1995). It is very sensitive. Allan confirmed this *in vitro* in the detailed fluence response curves which, using just one of the two LREs, showed a response threshold that agreed with the *in vivo* data within a factor of two to three (Froehlich et al 2002).

Van Gelder: You have shown sufficiency, which is fantastic, but to fully demonstrate equivalence of *in vitro* and *in vivo* responses, you need to show that the kinetics *in vitro* match the observed kinetics for photic responses *in vivo*.

Dunlap: Reflecting the stability of the light-induced change, the photoreceptor function in WC-1 seems to act like an integrator *in vitro* that senses or results in reciprocity between duration and intensity of the signal, at least within limits. This is what the fluence response does. The threshold of about 30 μM of photons per square metre, delivered in a short pulse or in a longer dimmer pulse, can trigger the gel shift response *in vitro* and then it's stable, as Allan showed. It can also trigger the phase shift response or *frq* induction response *in vivo* within a few minutes as Sue Crosthwaite showed in 1995 (Crosthwaite et al 1995). It seems likely though that additional factors present *in vivo* but not *in vitro*, like VVD, further modulate the response. But Sue used light pulses as short as 10 seconds.

Van Gelder: Was that the intensity of light used for the *in vitro* translated reconstitution experiment?

Dunlap: No, that was done with longer pulses to deliver known amounts of total energy.

References

Crosthwaite SC, Loros JJ, Dunlap JC 1995 Light-induced resetting of a circadian clock is mediated by a rapid increase in *frequency* transcript. Cell 81:1003–1012

Dragovic Z, Tan Y, Görl M, Roenneberg T, Merrow M 2002 Light reception and circadian behavior in 'blind' and 'clock-less' mutants of *Neurospora crassa*. EMBO J 21:3643–3651

Froehlich AF, Loros JJ, Dunlap JC 2002 WHITE COLLAR-1, a circadian blue light photoreceptor, binding to the *frequency* promoter. Science 297:815–819

Schwerdtfeger C, Linden H 2003 VIVID is a flavoprotein and serves as a fungal blue light photoreceptor for photoadaptation. EMBO J, in press

Expression of clock gene products in the suprachiasmatic nucleus in relation to circadian behaviour

M. H. Hastings, A. B. Reddy, M. Garabette*, V. M. King*, S. Chahad-Ehlers*, J. O'Brien and E. S. Maywood

*Neurobiology Division, Laboratory of Molecular Biology, MRC Centre, Hills Road, Cambridge CB2 2QH and *Department of Anatomy, University of Cambridge, Downing Street, Cambridge CB2 3DY, UK*

Abstract. Circadian timing within the suprachiasmatic nucleus (SCN) is modelled around cell-autonomous, autoregulatory transcriptional/post-translational feedback loops, in which protein products of canonical clock genes *Period* and *Cryptochrome* periodically oppose transcription driven by CLOCK:BMAL complexes. Consistent with this model, mCLOCK is a nuclear antigen constitutively expressed in mouse SCN, whereas nuclear mPER and mCRY are expressed rhythmically. Peaking in late subjective day, mPER and mCRY form heteromeric complexes with mCLOCK, completing the negative feedback loop as levels of *mPer* and *mCry* mRNA decline. Circadian resetting by light or non-photic resetting (mediated by neuropeptide Y) involves acute up- and down-regulation of *mPer* mRNA, respectively. Expression of *Per* mRNA also peaks in subjective day in the SCN of the ground squirrel, indicating common clock and entrainment mechanisms for nocturnal and diurnal species. Oscillation within the SCN is dependent on intercellular signals, in so far as genetic ablation of the VPAC2 receptor for vasoactive intestinal polypeptide (VIP) suspends SCN circadian gene expression. The pervasive effect of the SCN on peripheral physiology is underscored by cDNA microarray analysis of the circadian gene expression in liver, which involves ca. 10% of the genome and almost all aspects of cell function. Moreover, the same molecular regulatory mechanisms driving the SCN appear also to underpin peripheral cycles.

2003 Molecular clocks and light signalling. Wiley, Chichester (Novartis Foundation Symposium 253) p 203–222

The identification of putative mammalian clock genes by both homology and mutagenesis screening is described elsewhere in this volume. It represents an enormous achievement, and already it is possible to speak of the 'new' circadian biology in which molecular genetics can be used to understand and re-interpret more classical, black-box descriptions of mammalian circadian behaviour. A key component of this analysis is to understand the behaviour of the clock genes, and especially their protein products, in their 'native' environment, the suprachiasmatic nuclei (SCN). Specifically, this chapter will address questions regarding how well the behaviour of clock gene products in the SCN supports the autoregulatory feedback model of the clockwork, and how that behaviour maps to circadian behaviour of the whole organism. It will then consider how the SCN molecular cycle is entrained to solar time, without which the mechanism would have no adaptive relevance, and how it contributes to circadian physiology more widely in the periphery.

Clock proteins in the SCN

Elsewhere in this volume are descriptions of the 'interlocked molecular loops' model of the SCN clockwork, in which the protein products of the canonical clock genes *Period* and *Cryptochrome* feedback with a delay to suppress activation of their cognate (and other clock-controlled) genes by interfering with transcriptional drive mediated by CLOCK:BMAL complexes acting via E-box *cis*-regulatory sequences. Stability, high amplitude and precision are conferred on the system by a feed-forward loop, in which E-box dependent circadian expression of REV-ERBα drives, by a disinhibitory mechanism, an antiphasic cycle of *Bmal* expression (Preitner et al 2002). This ensures that even as the current cycle of gene expression is being terminated, events leading to the subsequent round of circadian gene expression are set in motion. This model makes strong predictions about the behaviour of CLOCK, PER and CRY proteins that by and large have been confirmed by experimental analyses. If transcriptional feedback is to be important in the SCN, the foremost observation is that mCLOCK, mPER1, mPER2 and mCRY1 and mCRY2 are nuclear antigens in the SCN neurons (Fig. 1) (Hastings et al 1999, Kume et al 1999, Field et al 2000, Maywood et al 2003). Expression of these proteins within the cytoplasm is below the level of detectability, indicating that newly synthesized protein is either rapidly transferred to the nucleus, and/or unstable in the cytoplasm. Consistent with constitutive expression at the mRNA level, the expression of mCLOCK in the mouse SCN, assessed by immunostaining or immunoblots, is constant in circadian time. In contrast, nuclear mPER- and mCRY-immunoreactivity (-ir) varies dramatically across the circadian cycle. For most of the SCN, it is low at the beginning of subjective day and peaks at the end of subjective day and into the first few hours of subjective night. This circadian change occurs synchronously across the two major SCN sub-divisions, distinguished by AVP-ergic and VIP-ergic neurons. There remains, however, a small sub-division of the central SCN in which protein (and mRNA) expression is antiphasic to the main body of the nucleus (Fig. 1). These cells are neither AVP- nor VIP-ir and their functional role is unclear (King et al 2003), although it should be noted that

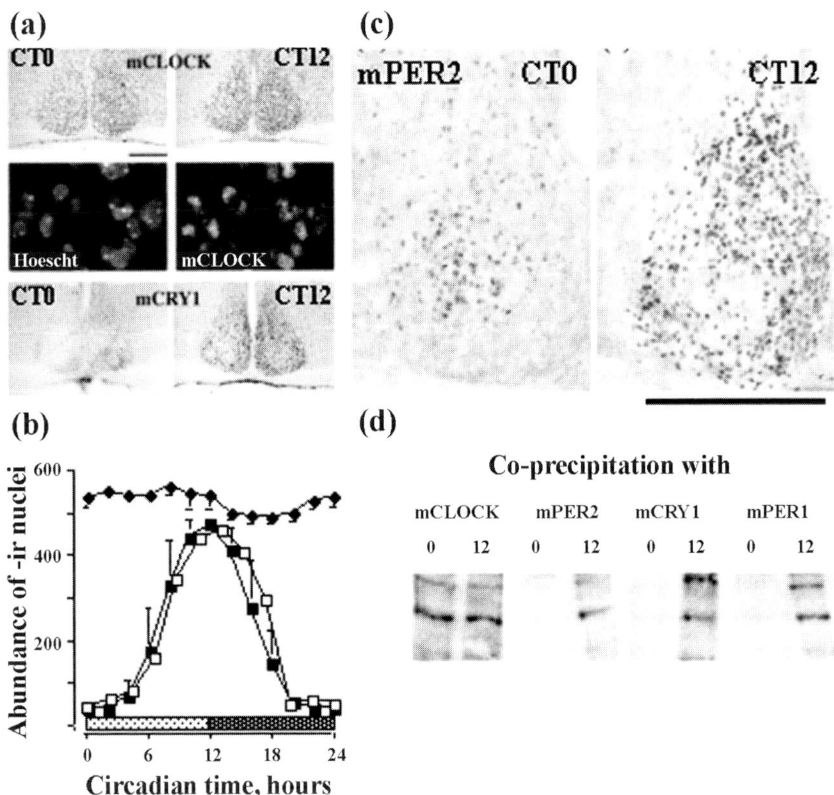

FIG. 1. Expression of clock proteins in mouse SCN. (a) Immunostaining of mouse SCN sampled at beginning (CT0) and end (CT12) of circadian day reveals constitutive expression of mCLOCK, and rhythmic expression of mCRY (scale bar 500 μm). Inset high power confocal views of mCLOCK-ir and Hoescht DNA stain confirm nuclear localization of mCLOCK-ir. (b) Quantitative analysis confirms constitutive expression of mCLOCK-ir (*closed circles*) and circadian expression of mPER1-ir (*closed squares*) and mPER2-ir (*open squares*) in adjacent sections of mouse SCN sampled on second cycle in continuous darkness (mean $\pm$ SEM, $n = 3$). (c) Regionally specific, antiphasic expression of nuclear mPER2-ir within SCN. At CT0 expression is confined to a central zone of SCN, whilst at CT12 expression is far more abundant across the SCN, but absent from the central region. (d) Immunoblots of mCLOCK-ir in SCN tissue punches collected at CT0 or CT12 following immunoprecipitation with anti-mCLOCK, anti-mPERs or anti-mCRY. Note constitutive expression of mCLOCK (*arrowed bands*), immunoprecipitated at both times and rhythmic occurrence of mCLOCK:mPER, and mCLOCK:mCRY complexes, precipitated only at CT12. (Data redrawn from Maywood et al 2003.)

in vitro electrophysiological recordings of hamster SCN have also revealed separable antiphasic components of the clock (Jagota et al 2000), and differentially phased expression of phosphorylated MAP kinase in local sub-regions also indicates the existence of separable oscillators in the SCN (Lee et al 2003).

The rise and fall in nuclear accumulation of mPER and mCRY across the bulk of the SCN follow the cycle of mRNA expression but with a lag of several hours, as predicted by the delayed feedback loop model of the oscillator. It also suggests that the factors are more stable in the nucleus than in the cytoplasm, and that nuclear entry can only occur once protein synthesis rates are sufficiently high to exceed cytoplasmic degradation. As a consequence of this periodic accumulation and disappearance of mPER and mCRY, the presence of mPER:mCLOCK and mCRY:mCLOCK complexes in the SCN is rhythmic, being undetectable by co-immunoprecipitation at the start of circadian day but present at its end (Fig. 1d) (Maywood et al 2003). Moreover, these associations are multimeric, probably involving a cocktail of mPERs, mCRYs, mBMAL1 and mCLOCK because at late circadian day, mPER can be immunoprecipitated with mCRY from SCN punches (Field et al 2000, Lee et al 2001). The formation of complexes between mCLOCK and the negative factors mCRY and mPER at this time coincides with the onset of negative transcriptional feedback. This supports the idea that the transcriptional actions of mCLOCK are interrupted once it comes into association with mPER and/or mCRY. A further function of heteromeric associations also appears to be stabilization of mPER and mCRY. In mice lacking mCRY proteins, mPER1 levels are markedly reduced, and mPER2 disappears from the SCN (Shearman et al 2000). Reciprocally, in mice lacking mPER1 or mPER2, mCRY levels in the SCN are low (Bae et al 2001). The delay in the cycles of protein abundance, relative to the SCN mRNA rhythm in wild-type animals may also be a consequence of this need for mutual stabilization. For example, even though *mPer1* mRNA expression in the SCN starts some hours before *mPer2* and *mCry1*, the three proteins rise in level simultaneously.

Circadian entrainment by light

To be of value to the organism, and to predict solar time, the biological clock of the SCN has to be entrained. In mammals, entrainment by light is mediated by direct retinal innervation to the SCN, with glutamate as the principal neurotransmitter (see other chapters in this volume for consideration of phototransduction mechanisms). Earlier studies had demonstrated a role for glutamatergic gene induction in photic resetting (Ginty et al 1993, Ebling 1996), and it has recently become clear that photic induction of *mPer1* and *mPer2* mRNA is a central event in entrainment (Shigeyoshi et al 1997, Albrecht et al 1997). The molecular basis to the photic response of *mPer1* and *mPer2* genes lies in their CRE regulatory sequences (Travnickova-Bendova et al 2002). Through these, the transcription factor pCREB, activated by glutamatergic retinal signals to the SCN, is able to drive *mPer* expression (Ginty et al 1993, Schurov et al 1999).

Brief light pulses delivered during early or late subjective night can, respectively, delay or advance the circadian clock. This resetting occurs when *Per* mRNA levels are approaching or are at their nadir. Induction on the declining phase will acutely reverse the fall in mRNA and delay the cycle, whereas induction at the nadir or just after will prematurely accelerate the rise and thereby advance the clockwork. Light pulses delivered when spontaneous *Per* expression is maximal during circadian day have little additional effect on mRNA levels and consequently map to the dead zone of the behavioural phase response curve. As noted by Shigeyoshi et al (1997), however, this qualitative model for resetting explains neither the magnitude nor the time course of behavioural resetting. Nocturnal light acutely induces *mPer* mRNA levels equivalent to the circadian peak, and so should sustain immediate phase shifts of up to 12 hours. Behavioural rhythms shift by a maximum of 3 hours, and take several cycles to be expressed in full, especially phase advances. This attenuation of the expected shift reflects the fact that peak levels of *mPer* mRNA are not translated into a peak abundance of mPER protein. Only with sustained illumination can mPER levels be held at the circadian maximum (Field et al 2000). This attenuation of resetting may occur because acute light pulses do not induce expression of *mCry* genes, which do not contain CREs and so must respond to photic cues indirectly. As a result *de novo* mPER lacks its stabilizing partner and will be rapidly degraded, limiting the magnitude of any consequent phase shift.

The resistance of *mCry* to photic cues may also contribute to the inertia in circadian resetting. When subject to a 6 h advance of the light:dark cycle, the cycle of expression of *mPer1* and *mPer2* in mouse SCN reacts rapidly and the advance is completed in 3 days (Reddy et al, 2002). In contrast, the cycle of *mCry1* expression has only achieved about 60% of the ultimate shift, the same degree to which the overt rhythm of rest/activity has also shifted by that point. When the behavioural shift is completed after 8 days, then so is the *mCry1* cycle fully advanced, and back in synchrony with *mPer*. In the opposite condition when the light:dark cycle is acutely delayed by 6 hours, the activity/rest cycle adjusts rapidly and is accompanied by immediate resetting of both the *mPer* and the *mCry* elements of the clockwork. These differential responses to time-zone transitions support the interpretation that *mCry1* expression defines the phase of rhythmic outputs dependent on the SCN, whilst *mPer* expression is more important as the entry point into the loop for resetting stimuli.

The biological relevance of this major 'fault-line' in the circadian system, the contrast between acutely responsive *Per* genes and non-responsive *Cry*, has been discussed extensively elsewhere (Hastings 2001). In particular, the expression of *mPer*/mPER in the SCN will be extended in the longer photoperiods of summer and the molecular clockwork will therefore have additional calendrical properties (Fig. 2) (Nuesslein-Hildesheim et al 2000). Consequently, PER-dependent outputs will be able to sculpt the profile of systems such as the secretion of melatonin by the pineal, matching them to daylength and thereby facilitating seasonal adaptation to natural habitats.

In addition to light, the SCN clockwork can also be reset by non-photic stimuli, mediated by a combination of serotonergic and neuropeptide Y (NPY)-positive afferents to the SCN from brain stem and thalamus, respectively, that reflect the animal's state of arousal. Whereas light shifts the clock by *up-regulating Per* expression when it is spontaneously low at night, non-photic cues reset the molecular cycle by *suppressing Per* levels when they are maximal in the middle of circadian day (Maywood et al 1999, 2002; Fig. 3a,b). This rapid suppression will lead to an acute advance of the SCN molecular loop, but what about expression of *Cry*/CRY? Although not tested directly, the strong prediction is that a premature decline in PER expression will destabilise CRY, accelerating the clock to its new phase. This model also explains why non-photic cues are without effect during subjective night. At this time the spontaneous expression of *Per* is basal and cannot be suppressed further.

The opposite molecular effects of light and non-photic cues explain, therefore, their contrasting phase response curves, with active zones in the subjective night (*Per* low) and subjective day (*Per* high), respectively. This highlights the role of *Per* as a point of convergence for multiple resetting cues, and indeed different cues may interact in their regulation of *Per*. For example, light delivered during subjective day reverses the acute suppression of *Per* levels in the SCN and blocks the resetting effect of non-photic stimuli (Maywood & Mrosovsky 2001, Maywood et al 2002). Equally, resetting by nocturnal light pulses can be blocked by NPY (Biello & Mrosovsky 1995), presumably because NPY blocks acute *Per* induction. Consideration of the particular phase-dependent consequences for *Per* expression of light and non-photic cues provides, therefore, a synthetic and mechanistic *explanation* for circadian resetting that replaces the earlier *descriptive* formalisms.

The question of diurnal versus nocturnal species

If the model for the oscillator is correct, *Per* and *Cry* expression will define solar/circadian time, driving the activity–rest cycle rather than just being a passive reflection of the activity cycle. Therefore, their expression patterns should exhibit the same phase in the SCN of nocturnal and diurnal species. This is confirmed by examination of *Per* expression in the SCN of the diurnally active ground squirrel, *Spermophilus* (Mrosovsky et al 2001). The rhythm of *Per1* and *Per2* expression in

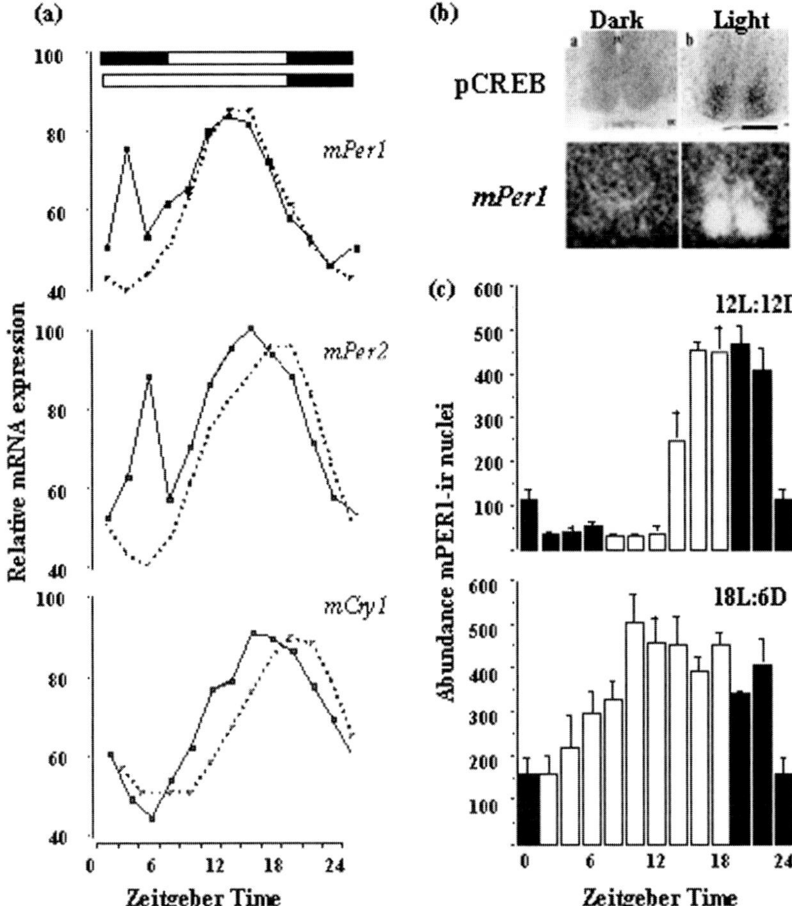

FIG. 2. Photic regulation of circadian gene products in the mouse SCN. (a) Comparison of mRNA expression levels for *mPer1*, *mPer2* and *mCry1* in mice held on 12L:12D (*dotted line*) or a 'super-long' day of 18L:6D (*solid line*). Note acute induction of both *mPer* genes at lights-on under 18L:6D, but no effect on *mCry* levels. All three genes peak later in the light phase, under the control of the circadian clock entrained to the light–dark schedule. Despite the absence of an acute response to light, the later peak of *mCry1* mRNA is phase advanced on 18L:6D compared to 12L:12D, indicating indirect photoperiodic regulation. ($n = 3$ mice per time point, error bars omitted for clarity, 12L:12D data redrawn from Field et al 2000). (b) Nocturnal expression of pCREB-ir and *mPer1* mRNA is very low in control mice in darkness. A brief light pulse rapidly induces *mPer1* mRNA directly in the retinorecipient region of the SCN, mapping directly to the region where light induces pCREB-ir (bar = 500 μm). (c) Photoperiod is reflected in the duration of elevated mPER1 expression in the mouse SCN (mean ± SEM, $n = 3$–6 per time point, 12L:12D data redrawn from Field et al 2000).

the SCN exhibits very high amplitude and peaks in subjective day, directly in register with the cycle of expression reported in nocturnal rodents such as the Syrian hamster, rat and mouse (Fig. 3c). The nature of diurnal and nocturnal species is determined therefore, by factors distal to the molecular loop of the SCN, and in the case of the squirrel and hamster leads to differences in the phase of *Per* expression in motor pathways, including the corpus striatum and motor cortex. In such sites, *Per* appears to behave more like an immediate early gene, reporting overall neuronal activity, and so expression is high at activity onset at dawn in the diurnal species and high at activity onset at dusk in nocturnal species. The common phase of the rhythm of SCN *Per* expression is consistent with the overlap of the nocturnal active zones of the photic phase response curves of diurnal and nocturnal species — in both cases resetting by light involves up-regulation of *Per* expression about its nocturnal nadir. This also suggests that diurnal species such as humans will be most responsive to non-photic cues during subjective day, even though this is their active phase, whereas in nocturnal species non-photic cues are effective during the inactive phase.

Peptidergic signalling and the molecular clockwork

As discussed above, the model of cell autonomous, interlocked feedback loops, implicates inter-neuronal signalling in entrainment via retinal and brain stem afferents, and circadian output to target structures. At some level there must also be signalling between SCN neurons to ensure their synchrony. One important synchroniser is GABA (Liu & Reppert 2000), a neurotransmitter common to all SCN neurons. A second is the peptide VIP, co-released with GABA by retinorecipient SCN neurons that project both within the SCN and to its targets. VIP acts through the VPAC2 receptor, which is highly expressed across the SCN and in its targets innervated by VIP-ergic efferents. Targeted genetic ablation of the receptor produces mice with a pronounced circadian disturbance (Harmar et al 2002). Whilst able to exhibit a coherent rest–activity cycle when on a light–dark cycle, this is primarily a masking response. The activity of the animals is suppressed by illumination, whereas exposure to darkness during the day immediately releases this suppression and the mice become active. In contrast, the circadian regulation to behaviour of wild-type animals prevents them becoming active during daytime dark pulses. When transferred to continuous darkness, the phenotype of the VPAC2 knockout mice is seen to be even more severe and their locomotor activity rhythm either breaks down completely, or adopts an antiphase pattern with poorly defined bouts of activity observed in subjective day (Fig. 4a). One interpretation of this phenotype is that in the absence of the receptor, the circadian signal of the SCN is not effectively recognised by target structures such as striatum and motor cortex that express the activity cycle. The phenotype is

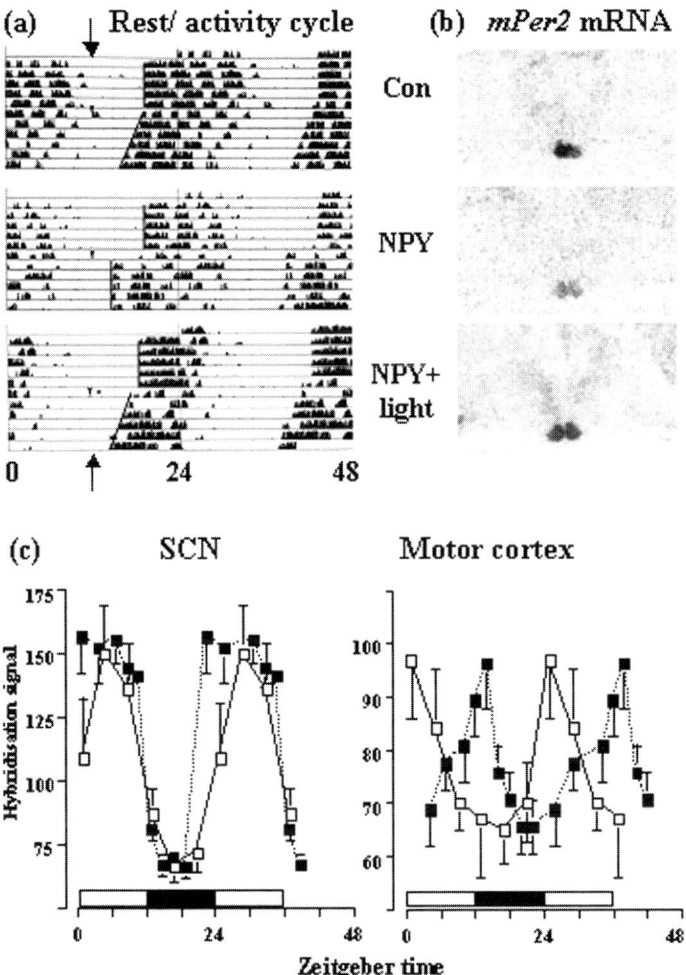

FIG. 3.   Regulation of *mPer* expression in SCN as a function of light, non-photic cues and behavioural profile. (a) Representative actograms of mice transferred to continuous darkness on day 6, and infused centrally with vehicle or NPY (*arrows*), or given NPY infusion plus a light pulse. Note non-photic, NPY-induced shift (*central panel*) is reversed by light presented immediately afterwards during circadian day. (b) *mPer2* mRNA expression in SCN of mice sampled immediately after treatments depicted in (a). Note suppression of *mPer2* mRNA by NPY, and reversal of this effect by light. (Data redrawn from Maywood et al 2002). (c) Expression profiles of *Per2* mRNA in nocturnal Syrian hamster (*dotted line*) and diurnal ground squirrel (*solid line*). Note synchronous cycles in SCN, with peaks in circadian day, but antiphasic cycles in motor cortex, reflecting oppositely phased behavioural profiles. (Data represent mean ± SEM, *n* = 3–6, redrawn from Mrosovsky et al 2001 and Maywood et al 1999.)

considerably more complex than this, however. The cycles of clock gene expression in the SCN of homozygous mutant mice held on a light–dark cycle are severely dampened, although *mPer* expression in the motor cortex and striatum shows a high amplitude cycle equivalent to that of wild-type controls and in register with the locomotor rhythm. In continuous darkness, the circadian rhythm of *Per* expression in the motor cortex and striatum is lost. Rhythmic gene expression is also completely lost in the SCN, not just for the *mPer* genes but also for *mCry* and the clock controlled gene encoding AVP (Fig. 4b–d). One possibility is that in the absence of the receptor, the SCN neurons remain rhythmic but immediately lose synchrony on the second cycle in continuous darkness. This is, however, unlikely because it would lead to an expression level across the SCN equivalent to the 24 h mean. The observed expression levels are far below this average, and emulsion autoradiographic analysis failed to identify hot spots indicating high levels of gene expression in individual (rhythmic) cells. The results indicate that in the absence of the VPAC2 receptor, the molecular cycles within the SCN neurons are either suspended, or at the least sustained with extremely low amplitude.

The cellular basis of this effect awaits clarification, but it is accompanied by both a loss of the neuronal firing rate rhythm in the mutant SCN (Cutler et al 2003) and suppression of the overall firing rate below the normal circadian nadir. One possibility is that the loss of VIP-ergic signalling in the SCN leaves unopposed the inhibitory action of GABA, the co-transmitter to VIP, and this in turn causes a suppression of electrical activity across the nucleus. A consequence of this is that the molecular loop is also suspended. This is not unprecedented because electrical silencing of clock neurons in *Drosophila* using a transgene encoding an ectopic channel protein not only renders the flies behaviourally arrhythmic, but also suspends molecular cycles of *dPer* expression in the clock neurons (Nitabach et al 2002).

FIG. 4.   Targeted deletion of VPAC2 receptor disrupts circadian timing in mouse SCN. (a) Representative actograms of wild-type and knockout mice, initially held on 12L:12D (darkness shaded) then subjected to forward and reverse phase shifts, and finally transferred to continuous dim light. (b) Representative images of *mPer2* mRNA, assessed by emulsion autoradiography, in SCN of wild-type and knockout mice sampled on second cycle of release into continuous darkness. The pronounced rhythm observed in wild-type mice is lost in the mutants with very low expression across the SCN. (c) Representative images of mPER2-ir on second cycle of release into continuous darkness. As for the mRNA, the pronounced rhythm observed in wild-type mice is lost in the mutants with very low expression across the SCN. (d) Circadian expression profiles of *mPer2* and *pre-pro A VP* in SCN of wild-type and mutant mice reveal loss of circadian modulation, and constitutive expression at basal levels for both the canonical clock gene and a clock controlled gene. (Data plotted as mean $\pm$ SEM, $n=3$.) (All data redrawn from Harmar et al 2002).

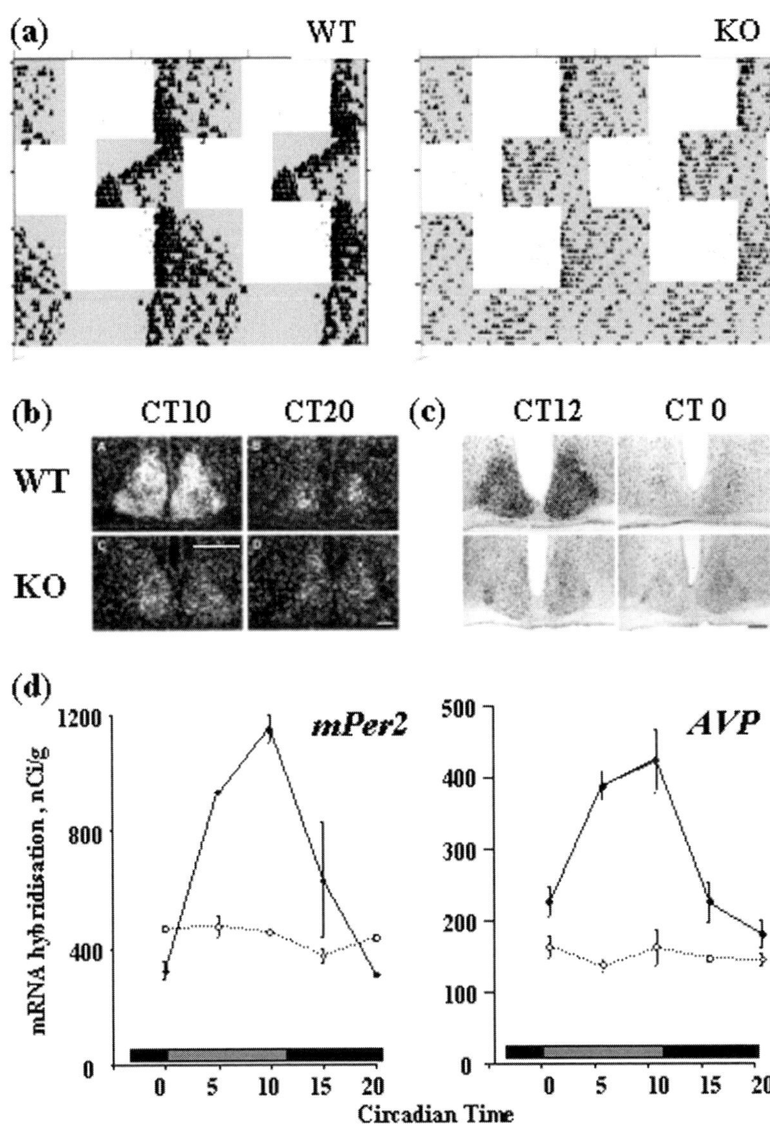
(a)
WT
KO
(b)
CT10
CT20
WT
KO
(c)
CT12
CT 0
(d)
mPer2
1200
800
400
0
mRNA hybridisation , nCi/g
AVP
500
400
300
200
100
0
0   5   10   15   20
Circadian Time

The SCN and control of the 'circadian transcriptome' in peripheral tissues

An important and unanticipated discovery arising from identification of the molecular clock of the SCN is that peripheral tissues not only express canonical clock genes, but they do so with a circadian pattern. Visceral organs and skeletal muscle, as well as extra-hypothalamic brain regions not only express the full complement of clock genes, they also have a limited ability to sustain circadian cycles of *Per* gene expression when isolated *in vitro* (see Menaker, this volume). This observation raises two important questions: how extensive is circadian gene expression, both across the genome and across peripheral tissues, and what is the relationship of such peripheral rhythms to the SCN? Using DNA microarrays to analyse gene expression profiles from liver tissues collected from mice over circadian time, we have been able to show that approximately 10% of the sampled genome is under circadian control (Akhtar et al 2002, and see chapters by Kramer et al 2003 and Panda et al 2003, this volume). The regulated genes contribute to most, if not all, cellular functions. Genes encoding cytoskeletal elements, vesicle recycling proteins and enzymes involved in carbohydrate metabolism that are particularly relevant to hepatic function, are under circadian regulation. The conclusion is that such transcriptional cycling adapts the liver to the circadian patterning of feeding and digestion, ultimately driven by SCN regulation of the sleep–wake cycle.

This adaptation is dependent on a precise temporal programme such that particular groupings of genes share a common phase of expression. In the case of *mPer1* and *mBmal1*, the anti-phasic relationship observed in the SCN is retained in the liver, albeit with a slightly delayed timing for both clusters (Fig. 5). Nevertheless, this peripheral programme is not robustly autonomous. Circadian gene expression persists in cultures of peripheral tissue for only a handful of cycles, dampening rapidly. Consistent with this, surgical ablation of the SCN either stops completely or severely dampens the amplitude of 95% of the rhythmic transcripts identified by cDNA microarray analysis, confirming the primacy of the SCN oscillator within circadian organisation (Akhtar et al 2002).

Conclusion

The behaviour of clock-gene products in the SCN provides strong support to the current model of the circadian clockwork as an autonomous intracellular mechanism based upon interlocked feedback and feed-forward loops. Intercellular signalling, more specifically convergent mechanisms for up- and down-regulation of *Per* expression, is critical to circadian entrainment, which can now be explained in mechanistic terms rather than described by formalisms.

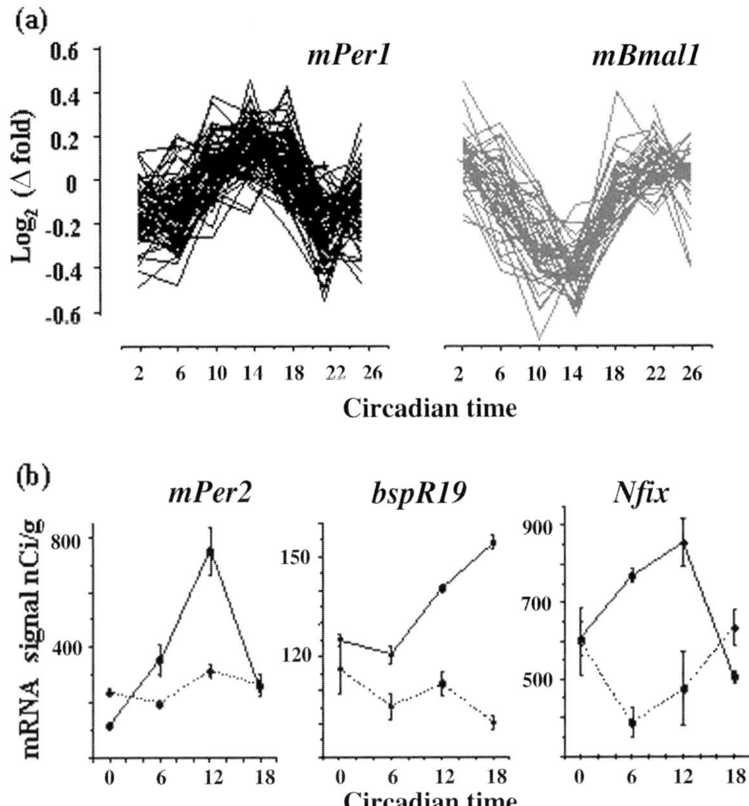

FIG. 5.   DNA microarray analysis of circadian gene expression in mouse liver. (a) Relative expression levels of representative circadian phase-specific gene-clusters, expressed predominantly in circadian day and associated with *mPer1* ($n=69$), or predominantly in circadian night and associated with *mBmal1* ($n=40$) (A.B. Reddy, unpublished data, based on Akhtar et al 2002). (b) Circadian gene expression cycles in liver of intact (solid line) and SCN-lesioned (dotted line) mice on second cycle of release to continuous darkness. Note high amplitude cycles of intact animals are lost, and expression of canonical gene (*mPer2*) and two representative novel circadian genes is constitutively low in SCN lesioned animals. (Redrawn from Akhtar et al 2002.)

Intercellular peptidergic signalling via the VPAC2 receptor is also necessary for circadian function in the SCN, a finding that implies the existence of reciprocal interplay between electrical firing and the core molecular loops. Finally, cDNA microarray analysis has demonstrated the remarkably diverse and pervasive effect of the SCN upon gene expression patterns in peripheral tissue. The observation that tissues undergo pronounced and widespread circadian changes in transcriptional activity will have far-reaching consequences for the medical application of circadian principles. Identification of the factors that confer on the SCN the remarkable ability to sustain circadian timing in itself, and thereby in other tissues, presents another major challenge to the field. Interdependence between electrical events at the neuronal membrane and the core molecular loop may be a fruitful avenue for such investigation.

Acknowledgements

Elements of the work described in this chapter were conducted in collaboration with Professors C. P. Kyriacou (University of Leicester), A. J. Harmar (University of Edinburgh), N. Mrosovsky (University of Toronto) and S. M. Reppert and D. R. Weaver (University of Massachusetts), and the work was funded by the Medical Research Council and the Biotechnology and Biological Sciences Research Council, UK.

References

Akhtar RA, Reddy AB, Maywood ES et al 2002 Circadian cycling of the mouse liver transcriptome, as revealed by cDNA microarray, is driven by the suprachiasmatic nucleus. Curr Biol 12:540–550

Albrecht U, Sun ZS, Eichele G, Lee CC 1997 A differential response of two putative mammalian circadian regulators, mper1 and mper2, to light. Cell 91:1055–1064

Bae K, Jin X, Maywood ES, Hastings MH, Reppert SM, Weaver DR 2001 Differential function of *mPer1*, *mPer2* and *mPer3* in the SCN circadian clock. Neuron 30:525–536

Biello SM, Mrosovsky N 1995 Blocking the phase-shifting effect of neuropeptide Y with light. Proc R Soc Lond B Biol Sci 259:179–187

Cutler DJ, Haraura M, Reed HE, et. al 2003 The mouse VPAC2 receptor confers suprachiasmatic nuclei cellular rhythmicity and responsiveness to vasoactive intestinal polypeptide in vitro. Eur J Neurosci. 17:197–204

Ebling FJ 1996 The role of glutamate in the photic regulation of the suprachiasmatic nucleus. Prog Neurobiol 50:109–132

Harmar AJ, Marston HM, Shen S et al 2002 The VPAC2 receptor is essential for circadian function in the mouse suprachiasmatic nuclei. Cell 109:497–508

Hastings MH 2001 Modelling the molecular calendar. J Biol Rhythms 16:117–123

Hastings MH, Field MD, Maywood ES, Weaver DR, Reppert SM 1999 Differential regulation of mPER1 and mTIM proteins in the mouse suprachiasmatic nuclei: new insights into a core clock mechanism. J Neurosci 19:RC11

Field MD, Maywood ES, O'Brien JA, Weaver DR, Reppert SM, Hastings MH 2000 Analysis of clock proteins in mouse SCN demonstrates phylogenetic divergence of the circadian clockwork and resetting mechanisms. Neuron 25:437–447

Ginty DD, Kornhauser JM, Thompson MA et al 1993 Regulation of CREB phosphorylation in the suprachiasmatic nucleus by light and a circadian clock. Science 260:238–241

Jagota A, de la Iglesia HO, Schwartz WJ 2000 Morning and evening circadian oscillations in the suprachiasmatic nucleus in vitro. Nat Neurosci 3:372–376

King VM, Chahad-Ehlers S, Shen S et al 2003 A hVIPR transgene as a novel tool for the analysis of circadian function in the mouse suprachiasmatic nucleus. Eur J Neurosci 17:822–832

Kramer A, Yiang F-C, Snodgrass P et al 2003 Regulation of daily locomotor activity and sleep by hypothalamic EGF receptor signalling. In: Molecular clocks and light signalling. Wiley, Chichester (Novartis Found Symp 253) p 250–266

Kume K, Zylka MJ, Sriram S et al 1999 mCRY1 and mCRY2 are essential components of the negative limb of the circadian clock feedback loop. Cell 98:193–205

Lee C, Etchegaray JP, Cagampang FR, Loudon AS, Reppert SM 2001 Posttranslational mechanisms regulate the mammalian circadian clock. Cell 107:855–867

Lee HS, Nelms JL, Nguyen M et al 2003 The eye is necessary for a circadian rhythm in the suprachiasmatic nucleus. Nat Neurosci 6:111–112

Liu C, Reppert SM 2000 GABA synchronizes clock cells within the suprachiasmatic circadian clock Neuron 25:123–128

Maywood ES, Mrosovsky N 2001 A molecular explanation of interactions between photic and non-photic circadian clock-resetting stimuli. Gene Expression Patterns, Brain Res 1:27–31

Maywood ES, Mrosovsky N, Field MD, Hastings MH 1999 Rapid down-regulation of mammalian *Period* genes during behavioural resetting of the circadian clock. Proc Natl Acad Sci USA 96:15211–15216

Maywood ES, Okamura H, Hastings MH 2002 Opposing actions of neuropeptide Y and light on the expression of circadian clock genes in the mouse suprachiasmatic nucleus. Eur J Neurosci 15:216–220

Maywood ES, O'Brien JA, Hastings MH 2003 Expression of mCLOCK and other circadian clock-relevant proteins in the mouse suprachiasmatic nuclei. J Neuroendocrinol 15:329–334

Mrosovsky N, Edelstein K, Hastings MH, Maywood ES 2001 Cycle of *period* gene expression in a diurnal mammal (*Spermophilus tridecemlineatus*): implications for nonphotic phase shifting. J Biol Rhythms 16:471–478

Nitabach MN, Blau J, Holmes TC 2002 Electrical silencing of *Drosophila* pacemaker neurons stops the free-running circadian clock. Cell 109:485–495

Nusslein-Hildesheim B, O'Brien JA, Ebling FJP, Maywood ES, Hastings MH 2000 The circadian cycle of mPER clock gene products in the suprachiasmatic nucleus of the Siberian hamster encodes both daily and seasonal time. Eur J Neurosci 12:2856–2864

Panda S, Hogenesch JB, Kay SA 2003 Circadian light input in plants, flies and mammals. In: Molecular clocks and light signalling. Wiley, Chichester (Novartis Found Symp 253) p 73–88

Preitner N, Damiola F, Lopez-Molina L et al 2002 The orphan nuclear receptor REV-ERBα controls circadian transcription within the positive limb of the mammalian circadian oscillator. Cell 110:251–260

Reddy AB, Field MD, Maywood ES Hastings MH 2002 Differential resynchronisation of circadian clock gene expression within the suprachiasmatic nuclei of mice subjected to experimental 'jet-lag'. J Neurosci 22:7326–7330

Schurov IL, McNulty S, Best JD, Sloper PJ, Hastings MH 1999 Glutamatergic induction of CREB phosphorylation and Fos expression in primary cultures of the suprachiasmatic hypothalamus in vitro is mediated by co-ordinate activity of NMDA and non-NMDA receptors. J Neuroendocrinol 11:43–51

Shearman LP, Sriram S, Weaver DR et al 2000 Interacting molecular loops in the mammalian circadian clock. Science 288:1013–1019

Shigeyoshi Y, Taguchi K, Yamamoto S et al 1997 Light-induced resetting of a mammalian circadian clock is associated with rapid induction of the mPer1 transcript. Cell 91:1043–1053

Travnickova-Bendova Z, Cermakian N, Reppert SM, Sassone-Corsi P 2002 Bimodal regulation of mPeriod promoters by CREB-dependent signaling and CLOCK/BMAL1 activity. Proc Natl Acad Sci USA 99:7728–7733

DISCUSSION

Menaker: Have you measured feeding behaviour in the *Vpac2* knockout mice?

Hastings: No. It's clearly something we have to do. We can make strong predictions about their feeding pattern based on their locomotor activity patterns, and we should be able to pull the liver cycle forwards and backwards by restricted feeding. We don't know the glucocorticoid profiles, which we will need to interpret the liver data.

Menaker: It is possible that this will explain it very neatly.

Schibler: The one thing it will not explain is why they are antiphasic. This is mysterious.

Hastings: I am not a geneticist, but it is safe to say that penetrance isn't complete. Some of the animals are completely arrhythmic, but most of them aren't. This is a phenotype we have seen in several different animal facilities. These mice are interesting at a number of levels. For example, they can be useful as a model to study the liver devoid of SCN control. As for the origin of the antiphasic behaviour, it is a systems neuroscience question. Recent work from Mike Menaker's lab (Abe et al 2002) has shown the existence of weak extra-SCN oscillators in the brain which may or not be involved in the antiphasic behaviour of *Vpac2* knockout mice under dark–dark conditions (DD).

Rosbash: Have you lesioned the mutant animals?

Hastings: No. We have not even given them amphetamine. If we were to do this, we might get much better definition to that behaviour.

Young: Are the phase differences in light–dark conditions (LD) so subtle that you can't try to map the motor cortex rhythms?

Hastings: I think we could do; this reflects the limits to the measurements. If we get a fine-scale resolution to feeding or drinking behaviour in the order of minutes, we should be able to see differences in what they do.

Weitz: It would be nice to see what the true phase relationship is to light in the entrained state. It looks like there is incomplete masking of what would be a lot of behaviour in the antiphase relationship.

Hastings: Absolutely. We did some probe tests that involved dropping brief intervals of darkness into the light phase. If you do this to the intact wild-type animals they won't run. Although darkness is a conditional factor it is not sufficient because the intact clock in the wild-types suppresses activity during subjective day. If you drop the probe dark pulse into the light phase with the mutants, they immediately become active.

Schibler: Did you look at the eye rhythms? They could feed on other brain regions.

Hastings: No.

Lee: Do these animals phase shift normally?

Hastings: Our simple interpretation is that they don't have a clock to shift. As we move the LD schedule they just go completely with it on first cycle. There isn't evidence for any transient re-setting which would indicate that there is inertia in the behaviour that is reflective of a clock mechanism (Reddy et al 2002). I didn't show the LD cycle data, but if you look at *Per* expression on an LD cycle it is extremely low. The retinal mechanism in these mice is also impaired and *Per* induction can't be driven with light.

Van Gelder: The VPAC2 receptor is also the PACAP receptor, and PACAP is the major peptide of RHT.

Hastings: That is right. The confounding possibility is that the receptor has knocked out some RHT sensitivity. But given that we know that *in vitro* you can put NMDA on a rat slice and get competent phase shifts in the absence of PACAP insensitivity to PACAP is unlikely to be the full reason for impaired circadian entrainment.

Van Gelder: Chris Colwell has data on the PACAP knockout suggesting that they had markedly reduced photosensitivity for phase shifting and entrainment. Although glutamate is sufficient for phase shifting *in vitro* it may be that *in vivo* to get normal entrainment or phase response PACAP release is also needed.

Hastings: It is a question of how abnormal these mice are. I would suggest that they are so abnormal in their light response that it wouldn't be possible to assign it all to a peptidergic dysfunction.

Weitz: In Hugh Piggin's electrophysiology experiments, did they try resetting manipulations in the dish that avoided PACAP receptors? Can the tissue be provoked into something rhythmic? This would get around the possibility. One outside possibility is that this is a SCN that had never seen light and had never been synchronized: it had never become competent to be rhythmic.

Van Gelder: In this case our *math5*$^{-/-}$ mutants should be arrhythmic. These are probably in their 14th generation of never having seen any light in the SCN and they are still rhythmic.

Hastings: They certainly put on VIP and showed that the electrical responses of the wild-type to VIP are absent in the mutants. What I didn't say is that the neurons of the SCN, in addition to having the peptide, also use GABA as a co-transmitter. In a simple model you might have VIP peptide as the afferent stimulatory link and GABA as a negative link in a synapse between a VIP neuron and another cell. My personal view at the moment is that the VPAC knockout phenotype is a neurochemical defect, and in some weird way the SCN network is reciprocally suppressing through enhanced GABA tone. We plan to put on GABA antagonists and we should be able to spring the thing back into life by blocking GABA if this model is correct.

Rosbash: In the simple experiments of the clock protein profiles in the wild-type, do you see PER build up cytoplasmically before you see a nuclear PER?

Hastings: With the antibodies as we use them, we never see the characteristic doughnut images. When we first started doing our work we were expecting to see them. Clearly, the protein has to be in the cytoplasm at some stage, but it is never at levels detectable by the antibodies as we use them.

Rosbash: So there is no evidence of any gating of nuclear entrance.

Hastings: No, and this would be consistent with the things entering as soon as they are stabilized by complex formation.

Rosbash: What you described for the *tau* mutant hamster, with an effect on the decay portion of the curve is exactly what we published for the two *doubletime* long alleles.

Hastings: You had a prolongation.

Rosbash: Only of the declining phase. There was no effect on the accumulating phase.

Young: In the original paper we saw evidence for a functional requirement for *dbt* in the nucleus. You refer to hypo-phosphorylated PER in the model, but casein kinase 1 delta is still there. I thought in *tau* mutants the phosphorylation patterns for PER were not really distinguishable from wild-type.

Hastings: In the *Cell* paper (Lee et al 2001) they used liver, and they were unable to show a difference in phosphorylation patterns with the PER protein between the two genotypes. These sorts of data don't tell us the phosphorylation state in the SCN. If it were possible, one would hope to see a difference in phosphorylation state in the mutants.

Young: I also wonder whether a part of the regulation in mammals might be a shift between these two kinases — a qualitative difference in phosphorylation patterns rather than a presence or absence of it.

Hastings: This could also be tissue specific. It may be that the contribution is different in the SCN from the liver.

Young: Is casein kinase 1 delta expressed in the SCN?

Hastings: I think it is at low levels.

Takahashi: We don't have good information but my feeling is that delta is in SCN. Epsilon is not in the liver.

Schibler: BMAL is low in the SCN, so if PER is low that is fine because there is no BMAL. However, CRY is high. What happens to CRY levels in the *Bmal* knockout?

Takahashi: We didn't look.

Schibler: This is very important. There is very little evidence that CRY is regulated by CLOCK and BMAL. Perhaps CRY is regulated completely differently from PER.

Van Gelder: Did your SCN lesions show masking?

Hastings: Yes.

Van Gelder: So the SCN-lesion animals mask and the VPAC2 receptor knockouts mask, and you can only drive peripheral gene expression rhythmically in the VPAC2 knockout. Yet you can't drive circadian gene expression in the SCN with light in the VPAC2 knockout. How do you put that together? We discussed earlier the notion of some masking going through the SCN and other masking going around it. Clearly in the lesion this is going around. What interpretation do you have for the masking that you see in the VPAC2 mutant, and particularly its ability to entrain peripheral oscillators?

Hastings: I would take that to be a consequence of feeding patterns.

Van Gelder: Presumably feeding patterns are also rhythmic in the SCN lesioned animals that are masked, yet they don't seem to synchronize peripheral gene expression.

Hastings: If there is an extra-SCN oscillator, presumably it is influenced by the SCN. In the two conditions it is getting a different type of perturbation. If we lesion the SCN this oscillator has no afferent input from the SCN. In the VPAC2 mutant the neurons are still there and the SCN appears fine anatomically. And so some residual communication between SCN and its targets is possible even if it is not rhythmic. It is not the same as complete deafferentation.

Weitz: The assumption is that an SCN-lesioned animal, monitored under an LD cycle and showing masking of locomotor activity, will have an altered feeding activity. Is this known?

Hastings: Yes. The lesion has an effect on feeding as well as behaviour.

Menaker: Masking isn't always seen with SCN lesions.

Hastings: The pathways that are necessary for masking to occur will run through the chiasm, even if they are not synapsing in the SCN. If you lesion other afferents to the midbrain or thalamus you may take away the neural substrate for masking with a large or misplaced SCN lesion.

Weitz: There are now many examples of SCN lesions showing arrhythmic locomotor activity that is masked by light.

Rosbash: So you would predict that if you lesion the SCN in these mutant animals that this peripheral gene expression would disappear and perhaps also some of the behavioural effects.

Hastings: The expectation would be that with a lesion to the SCN we would see arrhythmia to behaviour in DD, and then peripheral cycles would dampen out. I don't know how long it would take to damp out.

Schibler: We have always assumed that if we remove the SCN, the important thing to remove is the cycling system. But how about if the SCN is simply a relay station to other regions in the brain? In this case, when the SCN is removed, this relay activity is removed. If you lesion the SCN, you also lesion further connections to other areas.

Hastings: If we have a good SCN lesion that doesn't impinge deep into the chiasm, animals such as this will still mask their behaviour on a LD schedule. So there are routes avoiding the SCN that can control masking behaviour.

Menaker: I gather that no one has managed to lesion the SCN with kainic acid to spare the fibres that pass through it. It is highly resistant.

Hastings: Paradoxically, the whole photic cascade is dependent on NMDA and AMPA signalling, yet none of us have been able to use neurotoxins to kill it off.

Young: Is the same interdependence of CRY and PER stability that Steve Reppert described for the liver seen in the SCN?

Hastings: In the *Cry* double knockouts we have done immunostaining for PER2 and it is not present. Equally, in the *Per1* and *Per2* mutants, in the absence of PER1 and PER2 there are half-maximal levels of CRY immunostaining (Bae et al 2001). Part of this is because in the *Per2* mutant there is no *Cry* mRNA cycling, and in the *Per1* mutant rhythmic gene expression occurs but the protein is unstable in the absence of PER1.

References

Abe M, Herzog ED, Yamazaki S et al 2002 Circadian rhythms in isolated brain regions. J Neurosci 22:350–356

Bae K, Jin X, Maywood ES, Hastings MH, Reppert SM, Weaver DR 2001 Differential functions of mPer1, mPer2, and mPer3 in the SCN circadian clock. Neuron 30:525–536

Lee C, Etchegaray JP, Cagampang FR, Loudon AS, Reppert SM 2001 Posttranslational mechanisms regulate the mammalian circadian clock. Cell 107:855–867

Reddy AB, Field MD, Maywood ES, Hastings MH 2002 Differential resynchronisation of circadian clock gene expression within the suprachiasmatic nuclei of mice subjected to experimental jet lag. J Neurosci 22:7326–7330

Circadian rhythms in *Drosophila*

Michael Rosbash, Ravi Allada[1], Mike McDonald[2], Ying Peng and Jie Zhao

Howard Hughes Medical Institute and Department of Biology, Brandeis University, Waltham, MA 02454, USA

Abstract. We discuss some historical features of the circadian field in *Drosophila melanogaster*. We then describe some recent progress from our laboratory in three different areas. First, we discuss the regulation of circadian gene expression as assayed with microarrays. Results are discussed that verify and extend published data, both with respect to the previously identified cycling mRNAs as well as some clustering within the genome of some of the genes that give rise to these circadian transcripts. Also discussed are experiments that attempt to identify transcripts that are enriched in lateral neurons, the key circadian pacemaker cells in the *Drosophila* brain. Second, the issue of damping within the brain is addressed, by assaying molecular oscillations after many days in constant darkness. Third, the identification of a new circadian mutant is described, which is a fully recessive allele of the gene *Clock*. The previous allele in flies, as well as the single mutant allele in mice, is a dominant allele. This limits the conclusions that can be drawn from the genetic and molecular analyses in these mutant strains. Results with the new recessive allele not only support the notion that *Clock* is an important clock gene but also indicate that it contributes more to the amplitude of the rhythm rather than the period.

2003 Molecular clocks and light signalling. Wiley, Chichester (Novartis Foundation Symposium 253) p 223–237

The modern molecular genetic era of circadian rhythms arguably began more than thirty years ago when Konopka and Benzer published the results of a genetic screen, describing their three identified alleles of the period gene in *Drosophila melanogaster* (Konopka & Benzer 1971). This was not only a landmark achievement, which kick-started the molecular-circadian field, but it was also unusually prescient. This is because it was still several years before the first recombinant DNA technology was published and almost a decade before it became practical, even in the most sophisticated of laboratories. In other words,

---

Present addresses: [1]Department of Neurobiology and Physiology, Northwestern University, 2205 Tech Drive, #2-160, Evanston, IL 60208, USA and [2]University of Geneva, Department of Zoology and Biologie Animale, Quai Ernest-Ansermet 30, CH-1211 Geneve 4, Switzerland.

the period mutants were identified and characterized well before a molecular vision could be conceptualized let alone realized.

It was therefore not until 1984, 13 years later, that the period gene was cloned and used in what were arguably the first gene rescue experiments of behavioural import (Bargiello et al 1984, Zehring et al 1984). Yet even this achievement was not illuminating from the circadian point of view, because the function of the period protein (PER) was still unknown. During the next few years, we at Brandeis (Rosbash and Hall labs) and the Young laboratory at Rockefeller worked on the relationship of this gene to circadian rhythms, which resulted in several significant advances. Among these were the sequence of the complete protein and the location of the precise nucleotide changes responsible for the slow, fast and arrhythmic alleles (Baylies et al 1987, Yu et al 1987). However, not even the sequence was particularly clarifying. This is because it was a pioneer protein, with no known relatives. In those early days of DNA sequencing, it was much more frequent that a sequence did not reveal a protein's secrets. This was the situation until 1988, when there appeared the sequence of a *Drosophila* transcription factor with a clear relationship to the period protein (Crews et al 1988). Although the two proteins were not close relatives and the single motif in common was of uncertain function, the similarity was unambiguous and inspired us at Brandeis to pursue this hypothesis, namely, that PER was a transcription factor and that the regulation of transcription was central to circadian rhythms. In 1990, almost 20 years after the landmark Konopka and Benzer publication, we published the finding that period mRNA levels undergo circadian oscillations and that PER regulates the period and phase of its own mRNA cycling. In other words, the mRNA cycling was sensitive to the Konopka and Benzer mutations and paralleled the previously described changes in the behavioural cycling (Hardin et al 1990). We expanded on this observation over the next couple of years and showed that the regulation was transcriptional and almost certainly reflected a negative feedback loop, in which PER inhibits its own transcription (Hardin et al 1992, Zeng et al 1994). This feedback loop and transcriptional regulation have been cornerstones of the mammalian as well as the *Drosophila* circadian system, since the discovery in 1997 of the mammalian *period* genes.

In the decade or so since the publication of the second *Drosophila* circadian rhythm gene *timeless* by the Young laboratory, many fruit fly pacemaker components have been discovered. Like *period*, most of these are conserved in mammals, with similar if not identical functions (Allada et al 2001). Moreover, studies on the biochemistry of the timekeeping mechanism have continued to focus on transcriptional regulation. It is believed that the basic helix-loop-helix (bHLH) transcription factors CLOCK (CLK) and CYCLE (CYC) bind to upstream E-boxes (CACGTG) and directly activate transcription of the period (*per*) and timeless (*tim*) gene (Allada et al 1998, Darlington et al 1998, Hao et al

1997, Rutila et al 1998, Wang et al 2001). This view is based on strong biochemical evidence in both systems: in the *Drosophila* system, PER and TIM proteins subsequently feed back and inhibit transcriptional activation by CLOCK and CYCLE (Darlington et al 1998, Lee et al 1998, 1999). A similar focus on transcription and feedback loops exists in mammals, including humans. Of note, circadian transcription studies *in vivo* have relied heavily on two dominant negative (antimorphic) alleles of *Clock*, one in *Drosophila* (*Clk^{Jrk}*) and one in mouse (Allada et al 1998, King et al 1997a,b). In the fly system, transcription of the *per* and *tim* genes is incredibly low in the *Clk^{Jrk}* background.

Further studies have implicated a second feedback loop in circadian timing. Like *per* and *tim*, *Clk* and *cry* RNAs also oscillate with respect to time of day (Bae et al 1998, Darlington et al 1998, Emery et al 1998). However, these oscillations are antiphase to those of *per* and *tim*, suggesting that they are indirect targets of the *Clk–cyc* system. This is consistent with the levels of the *Clk* and *cry* RNAs in *Clk^{Jrk}* and *cyc^0* mutants; they are high, whereas the levels of *per* and *tim* RNAs are low (Emery et al 1998, Glossop et al 1999). It has been proposed that these genes, *per* and *tim* on the one hand and *Clk* on the other, define two interdependent transcriptional feedback loops. Transcriptional oscillations are thought to emerge from the dynamic interplay of these feedback loops, leading to behavioural and physiological rhythms.

Several aspects of circadian gene expression are also subject to post-transcriptional control, including RNA and protein stability as well as protein phosphorylation (Dembinska et al 1997, Kim et al 2002, So & Rosbash 1997). Protein levels and phosphorylation states of PER and TIM oscillate with time of day (reviewed in Allada et al 2001). Doubletime, a casein kinase I epsilon homologue; shaggy, a glycogen synthase kinase 3 homologue; and casein kinase 2, appear to phosphorylate PER and TIM (Kloss et al 1998, Kloss et al 2001, Lin et al 2002a, Atken et al 2003, Martinek et al 2001, Price et al 1998). These additional layers of feedback make it difficult to untangle the roles of different mechanisms in determining rhythm period, phase, and amplitude.

However, cycling RNAs are generally considered to be under transcriptional regulation. This is due in part to the fact that all RNA cycling is apparently eliminated in the *Clock* mutant *Clk^{Jrk}* (McDonald & Rosbash 2001). Of course many of these mRNAs could be regulated post-transcriptionally and only indirectly by the circadian transcription machinery, for example through the transcriptional regulation of a splicing factor. But the current view of the field is that most cycling mRNAs are regulated at the transcriptional level. This is also because in addition to *period* and *timeless*, the CLK–CYC heterodimer directly activates at least three additional transcription factor-encoding genes. Direct target genes of CLK–CYC have been defined in a microarray experiment with S2 tissue culture cells, in which the CLK–CYC heterodimer is able to activate target gene expression in the presence of cycloheximide. The activation takes place with dexamethasone and a CLK–GR (clock–glucocorticoid) fusion gene in the cells, so that the S2 cell-expressed endogenous CYC protein and the CLK–GR protein is activated by the addition of glucocorticoid and without a transcription or protein synthesis requirement. Only a handful of genes are activated in this system, including at least three transcription factors (McDonald & Rosbash 2001). These include VRI as well as PDP1, both of which have been subsequently shown to participate in clock gene regulation (Cyran et al 2003, Glossop et al 2003). Our current view is that the large number of cycling genes is the product of a transcriptional cascade and that the CLK–CYC heterodimer sits at the top of the pyramid.

At the base of that pyramid sit 134 mRNAs, which undergo circadian oscillations. Based on sequence criteria, these genes are grouped into pathways with different functions, many of which had already been discovered to have roles in circadian rhythms. There were also several novel pathways, suggesting that many different physiological systems are under clock control. Although the number and identity of cycling mRNAs was very different in the different studies published to date (Ceriani et al 2002, Claridge-Chang et al 2001, Lin et al 2002b, McDonald & Rosbash 2001, Ueda et al 2002), they identified many genes in related biochemical and metabolic pathways. A large fraction of the variation may come from biological differences; that is how the samples were collected and RNA harvested. Perhaps even more important is the fact that all groups used different methods of analysis and different thresholds for significance. In fact, we could never recapitulate any results with another method of analysis, either our results with another methodology or other results with our methods. In any case, it is presently uncertain whether the rather small overlap (for example, about 25% between McDonald & Rosbash 2001, Claridge-Chang et al 2001 and Etter & Ramaswami 2002) is due to a high fraction of false-positives in the various studies or a high fraction of false negatives. Although it has been suggested that small overlap is due to false positives, we prefer the false negative explanation. This fits with the fact that all groups used different methods and criteria to define their cycling mRNA subpopulations. Also, it should be easier to disprove the 'high fraction of false positives' hypothesis.

To begin an examination of the fraction of false positives, we chose 14 of our cycling mRNAs at random and examined their cycling by real-time PCR. We could clearly confirm the cycling and microarray patterns for 10 of the 14 mRNAs, and the cycling was likely positive for two more. Only in two of the 14 cases was circadian cycling unlikely, based on the real-time results. We conclude that most of our 134 mRNAs are real cyclers and that false positives constitute only a minority of the 134 mRNAs. The number 134 is probably a gross underestimate.

Another finding from our microarray paper was the fact that there are clusters of cycling genes, closely spaced within a single chromosomal region. To verify and extend this observation, we took the same real-time PCR approach and examined every open reading frame within the takeout-Duf 227 cluster. The results verified the cycling mRNAs originally identified and also identified several new cycling genes within this cluster. Moreover, expression of many more genes from this region were identifiable by real-time criteria than by microarray criteria. It is unclear at present why this is the case, i.e. whether these genes were just expressed at levels too low to detect or whether there is some other obstacle that limits the sensitivity or the generality of the microarray approach.

A third approach we are taking with microarrays is to identify mRNAs that are highly expressed in the brain neurons most important for locomotor activity rhythms in flies. Because there are no available techniques for sorting or enriching adult brain neurons from *Drosophila*, we have taken an ablation approach and eliminated brain neurons by specifically expressing cell-death genes in these cells. This approach has been previously used in behavioural studies, and these neurons, the sLN$_v$s and the lLN$_v$s, can be killed without any adverse effects other than a loss of circadian rhythms (Renn et al 1999). Head microarrays from these strains identify a number of genes that are low in the ablation strains compared to a wild-type strain. Because the neuropeptide PDF is specifically expressed in the LN$_v$s, this gene serves as a positive control; *pdf* mRNA is indeed present only at low levels in these strains. These studies identify a number of mRNAs that behave like *pdf* and are present at low levels in the cell-ablated strain. We have tested three of them by *in situ* hybridization, and all three give strong signals in both groups of LN$_v$s; this is consistent with the notion that they are highly expressed in those cells. More work needs to be done on these genes to verify that they make a contribution to circadian rhythms.

We have also been interested in the function of these few clusters of brain neurons, i.e. how they contribute to circadian gene expression. The issue at hand is damping, the fact that the amplitude of gene expression oscillations decreases as a function of the time that the animals are in constant conditions (constant darkness). Damping in the fly system was observed in the original cycling gene expression observations from the early 90s: after several days in constant darkness, the amplitude of gene expression cycling is very modest compared to LD (light–dark) conditions or compared to the first day in constant darkness. More recently, immunohistochemical experiments suggest that molecular rhythms in the eyes and even in some of the circadian brain neurons undergo dramatic damping in constant darkness. Importantly, locomotor activity rhythms persist in constant conditions for at least two weeks with no detectable damping. Because of this conflict (molecular damping vs. no behavioural damping), we re-examined molecular cycling in the brain neurons by *in situ* hybridization with a *tim* antisense probe. Although this assay has been used often to examine brain neurons, it has not been applied to flies maintained for a long time in constant darkness. We examined flies after four and eight days in constant darkness and observed robust transcriptional oscillations, undiminished from what is observed in a light–dark cycle. We conclude that there is no conflict for the brain, in which the robust molecular oscillations match the robust behavioural oscillations. The damping of molecular rhythms must come principally from other head tissues like the eye, or the molecular damping that has been observed is probably due to the short-term adjustment in going from LD to constant darkness conditions.

I now want to return to the transcriptional cascade and the *Clk* gene. As mentioned previously, *Drosophila* circadian transcription studies *in vivo* have relied heavily on a single dominant negative allele of *Clock*, (*Clk^{Jrk}*) and one in mouse (Allada et al 1998). This is a precarious situation for the field, because some of the mutant phenotypes could be due to effects on other transcription factors and systems rather than just to low activity of the CLK–CYC complex. For this reason, we characterized a second allele of the *Clk* gene, which turned out to be a real recessive allele.

The mutant gene was found in our search for novel genes involved in circadian rhythmicity. We were screening ethyl methane-sulfonate (EMS) mutagenized flies for alterations in circadian locomotor activity (Rutila et al 1996). One line homozygous for a mutagenized third chromosome was arrhythmic. The phenotype mapped to the third chromosome, and homozygotes do not exhibit robust rhythms; in contrast, heterozygotes are virtually indistinguishable from wild-type. All other genetic and phenotypic characterization also indicated that the mutant, called *Clk^{ar}*, is fully recessive (Allada et al 2003).

Expression of *Clk* by *pdfgal4* in a *Clk^{ar}* background did not result in significant rescue of rhythmicity. On the other hand, *crygal4*-driven expression of *Clk* resulted in rescue in the rhythmicity of a majority of these flies. The rescued flies exhibited a slightly short period, similar to periods in flies with *crygal4*-driven expression of *Clk* in a wild-type background. The period shortening with increased *Clk* expression is consistent with the long periods of flies with only a single dose of *Clk*. BAC transgenic mice containing extra copies of *Clock* also exhibit short periods (Antoch et al 1997). We obtained similar results in a *Clk^{Jrk}* background: *crygal4*-driven *Clk* expression was able to rescue the rhythmicity of *Clk^{Jrk}* (18% rhythmic), although more weakly than *Clk^{ar}* (60% rhythmic), consistent with the antimorphic effects of *Clk^{Jrk}*.

We searched coding exons and exon–intron boundaries for EMS-induced base changes, comparing *Clk^{ar}* with sibs. We identified a single mutation at the 5′ splice site of the second intron, destroying the GT dinucleotide required for efficient splicing. The mutation is a G to A transition classically found in EMS-induced alleles. We examined *Clk* splice forms across the second intron in the *Clk^{ar}* mutant using reverse transcriptase-polymerase chain reaction (RT-PCR). RT-PCR across this intron identified a single band of the appropriate size in wild-type flies. In *Clk^{ar}*, multiple bands are observed, none of which correspond by electrophoretic migration to that seen in wild-type, consistent with the observed splice site mutation. Splice junctions between other coding exons were not grossly perturbed as assayed by RT-PCR. Exon 2 encodes for the N-terminal 13 amino acids, including the first two amino acids of the basic region. The exons beyond exon 2 encode the remainder of the CLK protein, including most of the basic region, the PAS dimerization motif and the glutamine-rich activation domain (Allada et al 1998). To determine whether these altered *Clk^{ar}* transcripts can produce functional CLK protein, we sequenced *Clk^{ar}* cDNAs. In only 4/22 clones, an upstream methionine codon is in frame with the remainder of the *Clk* gene. Assuming initiation from this methionine, translation of these transcripts would result in a CLK protein with novel N-termini: two of 15 amino acids and two of 28 amino acids. In all four cases, only the first two amino acids of the basic DNA binding domain are altered. Based on this analysis, we believe that there is a low level of CLK activity in the *Clk^{ar}* strain, which comes from a small fraction of aberrantly spliced mRNAs.

The molecular assays in *Clk^{ar}* indicate *bona fide* rhythms with a predominant effect on circadian rhythm amplitude and no more than a modest effect on phase or period. With circadian *per* and *tim* enhancers, we observed reduced enhancer activity and a reduced cycling amplitude in a *Clk^{ar}* background, consistent with the role of *Clk* in regulating these enhancers. Nonetheless, the phase of oscillating bioluminescence is similar to that of wild-type flies. The presence of molecular rhythms contrasts with the absence of detectable behavioural rhythms. We favour the notion that this reflects a level or amplitude reduction below a critical threshold for behavioural rhythmicity. The absence of anticipation of light–dark transitions makes it very unlikely that an effect restricted to the lateral neurons—the absence of the neuropeptide PDF, for example—is primarily responsible for the behavioural phenotypes. This is also because LD behavioural rhythms are largely normal in flies devoid of PDF or the pacemaker lateral neurons (Renn et al 1999). However, we cannot exclude the possibility of selective effects of *Clk^{ar}* on other behaviourally relevant neurons.

Previous results with *Clk^{Jrk}* also support a role for *Clk* in defining rhythmic amplitude. *Clk^{Jrk}* heterozygotes reveal a dominant reduction in the amplitude of molecular rhythms with little apparent change in phase (Allada et al 1998). These heterozygotes also exhibit reductions in rhythmic behaviour with only slightly long periods. Indeed, *Clk* over-expression results in a selective increase in the amplitude of *per* RNA oscillations (Kim et al 2002). This modest effect of varying *Clk* activity on period is similar to the phenotype of transgenic strains missing the *per* promoter or expressing *per* and *tim* from constitutive promoters

(Frisch et al 1994, Yang & Sehgal 2001). These strains also have reasonable periods (22–26 h) with poor rhythm amplitudes, as evidenced by the poor penetrance of rhythmicity. Taken together, these data suggest that changes in clock gene transcription have limited effects on circadian period. Separate control of circadian rhythm amplitude on the one hand and period (or phase) on the other is also consistent with anatomical experiments in both the fly and mammalian system (Liu et al 1991, Low-Zeddies & Takahashi 2001).

We propose that the post-transcriptional phosphorylation turnover feedback loop involving several clock components (e.g. *per*, *tim* and the protein kinase *Dbt*) is predominantly responsible for period determination. Excluding null alleles that are either arrhythmic or lethal, Flybase lists mutant alleles of *per*, *tim* and *Dbt* which exhibit period alterations ranging from 16–30 h for *per* (8 mutant alleles), 21–33 h for *tim* (8 mutant alleles) and 18–29 h for *Dbt* (5 mutant alleles; Flybase 2002). Indeed, the only *Dbt* allele that fails to exhibit rhythmicity as a homozygote, displays a potent period-altering phenotype as a heterozygote (Rothenfluh et al 2000). More recent additions to this list are the protein kinases *shaggy* (Martinek et al 2001) and *CK2*. Indeed, one mutant allele of *CK2*, *CK2^{Tik}*, exhibits one of the strongest dominant period effects of any rhythm mutant (Lin et al 2002a). These large period effects contrast with the transcriptional factor mutants of *Clk* and *cyc*. Their phenotypes indicate that near-normal periods are maintained despite large protein level changes.

Acknowledgements

This work was supported by the NIH and the Howard Hughes Medical Institute.

References

Allada R, White NE, So WV, Hall JC, Rosbash M 1998 A mutant *Drosophila* homolog of mammalian *Clock* disrupts circadian rhythms and transcription of *period* and *timeless*. Cell 93:791–804

Allada R, Emery P, Takahashi JS, Rosbash M 2001 Stopping time: the genetics of fly and mouse circadian clocks. Annu Rev Neurosci 24:1091–1119

Allada R, Kadener S, Nandakumar N, Rosbash M 2003 A recessive mutant of *Drosophila Clock* reveals a role in circadian rhythm amplitude. EMBO J 22:3367–3375

Antoch MP, Song E-J, Chang A-M et al 1997 Functional identification of the mouse circadian *clock* gene by transgenic BAC rescue. Cell 89:655–667

Atken B, Javch E, Genova GK et al 2003 A role for CK2 in the *Drosophila* circadian oscillator. Nat Neurosci 6:208–210

Bae K, Lee C, Sidote D, Chuang KY, Edery I 1998 Circadian regulation of a *Drosophila* homolog of the mammalian Clock gene: PER and TIM function as positive regulators. Mol Cell Biol 18:6142–6151

Bargiello TA, Jackson FR, Young MW 1984 Restoration of circadian behavioural rhythms by gene transfer in *Drosophila*. Nature 312:752–754

Baylies MK, Bargiello TA, Jackson FR, Young MW 1987 Changes in abundance and structure of the *per* gene product can alter periodicity of the *Drosophila* clock. Nature 326:390–392

Ceriani MF, Hogenesch JB, Yanovsky M, Panda S, Straume M, Kay SA 2002 Genome-wide expression analysis in *Drosophila* reveals genes controlling circadian behavior. J Neurosci 22:9305–9319

Claridge-Chang A, Wijnen H, Naef F, Boothroyd C, Rajewsky N, Young MW 2001 Circadian regulation of gene expression systems in the *Drosophila* head. Neuron 32:657–671

Crews ST, Thomas JB, Goodman CS 1988 The *Drosophila single-minded* gene encodes a nuclear protein with sequence similarity to the *per* gene product. Cell 52:143–152

Cyran SA, Buchsbaum AM, Reddy KL et al 2003 vrille, Pdp1, and dClock form a second feedback loop in the *Drosophila* circadian clock. Cell 112:329–341

Darlington TK, Wager-Smith K, Ceriani MF et al 1998 Closing the circadian loop: CLOCK-induced transcription of its own inhibitors *per* and *tim*. Science 280:1599–1603

Dembinska ME, Stanewsky R, Hall JC, Rosbash M 1997 Circadian cycling of a *period-lacZ* fusion protein in *Drosophila*: evidence for an instability cycling element in PER. J Biol Rhythms 12:157–172

Emery P, So WV, Kaneko M, Hall JC, Rosbash M 1998 CRY, a *Drosophila* clock and light-regulated cryptochrome, is a major contributor to circadian rhythm resetting and photosensitivity. Cell 95:669–679

Etter PD, Ramaswami M 2002 The ups and downs of daily life: profiling circadian gene expression in *Drosophila*. BioEssays 24:494–498

Frisch B, Hardin PE, Hamblen-Coyle MJ, Rosbash M, Hall JC 1994 A promoterless DNA fragment from the *period* locus rescues behavioral rhythmicity and mediates cyclical gene expression in a restricted subset of the *Drosophila* nervous system. Neuron 12:555–570

Glossop NR, Houl JH, Zheng H, Ng FS, Dudek SM, Hardin PE 2003 VRILLE feeds back to control circadian transcription of clock in the *Drosophila* circadian oscillator. Neuron 37:249–261

Glossop NR, Lyons LC, Hardin PE 1999 Interlocked feedback loops within the *Drosophila* circadian pacemaker. Science 286:766–768

Hao H, Allen DL, Hardin PE 1997 A circadian enhancer mediates PER-dependent mRNA cycling in *Drosophila melanogaster*. Mol Cell Biol 17:3687–3693

Hardin PE, Hall JC, Rosbash M 1990 Feedback of the *Drosophila period* gene product on circadian cycling of its messenger RNA levels. Nature 343:536–540

Hardin PE, Hall JC, Rosbash M 1992 Circadian oscillations in *period* gene mRNA levels are transcriptionally regulated. Proc Natl Acad Sci USA 89:11711–11715

Kim EY, Bae K, Ng FS, Glossop NR, Hardin PE, Edery I 2002 *Drosophila* CLOCK protein is under posttranscriptional control and influences light-induced activity. Neuron 34:69–81

King DP, Vitaterna MH, Chang A-M et al 1997a The mouse *clock* mutation behaves as an antimorph and maps within the W^{19H} deletion, distal of *kit*. Genetics 146:1049–1060

King DP, Zhao Y, Sangoram AM et al 1997b Positional cloning of the mouse circadian *clock* gene. Cell 89:641–653

Kloss B, Price JL, Saez L et al 1998 The *Drosophila* clock gene *double-time* encodes a protein closely related to human casein kinase Iε. Cell 94:97–107

Kloss B, Rothenfluh A, Young MW, Saez L 2001 Phosphorylation of period is influenced by cycling physical associations of double-time, period, and timeless in the *Drosophila* clock. Neuron 30:699–706

Konopka RJ, Benzer S 1971 Clock mutants of *Drosophila melanogaster*. Proc Natl Acad Sci USA 68:2112–2116

Lee C, Bae K, Edery I 1998 The *Drosophila* CLOCK protein undergoes daily rhythms in abundance, phosphorylation and interactions with the PER-TIM complex. Neuron 4:857–867

Lee C, Bae K, Edery I 1999 PER and TIM inhibit the DNA binding activity of a *Drosophila* CLOCK-CYC/dBMAL1 heterodimer without disrupting formation of the heterodimer: a basis for circadian transcription. Mol Cell Biol 19:5316–5325

Lin JM, Kilman VL, Keegan K et al 2002a A role for casein kinase 2alpha in the *Drosophila* circadian clock. Nature 420:816–820

Lin Y, Han M, Shimada B et al 2002b Influence of the period-dependent circadian clock on diurnal, circadian, and aperiodic gene expression in *Drosophila melanogaster*. Proc Natl Acad Sci USA 99:9562–9567

Liu X, Yu Q, Huang Z, Zwiebel LJ, Hall JC, Rosbash M 1991 The strength and periodicity of *Drosophila melanogaster* circadian rhythms are differentially affected by alterations in period gene expression. Neuron 6:753–766

Low-Zeddies SS, Takahashi JS 2001 Chimera analysis of the Clock mutation in mice shows that complex cellular integration determines circadian behavior. Cell 105:25–42

Martinek S, Inonog S, Manoukian AS, Young MW 2001 A role for the segment polarity gene *shaggy*/GSK-3 in the *Drosophila* circadian clock. Cell 105:769–779

McDonald MJ, Rosbash M 2001 Microarray analysis and organization of circadian gene expression in *Drosophila*. Cell 107:567–578

Price JL, Blau J, Rothenfluh-Hilfiker A, Abodeely M, Kloss B, Young MW 1998 *double-time* is a novel *Drosophila* clock gene that regulates PERIOD protein accumulation. Cell 94:83–95

Renn SC, Park JH, Rosbash M, Hall JC, Taghert PH 1999 A pdf neuropeptide gene mutation and ablation of PDF neurons each cause severe abnormalities of behavioral circadian rhythms in *Drosophila*. Cell 99:791–802

Rothenfluh A, Abodeely M, Young MW 2000 Short-period mutations of *per* affect a *double-time*-dependent step in the *Drosophila* circadian clock. Curr Biol 10:1399–1402

Rutila JE, Zeng H, Le M, Curtin KD, Hall JC, Rosbash M 1996 The *tim^{SL}* mutant of the *Drosophila* rhythm gene *timeless* manifests allele-specific interactions with *period* gene mutants. Neuron 17:921–929

Rutila JE, Suri V, Le M, So WV, Rosbash M, Hall JC 1998 CYCLE is a second bHLH-PAS protein essential for circadian transcription of *Drosophila period* and *timeless*. Cell 93:805–814

So WV, Rosbash M 1997 Post-transcriptional regulation contributes to *Drosophila* clock gene mRNA cycling. EMBO J 16:7146–7155

Ueda HR, Matsumoto A, Kawamura M, Iino M, Tanimura T, Hashimoto S 2002 Genome-wide transcriptional orchestration of circadian rhythms in *Drosophila*. J Biol Chem 277:14048–14052

Wang GK, Ousley A, Darlington TK et al 2001 Regulation of the cycling of timeless (tim) RNA. J Neurobiol 47:161–175

Yang Z, Sehgal A 2001 Role of molecular oscillations in generating behavioral rhythms in Drosophila. Neuron 29:453–467

Yu Q, Jacquier AC, Citri Y, Hamblen M, Hall JC, Rosbash M 1987 Molecular mapping of point mutations in the period gene that stop or speed up biological clocks in *Drosophila melanogaster*. Proc Natl Acad Sci USA 84:784–788

Zehring WA, Wheeler DA, Reddy P et al 1984 P-element transformation with *period* locus DNA restores rhythmicity to mutant, arrhythmic *Drosophila melanogaster*. Cell 39:369–376

Zeng H, Hardin PE, Rosbash M 1994 Constitutive overexpression of the *Drosophila period* protein inhibits *period* mRNA cycling. EMBO J 13:3590–3598

DISCUSSION

Weitz: I have a question regarding the ectopic clocks. Is PDF driven in those cells, or are there any other PDF-like transcripts?

Rosbash: I mentioned in passing that when you double stain for PDF, it is restricted to its original homes. There is no PDF expression in any of these ectopic locations. Therefore we don't know what connects from these locations to behavioural outputs.

Kyriacou: The implications of that experiment with *dClock* are that in the more primitive insects, where the oscillators are located more out in the optic lobes, *dClock* may have had a direct effect on placing those cells there, in the same way that homeotic master control genes can put *Drosophila* eyes anywhere.

Rosbash: That is the implication. I think a prediction and a line of experimentation is called for that is simply to look at clock expression in development in different animals.

Sehgal: What you find for *Clock* rescue by *pdf* — that it is really weak — is also true for *per* and *tim*.

Rosbash: That is correct. The point here is that expression in the lateral neurons (by other people's experiments) is insufficient. So there is nothing weird about that *Clk* result. What hadn't been done before is the *cry* rescue, which is much better. My guess would be that if we use *per* or *tim*, we would get the same result.

Sehgal: What you call 'ectopic *Clk*' expression is not necessarily ectopic *Clk*. We don't know what the expression pattern of *Clk* is. It is ectopic on the basis of the fact that it is not in the subsets of cells that you think are circadian-relevant.

Rosbash: To put your question another way, is the rest of the brain really negative for *Clk* expression?

Sehgal: Or even for *per* and *tim* expression.

Rosbash: In the 10 years that this has been described, no one has seen PER or TIM protein or mRNA expression outside of these five centres. When you use *pergal4* or *timgal4* drivers, it is indeed present elsewhere.

Sehgal: It is everywhere.

Rosbash: It is not everywhere. It is in a lot of other places. These places don't correspond to these five centres. I'm sceptical that these are locations of *bona fide Clk* gene expression.

Sehgal: According to some people, the *per* and *timgal4* drivers are markers for all the neurons, or maybe even all the cells in the adult. I agree that they are not found everywhere, but the expression is widespread. One explanation is that the drivers are promiscuous. The other is that there are low levels of expression in these places that we don't detect any other way. I would think that this is kind of supported by some of the *doubletime* data, where, in a *doubletime* mutant we get *per* expression in ectopic locations. This would argue that *per* is synthesized in a lot of locations where you normally don't detect it because it is destabilized by *doubletime*.

Young: This could be an amplifier.

Sehgal: That is what I am thinking.

Rosbash: This is part and parcel of the question as to whether there is something special about the cells which now appear in this experiment to have properties which allow them to show cycling clock gene expression, and even connect up with behaviour. This is entirely possible. One of these properties could be some low level of clock gene expression which is there normally.

Stanewsky: In addition to these five groups of cells there are all the glial cells in the brain.

Rosbash: Definitely; I was just referring to neurons.

Stanewsky: Do you know what type of cells these are? John Ewer showed that mosaics which only have glial expression can rescue behaviour. It could be that using this promoter you get up-regulation in glial cell expression.

Rosbash: To put it more generally, if you do a misexpression experiment and get a behavioural consequence with this vector, it is impossible to distinguish whether it is because of up-regulation in the usual locations or because of ectopic expression in the new locations. This is due to the fact that the construct also expresses more heavily in the traditional clock cells. Therefore it could be the up-regulation in the traditional clock cells that leads to the behavioural consequences.

Stanewsky: Did you do a double stain against ELAV (this is a protein expressed in all neurons), for example?

Rosbash: No, but you can see that many of these cells are gigantic, i.e. almost certainly neurons.

Hardin: Have you used either a *per* or *timgal4* driver to do the rescue with *Clk*?

Rosbash: This was the first thing we tried and they are both lethal. And we don't get any aberrant behaviour with *pdfgal4* or *crygal4*, which both over-express in those lateral neurons.

Young: Your interpretation of the *cry* versus *pdf* promoter is that it is a cell-type pattern difference. But CRY is going to oscillate with the same phase as CLK in wild-type flies. Is this necessarily a cell-type difference, or could it be the fact that you are supplying a cycling promoter?

Rosbash: We assume that when we use GAL4, because of its stability, there is little or no cycling left because that protein is so stable. Without an unstable protein it doesn't matter what happens at the RNA level. Secondly, I should have made the point that swapping the promoters on *Clk* has only very modest behavioural consequences. The general sense and feel is that fooling around with promoters doesn't do a great deal, at least from the behavioural point of view. The cell-type issue is legitimate, and although it is very hard to draw a relationship between the cell types and behaviour, I would be shocked if the promoters were doing much here.

Young: We have used the same argument about GAL4. You can make sure by using another route, but that is probably it.

Sassone-Corsi: I was interested in the *Clk* recessive allele. I was wondering about the expression profile. You mentioned some alternative splicing.

Rosbash: One splice doesn't occur, but there is probably some re-initiation of something to provide a little bit of transcript that gives rise to either a bit of normal protein or almost normal protein. There are two reasons for drawing this conclusion. One is that by PCR analysis of the transcripts there is still some stuff there that could give rise to normal protein. Second, there is low but not super-low *per* and *tim* expression, and there is still weak amplitude RNA cycling. In other words, these proteins are made in greatly increased amounts compared with the dominant *Jrk* mutation. We infer that the amplitude of cycling is so low relative to wild-type that the flies are behaviourally arrhythmic. The period is only marginally affected. By and large this is an amplitude mutant and not a period mutant.

Sassone-Corsi: Is the alternative splicing present in the same cells?

Rosbash: Who knows. No one has got good *Clk in situ* hybridization so far.

Van Gelder: I'd like to switch to discussing the microarray experiments. There is a question about the relatively low concordance between the different groups who have done microarrays in what is a very similar experimental paradigm. 10% of these genes show up on multiple lists but that majority show up on only a single list. There are several possible reasons for this: statistics, how one decides whether something is oscillating or not, environmental conditions and day 1 DD versus day 3 DD. But this is a really important question because at a genomic level, what does the clock do? Is transcription the output or is it not?

Rosbash: You have touched on two different issues. I would leave aside the question of whether transcription is the output. Let's talk about why the numbers are so different. We have thought a lot about this. One big factor is indeed the method of analysis. What we have done is taken our method of analysis. We had access to Mike Young's raw data and we carried out our method of analysis on both data sets. We also got Straume's method of analysis. The different methods of analysis gave completely different results on the same data set. Then we tried one method of analysis on two data sets, but we still got different results. In other words, we still don't know why the conclusions are so different, but it is at least due to two differences between laboratories.

Van Gelder: I have a problem with the data you showed. There were genes with 30% peak to trough amplitude by RT-PCR. There is no way that you can distinguish less than a twofold change with any reliability on RT-PCR.

Rosbash: I disagree. You can pick out any one graph or gene and we can have a lengthy discussion, but I don't agree with that characterization. Most of the curves are very similar with the two methods. In other words, with real-time analysis and internal standard curves that is just incorrect. But if 70% of the data depend on 30% amplitudes, then things are indeed dicey. In our paper we set a minimal criterion of 50% amplitude change. The majority of the things we have tested with RT-PCR are greater than this.

Van Gelder: Mike Young's group validated their targets by Northern blot or RNAse protection. Yet those still don't show up in your list.

Rosbash: Our methods were too stringent.

Weitz: They had different methods for selecting waveform; it is not just a matter of stringency.

Rosbash: I would have posed the question a slightly different way at the outset. Everyone has different numbers and the overlap is small. You could break down the question even further: do we have large numbers of false negatives, or large numbers of false positives?

Van Gelder: We know that we don't have that many false negatives because we all found the canonical cycling genes. If the specificity had been very low we wouldn't have recovered those genes.

Rosbash: That is a little facile. We found them all too. In fact, in attempting to get the methods to do this we changed the parameters so that we could find them all. You in fact didn't find them all.

Van Gelder: Actually, we did find them all with the exception of *Takeout*.

Rosbash: Nor did Mike Young find them all.

Young: The list we gave had a particular cut-off, and they were recorded further down the list.

Rosbash: This gets very complicated. How far down the list do you go? This is a work in progress and most people are continuing to do other kinds of experiments.

Young: You can ask not only whether you have oscillation, but also whether the phase fits what you see in your microarray. And you can use the mutants to gain insight into the problem. You can ask what the response of the mutants is. As you gather additional pieces of information, your case gets either stronger or weaker.

Van Gelder: The reason I bring this up gets back to this question of output. How much cycling gene expression is facultative, and how much is mandatory for the organism? Our feeling from doing the analysis and doing the overlap analysis from the other groups' data is that there is only a core set of 25–30 genes that show up repeatedly as oscillating. These appear to be necessarily oscillating: the remainder are either false positives or facultatively oscillating genes that in one particular set-up will show oscillation and in another will disappear.

Rosbash: I think that will be a minor part of the explanation.

Van Gelder: I have a second point. The one thing we have all found, which is stunning and sometimes gets lost in the discussion of the oscillating genes, is how many genes there are whose constitutive level of expression is markedly affected by the clock genes. We don't have an explanation for this. We found over 400 genes showing markedly different basal levels of expression between wild-type and *per*[0] strains; and the statistical significance of these effects is phenomenal.

Rosbash: That has been brought up at this meeting several times. But there are simple explanations: these could be RNAs with long half-lives.

Van Gelder: If you are asking from both an evolutionary and genomic perspective what the clock genes do, and we took the naïve approach that we didn't really understand that clock genes oscillated, we would do chip experiments and conclude that clock genes are master regulators of static gene expression levels of a huge array of functions. We need to challenge the output model that the sole or major function of these genes is to drive transcriptional rhythmicity, and to also consider the possibility that one of the things that they are doing is setting static levels of gene expression and varying them with environmental conditions.

Menaker: You are focused on methods of analysis here. But it seems to me that if you are going to try to get concordance among experiments from different laboratories, you really have to worry about the conditions of the experiments. Michael Rosbash showed us nice data indicating that damped oscillations can grow in constant darkness in some of the cells. If you take this seriously, and you are trying to compare data sets that were generated after 1 d DD with those generated after 3 d DD, there is a problem.

Rosbash: I don't think of that initial response as damping. I think those cells freak out because of the lack of light cues at the appropriate time. There are enough data from the different groups for us to take comparable data and for this no longer to be a confounding feature.

Kyriacou: Would you then apply these 30 core genes to the mammalian work, where the correspondence is actually much greater among the various studies?

Rosbash: Liver gene expression is a much simpler situation than the fly head.

The role of phosphorylation and degradation of hPER protein oscillation in normal human fibroblasts

Koyomi Miyazaki*, Miho Mezaki* and Norio Ishida*†‡[1]

Clock Cell Biology, National Institute of Advanced Industrial Science and Technology (AIST), IMCB 6-5, 1-1-1, Higashi, Tsukuba, 305-8566, †Department of Biomolecular Engineering, Tokyo Institute of Technology, 4259 Nagatsuta, Yokohama, 226-8501 and ‡Department of Applied Biological Chemistry, University of Tsukuba, Tsukuba, Japan

Abstract. The circadian expression in *Drosophila* of clock gene products, such as PER and TIM, is thought to be important for driving overt rhythms. The constitutive expression of *per* by the heat-shock or rhodopsin promoters restores rhythmicity of the null allele of *per*, suggesting that *per* mRNA cycling may not be required for protein cycling or for locomotor rhythms. Furthermore, the constitutive expression of *tim* mRNA also supports protein cycling and behavioural rhythms in *tim* mutant flies. Other reports have also shown that eliminating the oscillations of PER and TIM proteins by their over-expression abrogated circadian rhythmicity. These data indicate that the circadian rhythmic expression of PER and TIM proteins is also important like their rhythmic mRNA expression in *Drosophila*. To compare the molecular mechanism of circadian clocks in divergent species, we report here cloning circadian mRNA and protein expression profiling of human clock genes in normal human fibroblasts. Circadian oscillations of *hPer1*, *hPer2*, *hPer3*, *hBMAL1* and *hCry2* mRNA expression were observed in serum-stimulated normal human fibroblasts. The serum shock of human fibroblasts also caused daily oscillations in the amount and size of human PER proteins as was shown using our novel antibodies. Inhibitor studies indicate that phosphorylation and degradation of PER proteins is an important process in the human molecular clock.

2003 Molecular clocks and light signalling. Wiley, Chichester (Novartis Foundation Symposium 253) p 238–249

The behaviour and physiology of most organisms shows circadian, 24 h rhythmicity. Negative feedback loops in clock genes are thought to control circadian oscillators in all organisms from bacteria to mammals (Dunlap 1998). Mammalian clock regulating genes are involved in a negative autoregulatory feedback loop that underlies overt rhythm generation (Ishida et al 1999).

---

[1]This paper was presented at the symposium by Norio Ishida, to whom correspondence should be addressed.

Oscillating molecules that control their own expression in a circadian fashion seem to be very important for generating circadian rhythms.

The circadian expression of clock gene products such as PER and TIM in *Drosophila* is thought to be important for driving overt rhythms. Several reports showed that the constitutive expression of *per* or *tim* restores locomotor rhythm and protein cycling of the null allele of these genes (Ishida et al 2001). It has been shown that eliminating the oscillations of PER and TIM proteins by their over-expression abrogates circadian rhythmicity (Yang & Sehgal 2001). These data indicate that the circadian rhythmic expression of PER and TIM proteins is also important like their rhythmic mRNA expression for maintenance of locomotor rhythm in *Drosophila* (Fig. 1).

Post-translational modification, including phosphorylation and protein degradation of clock gene products, underlies the mechanism of circadian rhythm generation in *Drosophila*. The *doubletime* (*dbt*) gene product phosphorylates PER and causes protein degradation (Kloss et al 1998, Price et al 1998). *Drosophila* TIM is degraded by a photic entrainment cue. Proteasome inhibitors block tyrosine phosphorylation-dependent dTIM degradation *in vitro* (Naidoo et al 1999). Thus, dTIM is degraded through the ubiquitin–proteasome pathway. A new clock gene, *shaggy*/glycogen synthase kinase 3 (GSK3) might also play a role in TIM phosphorylation (Martinek et al 2001). Data suggest that *shaggy*-dependent TIM phosphorylation increases PER/TIM heterodimerization or promotes the nuclear translocation of PER/TIM complexes in wild-type flies.

The phosphorylation and degradation mechanism should also be a critical regulation step of rhythm generation in mammals. Positional cloning has revealed that the *tau* locus (which shortens circadian rhythm) in hamsters is encoded by casein kinase Iε (CK1ε) (Lowrey et al 2000), a homologue of *dbt*. CK1ε phosphorylates PER1, PER2 and PER3, then renders them unstable (Camacho et al 2001). Recent findings indicate that the human PER2 site phosphorylated by CK1ε is mutated in familial advanced-sleep-phase syndrome (Toh et al 2001), which affects individuals who are of the 'morning type' with a 4 h advance of sleep, body temperature and melatonin rhythms. A recent mouse liver study showed that mPER1, mPER2, CLOCK and BMAL1 undergo circadian changes in terms of phosphorylation and abundance, which may play a role in maintenance of the rodent clock (Lee et al 2001). Co-overexpressed mPER1 and mPER3 with CKIε are phosphorylated and degraded by the ubiquitin–proteasome pathway, while phosphorylation in co-expressed PERs with loss-of-function CKIε is inhibited, leading to protection from degradation (Akashi et al 2002). However, the relationship between cycling of the mammalian clock protein and its post-translational modification in native systems is not yet thoroughly understood.

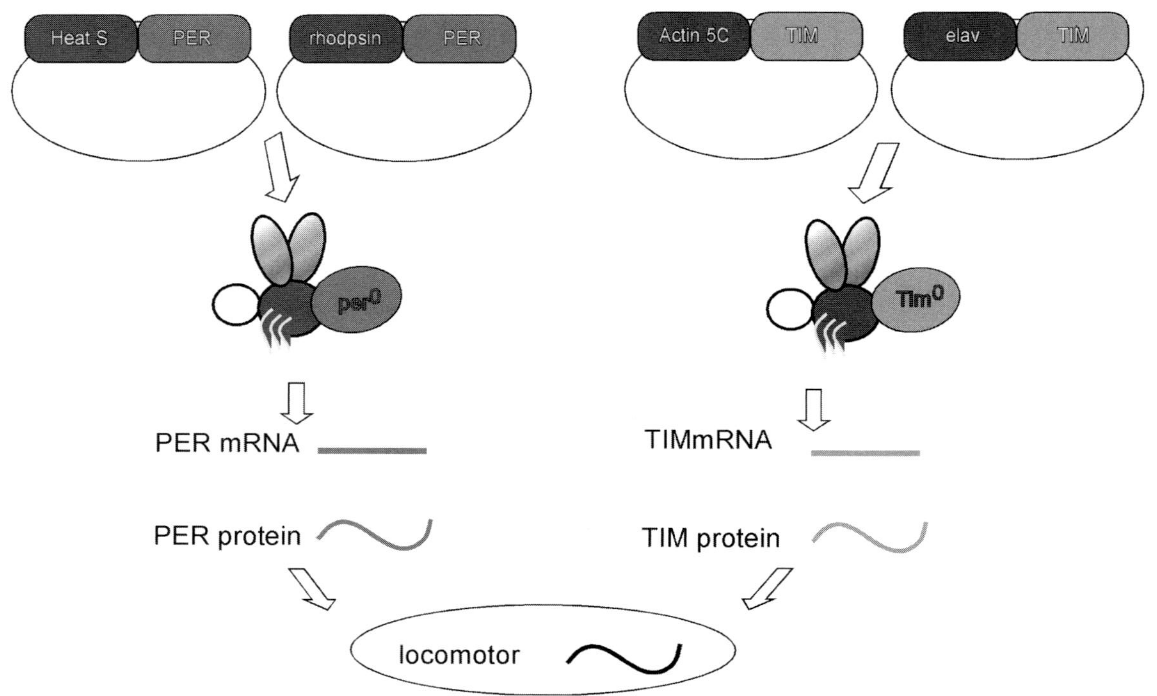

FIG. 1.　Rhythmic CLOCK protein is also important like its rhythmic mRNA expression. The circadian expression of clock gene products such as PER and TIM in *Drosophila* is thought to be important for driving overt rhythms. Several reports showed that the constitutive expression of *per* or *tim* restores locomotor rhythm and protein cycling of the null allele of *per* or *tim*, respectively (Ishida et al 2001). Eliminating the oscillations of PER and TIM proteins by over-expression abrogates circadian rhythmicity (Young & Sehgal 2001).

Familial advanced and delayed sleep phase syndromes can both be attributed to mammalian *per* mutations in a region of the PER protein, presumably a CKIε binding domain (Ebisawa et al 2001, Toh et al 2001). Because a phosphorylation disorder might cause some sleep syndromes, the establishment of a system with which to assay post-translational modifications of PER protein has recently received considerable focus.

To compare the molecular mechanisms of circadian clocks in divergent species, we cloned and analysed circadian mRNA expression of human clock genes in serum shocked fibroblasts. We also developed novel antisera and examined the temporal expression of three PER proteins (hPER1, hPER2, and hPER3). The results showed robust circadian profiles of hPER protein abundance, phosphorylation and degradation.

Results

Cloning and sequencing of human period genes

Human *PER3* (*hPER3*) cDNA encodes the predicted coding regions of 1210 amino acids. The sequence similarity of hPER3 to mouse, quail and zebrafish homologues (mPER3, qPER3 and zPER3) is 77, 34 and 38%, respectively (Fig. 2) (Delaunay et al 2000, Yoshimura et al 2000, Zylka et al 1998). The sequence of hPER3 exhibits overall identity of 39% and 38% to hPER1 and hPER2, respectively (Nagase et al 1998, Tei et al 1997). Several regions are conserved in the three predicted human PERs in overall sequences (Fig. 2). In particular, the PAS domains consisting of PAS A (residues 126–175 of hPER3) and PAS B (residues 264–316 of hPER3) motifs, which are protein–protein interaction domains with PAS-containing protein, are highly conserved. Figure 2 shows a domain structure of hPER3, including a nuclear localization signal (NLS; residues 744–752), a casein kinase binding site (residues 630–643), a nuclear export signal (NES; residues 401–413) and a CRY binding site (residues 1135–1188) predicted from sequence similarity with mammalian PER1, PER2 and PER3. These motifs are also highly conserved among mouse, quail and zebrafish PER3 (Delaunay et al 2000, Yoshimura et al 2000, Zylka et al 1998). In contrast to these conserved domains, human-specific repeats were located at residues 991–1064 of hPER3 (Fig. 2). The repetitive sequence in hPER3 has no homology with known motifs and its function remains to be elucidated.

Serum shock-induced circadian expression of clock genes in human fibroblasts

After cloning of human clock genes *PER1, PER2, PER3, TIM, CLK, BMAL1* and *CRY2,* we examined the tissue distribution of their transcripts in 14 tissues and three cell lines using RT-PCR. Although steady-state levels of the transcripts differed among tissues, these genes are expressed in all tissues including heart, brain, placenta, lung, liver, skeletal muscle, kidney, pancreas, spleen, thymus, prostate, testis, ovary and small intestine (data not shown). In cultured cells including the embryonic lung diploid fibroblast line WI-38, the immature myeloid cell line KG-1 and the epitheloid carcinoma cell line HeLa S3, we detected all clock gene transcripts that we examined. However, expression levels differed among the cell lines (data not shown). For example, less *hBMAL1* was expressed in KG-1 and HeLa S3 than in WI-38 cells. The expression of all clock genes except for *TIM* was quite low in HeLa S3 cells. Such ubiquitous expression suggests that clock genes play important roles in the circadian rhythms of many peripheral tissues.

Circadian oscillation profiles of clock genes are induced in several mammalian peripheral culture cells by serum shock (Balsalobre et al 1998). To elucidate whether serum induces the circadian expression of human clock genes in normal human diploid fibroblasts, we applied RT-PCR ELISA methods to detect RNA levels of clock genes in WI-38 cells after serum stimulation (Fig. 3). Since WI-38 cells in culture invariably undergo senescence after a finite number of doublings, we selected young WI-38 cells. The RT-PCR-ELISA data are expressed as amounts (Moles) of corresponding cDNA plasmids in 3 ng of starting total RNAs. The expression of c-*Fos* transcripts was transiently induced over 20-fold at 1 h after serum addition as expected (Fig. 3). Serum also stimulated the immediate expression of the *hPER1* and *hPER2* genes (Balsalobre et al 1998). Messenger RNA levels of these genes reached maximal levels at 1 and 4 h, respectively, and minimal levels at 12 h for both *PER* genes. Although *hPER3* was also expressed in a circadian fashion after 24 h, *hPER3* was not immediately expressed like *hPER1* and *hPER2*. The expression profile of *hCRY2* was similar to that of *hPER3*. The peak expression of *hBMAL1* transcripts was a reciprocal fluctuation compared to that of *hPER* as it is *in vivo* (Oishi et al 1998). On the other hand, the expression of *hCLK* did not significantly differ from that of *G3PDH* (data not shown).

To confirm immediate-early gene induction of *hPER1* and *hPER2*, the accumulation of *hPER*s was recorded for up to 4 h after serum shock in the presence of cycloheximide, an inhibitor of protein synthesis. Ongoing protein synthesis was not required for the immediate-early expression of *hPER1* and *hPER2* mRNAs in human fibroblasts as well as rodent fibroblasts (Balsalobre et al 1998). Moreover, the mRNA stability of *hPER1* and *hPER2* increased in CHX-treated cells. The regulatory profiles of various immediate-early genes including c-*Fos* and c-*Myc* are similar to those of *hPER1* and *hPER2* (Lau & Nathans 1987). These findings suggest that *hPER1* and *hPER2* have a role as immediate early genes. Figure 3 shows the circadian expression of three *PER* genes in serum-stimulated human WI-38 cells.

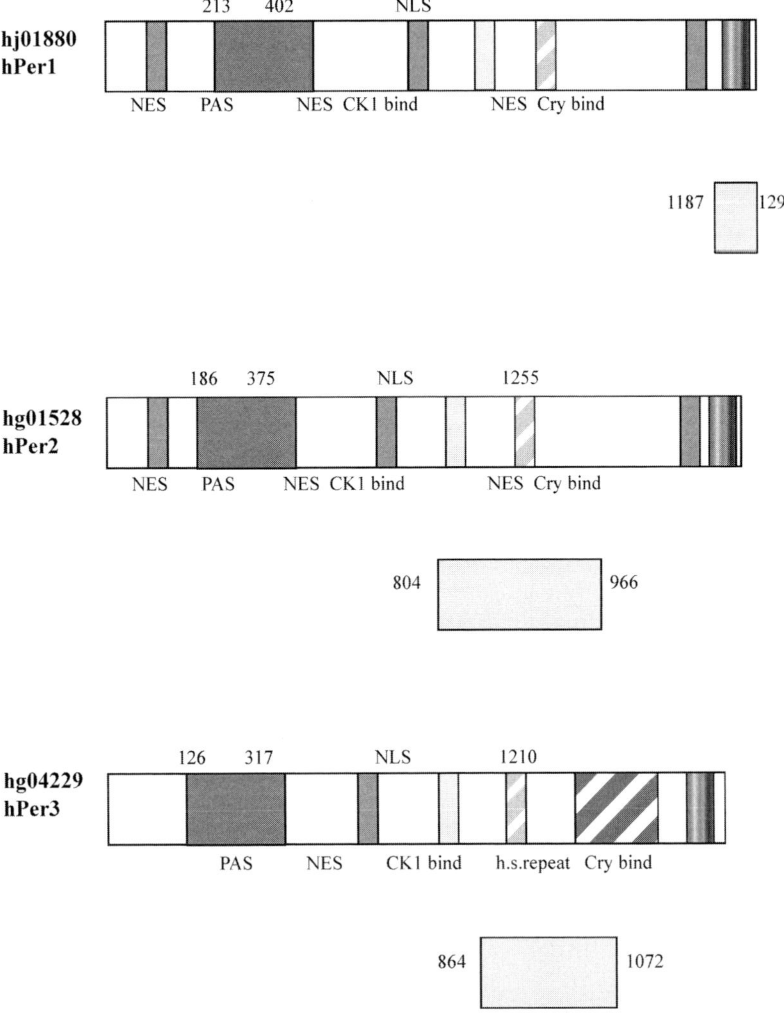

FIG. 2.   The structure of human PER proteins. Pale grey bars indicate the region of expressed protein in *E. coli* for making antibodies. NES, nuclear exporting sequence; CK1 bind, casein kinase 1 binding region; Cry bind, Cry binding region; NLS, nuclear localization sequence; PAS, period, arnt, single mind share common region; h.s., human specific.

Circadian oscillation and temporal change in apparent
size of PERIOD proteins in WI-38 cells stimulated with serum

To understand whether the circadian expression of *PER* genes leads to the expression of human PERIOD proteins (hPERs), we raised antisera against human PER1, PER2 and PER3 expressed in *Escherichia coli* (Fig. 4). Proteins were extracted from serum-stimulated WI-38 and analysed by immuno-precipitation or by Western blotting against anti-hPER antisera. We detected specific immunoreactive bands migrating at 183, 185 and 165/190 kDa. The protein accumulation profiles recorded after serum shock showed that the abundance of hPER1 and hPER2 oscillated in serum-stimulated WI-38 (Fig. 4A, top). The accumulation of hPER1 expression initially peaked at 6 h, then gradually fell for 20 h (Fig. 4A,B). After 23 h, the levels of hPER1 were not so high for the first peak but definitely increased once again to reach a second peak at 32 h (Fig. 4A,B). The accumulation of hPER2 started to increase at 3 h and reached a peak at 6 h (Fig. 4A, middle). The changes in hPER3 expression were not any more significant than those of hPER1 and hPER2 (Fig. 4A).

In addition to the oscillation of the amount of hPER1, the apparent size of hPER1 increased between 3 and 12 h from an apparent molecular mass of 188 kDa to 204 kDa (Fig. 4A). This indicates that hPER1 undergoes significant post-translational modifications as a function of time after serum stimulation. Edery et al (1994) reported that *Drosophila* PER protein in fruit fly head extracts undergoes daily oscillations in terms of apparent molecular mass as well as in abundance by phosphorylation. Furthermore mPERs in the liver are phosphorylated in a circadian manner (Lee et al 2001). To test whether the slowly migrating hPER1 in serum-stimulated cells is indeed caused by phosphorylation, we incubated the cell lysate at 12 h after serum stimulation with bacterial alkaline phosphatase (BAP). The mobility of BAP-treated hPER1 protein shifted to the level of the mock-treated sample and was essentially indistinguishable from hPER1 that initially appeared at 3 h (data not shown). This indicated that most of the time-dependent size increases in hPER1 after serum-stimulation is due to phosphorylation. The mobility change of hPER2 (apparent molecular mass) was less than that of hPER1, suggesting that phosphorylation for hPER2 is not as much as for hPER1.

The ubiquitin–proteasome pathway plays a key role in a variety of cellular processes, including cell cycle and transcriptional regulation. This pathway is also the route of dTIM and dCRY degradation (Lin et al 2001, Naidoo et al 1999). HA-tagged ubiquitin is incorporated into over-expressed exogenous mPERs and degraded on proteasomes (Akashi et al 2002, Yagita et al 2002). These data suggest that endogenous hPER1 may be also degraded through the ubiquitin–proteasome pathway after serum stimulation for protein cycling

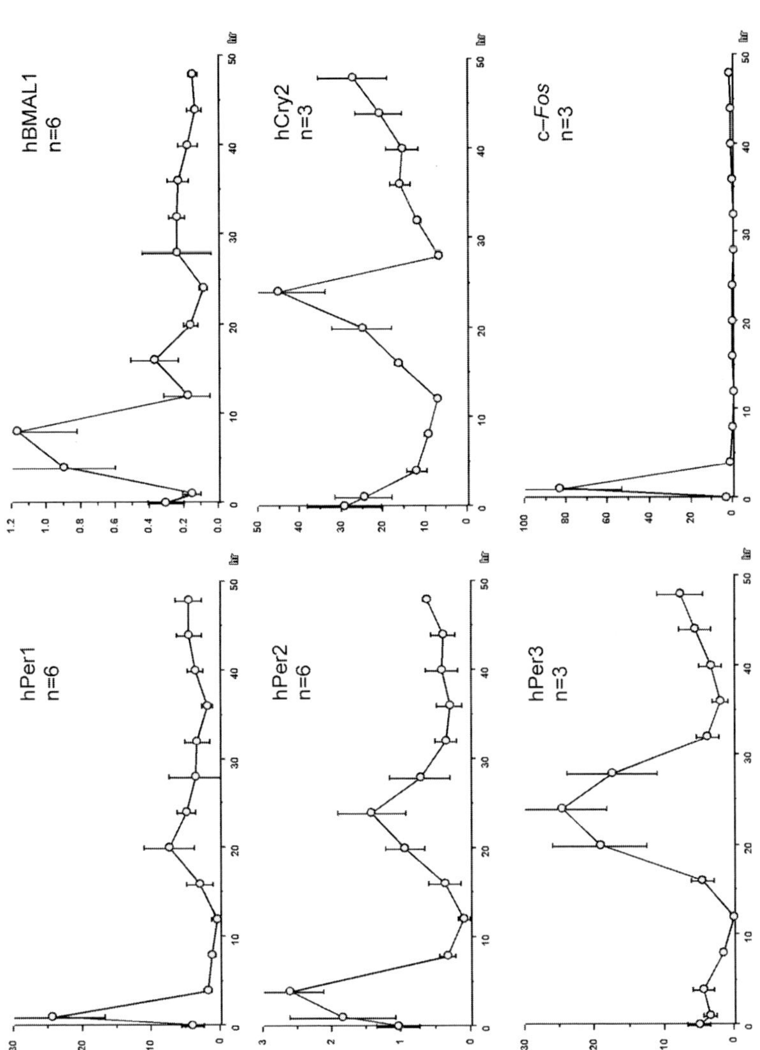

FIG. 3. Circadian expression profile and transient stimulation of human clock genes in serum-stimulated WI-38 cells. Expression of each gene was analysed by RT-PCR. Quiescent WI-38 cells were grown in BME-10+PSA to confluence then incubated in BME-0.5+PSA for 2 days. Quiescent cells were stimulated by changing the medium to BME-50+PSA for 2 h, then replacing this medium with serum-free BME+PSA. Total RNAs were prepared at various times (top of figure) after exposure to high serum concentration. Complementary DNA templates corresponded to 15 ng of starting total RNAs except that for G3PDH (1.5 ng). PCR products were quantified by ELISA. Each value is mean ± SD of two to six independent PCR experiments. Relative RNA levels were determined from ratios of each template cDNA used in PCR. Data from RT-PCR ELISA are expressed as amounts (moles) of corresponding cDNA plasmids in 3 ng of starting total RNAs.

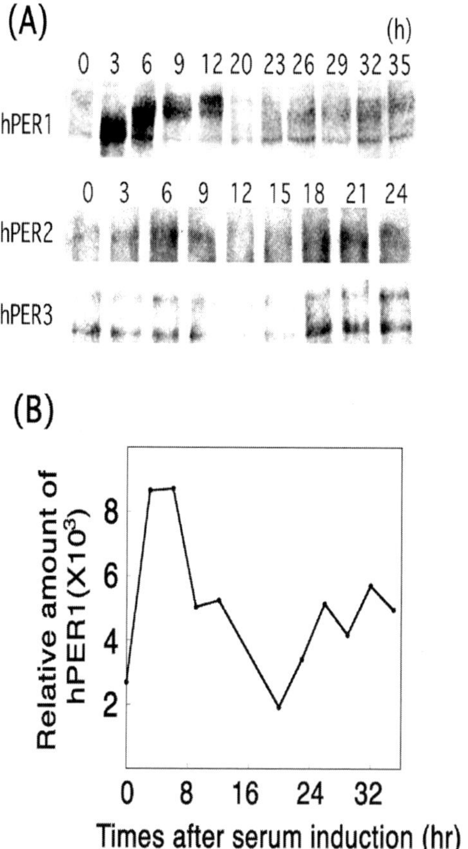

FIG. 4.   Temporal change of hPER1 accumulation in serum-stimulated WI-38 cells. (A) WI-38 cells were grown to confluence then starved of serum for 24 h. Cells were washed and shifted to a medium containing 50% calf serum. At various times (top of figure) after serum stimulation, cells were lysed and Western blotted (hPER1) or immunoprecipitated (hPER2 and hPER3). Lanes show 70 μg of extracted proteins. (B) Signals obtained in Western blotting shown in panel (A) for hPER1 protein were quantified and are indicated as a relative amount to the value at 0 h.

(Fig. 5). The phosphorylation and proteasome degradation of circadian clock proteins may play an important role in maintaining the circadian clock even in humans.

Discussion

Familial advanced (Toh et al 2001) and delayed (Ebisawa et al 2001) sleep phase syndrome is caused by disorders of the molecular circadian clock. To verify the

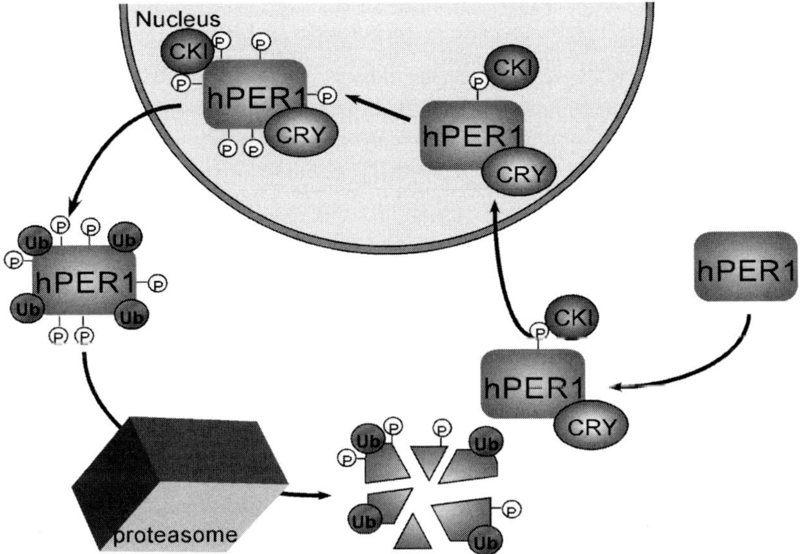

FIG. 5. Role of phosphorylation and degradation of clock protein PER in human normal fibroblasts. After the phosphorylation of hPER1 by casein kinase, the ubiquitin–proteosome pathway may be involved in its degradation in human cells.

molecular mechanism of human sleep disorders, an *in vitro* system is required with which to assay the human molecular clock. We developed a system for evaluating the post-translational modification of human clock molecules using normal diploid WI-38 cells, which are derived from human foreskin. Thus, disorders in post-translational regulation systems can be examined using skin fibroblasts cultured from patients. Since PER protein does not show circadian oscillation in aged WI-38 cells that have exceeded 40 population doublings (K. Miyazaki and N. Ishida, unpublished data), this could become a good model of insensitivity to circadian rhythms in aged people. Further investigation of post-translational regulating systems in human clock protein will lead to the understanding of rhythm disorders and their associated molecular mechanisms.

References

Akashi M, Tsuchiya Y, Yoshino T, Nishida E 2002 Control of intracellular dynamics of mammalian period proteins by casein kinase I epsilon CKIepsilon and CKIdelta in cultured cells. Mol Cell Biol 22:1693–1703

Balsalobre A, Damiola F, Schibler U 1998 A serum shock induces circadian gene expression in mammalian tissue culture cells. Cell 93:929–937

Camacho F, Cilio M, Guo Y et al 2001 Human casein kinase Idelta phosphorylation of human circadian clock proteins period 1 and 2. FEBS Lett 489:159–165

Delaunay F, Thisse C, Marchand O, Laudet V, Thisse B 2000 An inherited functional circadian clock in zebrafish embryos. Science 289:297–300

Dunlap J 1998 Circadian rhythms. An end in the beginning. Science 280:1548–1549

Ebisawa T, Uchiyama M, Kajimura N et al 2001 Association of structural polymorphisms in the human period3 gene with delayed sleep phase syndrome. EMBO Rep 2:342–346

Edery I, Zwiebel LJ, Dembinska ME, Rosbash M 1994 Temporal phosphorylation of the *Drosophila* period protein. Proc Natl Acad Sci USA 91:2260–2264

Ishida N, Kaneko M, Allada R 1999 Biological clocks. Proc Natl Acad Sci USA 96: 8819–8820

Ishida N, Miyazaki K, Sakai T 2001 Circadian rhythm biochemistry: from protein degradation to sleep and mating. Biochem Biophys Res Commun 286:1–5

Kloss B, Price JL, Saez L et al 1998 The *Drosophila* clock gene double-time encodes a protein closely related to human casein kinase Iepsilon. Cell 94:97–107

Lau LF, Nathans D 1987 Expression of a set of growth-related immediate early genes in BALB/ c 3T3 cells: coordinate regulation with c-fos or c-myc. Proc Natl Acad Sci USA 84: 1182–1186

Lee C, Etchegaray JP, Cagampang FR, Loudon AS, Reppert SM 2001 Posttranslational mechanisms regulate the mammalian circadian clock. Cell 107:855–867

Lin FJ, Song W, Meyer-Bernstein E, Naidoo N, Sehgal A 2001 Photic signaling by cryptochrome in the *Drosophila* circadian system. Mol Cell Biol 21:7287–7294

Lowrey PL, Shimomura K, Antoch MP et al 2000 Positional syntenic cloning and functional characterization of the mammalian circadian mutation tau. Science 288:483–492

Martinek S, Inonog S, Manoukian AS, Young MW 2001 A role for the segment polarity gene shaggy/GSK-3 in the *Drosophila* circadian clock. Cell 105:769–779

Nagase T, Ishikawa K, Suyama M et al 1998 Prediction of the coding sequences of unidentified human genes. XI. The complete sequences of 100 new cDNA clones from brain which code for large proteins *in vitro*. DNA Res 5:277–286

Naidoo N, Song W, Hunter-Ensor M, Sehgal A 1999 A role for the proteasome in the light response of the timeless clock protein. Science 285:1737–1741

Oishi K, Sakamoto K, Okada T, Nagase T, Ishida N 1998 Antiphase circadian expression between BMAL1 and period homologue mRNA in the suprachiasmatic nucleus and peripheral tissues of rats. Biochem Biophys Res Commun 253:199–203

Price JL, Blau J, Rothenfluh A, Abodeely M, Kloss B, Young MW 1998 double-time is a novel *Drosophila* clock gene that regulates PERIOD protein accumulation. Cell 94: 83–95

Tei H, Okamura H, Shigeyoshi Y et al 1997 Circadian oscillation of a mammalian homologue of the *Drosophila* period gene. Nature 389:512–516

Toh KL, Jones CR, He Y et al 2001 An hPer2 phosphorylation site mutation in familial advanced sleep phase syndrome. Science 291:1040–1043

Yagita K, Tamanini F, Yasuda M, Hoeijmakers JH, van Der Horst GT, Okamura H 2002 Nucleocytoplasmic shuttling and mCRY-dependent inhibition of ubiquitylation of the mPER2 clock protein. EMBO J 21:1301–1314

Yang Z, Sehgal A 2001 Role of molecular oscillations in generating behavioral rhythms in *Drosophila*. Neuron 29:453–467

Yoshimura T, Suzuki Y, Makino E et al 2000 Molecular analysis of avian circadian clock genes. Brain Res Mol Brain Res 78:207–215

Zylka MJ, Shearman LP, Weaver DR, Reppert SM 1998 Three period homologs in mammals: differential light responses in the suprachiasmatic circadian clock and oscillating transcripts outside of brain. Neuron 20:1103–1110

DISCUSSION

Hastings: If I understood correctly, this is opposite from what I would have expected on the basis of the *tau* mutant hamster. If you are suggesting that phosphorylation promotes degradation, how would you suggest that hyper-phosphorylation leads to a longer period phenotype?

Ishida: It is possible in the case of the *tau* mutants that there is a mutation in a serine residue of the PER2 protein, which is considered to be a targeted sequence for CK1ε. It causes phase-advanced-syndrome in humans. This is clearly affected by mutation of the serine. First, our case is hPER1 instead of hPER2. Thus, I cannot compare both events directly. Furthermore, our inhibitor is targeted mainly to CK1ε and δ, so we can't conclude from these experiments which one of them (or indeed both) is important for this phosphorylation.

Young: I was going to ask whether both of the kinases were hit by your inhibitor. If they are, then it might be the case that one class of phosphorylated protein is stabilized but an alternatively phosphorylated protein isn't.

Regulation of daily locomotor activity and sleep by hypothalamic EGF receptor signalling[1]

Achim Kramer*, Fu-Chia Yang*[2], Pamela Snodgrass†[2], Xiaodong Li†[2],
Thomas E. Scammell‡, Fred C. Davis†, Charles J. Weitz*[3]

*Department of Neurobiology, Harvard Medical School, Boston, MA 02115, †Department of
Biology, Northeastern University, Boston, MA 02115, and ‡Department of Neurology, Beth
Israel Deaconess Medical Center and Harvard Medical School, Boston, MA 02115, USA

Abstract. The circadian clock in the suprachiasmatic nucleus (SCN) is thought to drive
daily rhythms of behaviour by secreting factors that act locally within the hypothalamus.
In a systematic screen, we identified transforming growth factor (TGF)α as a likely SCN
inhibitor of locomotion. TGFα is expressed rhythmically in the SCN, and when infused
into the 3rd ventricle it reversibly inhibits locomotor activity and disrupts circadian
sleep–wake cycles. These actions are mediated by epidermal growth factor (EGF)
receptors, which we identified on neurons in the hypothalamic subparaventricular zone.
Mice with a hypomorphic EGF receptor mutation exhibit excessive daytime locomotor
activity and fail to suppress activity when exposed to light. These results implicate EGF
receptor signalling in the daily control of locomotor activity, and they identify a neural
circuit in the hypothalamus that likely mediates the regulation of behaviour both by the
SCN and the retina.

*2003 Molecular clocks and light signalling. Wiley, Chichester (Novartis Foundation Symposium
253) p 250–266*

Circadian rhythms of behaviour in mammals are robust and precise. For example,
in constant darkness and temperature, the circadian rhythm of locomotor activity
in laboratory rodents persists indefinitely (Pittendrigh 1993) and is accurate to
within a few minutes per day (Pittendrigh & Daan 1976a, Vitaterna et al 1994).

---

[1]Abstracted from Kramer et al 2001 with permission from the American Association for the
Advancement of Science.
[2]These authors contributed equally.
[3]This paper was presented at the symposium by Charles J. Weitz to whom correspondence
should be addressed.

The circadian clock driving locomotor activity and other circadian behaviours, such as the sleep–wake cycle, is located within the suprachiasmatic nucleus (SCN) of the hypothalamus (Klein et al 1991).

The molecular mechanisms by which the SCN drives circadian rhythms of locomotor activity and other behaviours are unknown. Intriguing clues, however, have come from SCN transplant studies. In animals made arrhythmic by SCN lesions, SCN grafts drive circadian rhythms of locomotor activity (Ralph et al 1990), even if the grafts are encapsulated, thereby preventing extension of axons but allowing diffusion of secreted factors (Silver et al 1996). A study of 'temporal chimeras' (Vogelbaum & Menaker 1992), hamsters with functional SCN tissue of both wild-type and short-period mutant genotypes, indicated that the SCN inhibits locomotor activity at one phase and promotes it at another, inhibition apparently dominating when the two influences coincided. Together, these and related studies (Davis & Menaker 1980, Earnest et al 1999) suggest that the SCN drives circadian rhythms of locomotor activity by secreting at least one 'locomotor inhibitory factor' at one phase and at least one 'locomotor activating factor' at another. Although the effects of SCN grafts are mediated by factors secreted into the 3rd ventricle in a paracrine fashion, in the intact animal it is possible that the secreted SCN factors act synaptically (Silver et al 1996).

Transplant experiments indicate that the receptors for the secreted SCN factors are located near the 3rd ventricle (LeSauter & Silver 1998). The major projection of the SCN is to the subparaventricular zone (SPZ) (Watts & Swanson 1987), a little understood hypothalamic region flanking the 3rd ventricle. Lesions of the SPZ disrupt circadian regulation of locomotor activity (Lu et al 2001), making the SPZ the likely location of receptors for secreted SCN locomotor factors, whether synaptic or paracrine.

Under a 24 h light–dark cycle, the daily timing of locomotor activity depends on both light and the circadian clock. Light influences locomotor behaviour in two ways. First, it resets the circadian clock (Pittendrigh & Daan 1976b) via the retinohypothalamic tract (RHT), a direct projection from the retina to the SCN (Klein et al 1991) and other hypothalamic sites (Kita & Omamura 1982, Johnson et al 1988). Second, it acts acutely in an effect termed 'masking'. In nocturnal animals like hamsters and mice, light suppresses locomotor activity (Mrosovsky 1999) independently of the circadian clock, requiring neither a genetically functional clock (van der Horst et al 1999) nor an intact SCN (Redlin & Mrosovsky 1999). Nevertheless, both masking (Mrosovsky et al 1999) and circadian clock resetting (Freedman et al 1999) involve similar or identical novel photoreceptors in the inner retina, raising the possibility that masking, like clock resetting, is mediated by the RHT, which is known to make a direct projection to the SPZ (Johnson et al 1988). The molecular basis of masking is unknown.

Screen for secreted SCN 'locomotor factors'

We performed a systematic molecular and behavioural screen to identify
locomotor factors secreted by the SCN. To find secreted factors not
previously documented in the SCN, we screened a hamster SCN cDNA library
in a yeast secretion-trap system (Klein et al 1996). We then carried out a
behavioural screen in which newly identified and previously documented
(Earnest et al 1999, Miller et al 1996, Ma et al 1992) SCN factors were tested
for an effect on circadian locomotor activity by constant infusion into the 3rd
ventricle of hamsters for 2 to 3 weeks (Kramer et al 2001). In general, constant
infusion of a SCN locomotor factor should alter locomotor activity reversibly
without affecting the underlying SCN circadian clock. A locomotor inhibitory
factor, for example, should block locomotor activity for the duration of the
infusion. Because the SCN clock should not be affected, the circadian rhythm of
locomotor activity should reappear with its expected phase and period upon
cessation of the infusion. In contrast, constant infusion of SCN factors involved
only in outputs other than locomotor activity should have no effect on locomotor
behaviour.

Chronic infusions of artificial cerebrospinal fluid (aCSF) into the 3rd ventricle
had little effect on the circadian rhythm of running-wheel behaviour, causing at
most a modest reduction in overall activity without affecting the period, phase,
or precision of the rhythm (Fig. 1, upper left). Altogether, we tested 32 secreted
peptide or protein factors at least twice each (singly or in pools), of which 11 were
among the newly identified SCN factors and the rest were previously documented.
Most had little or no effect on locomotor behaviour. For example, co-infusion of
neuropeptides thought to be co-released from SCN neurons (vasoactive intestinal
polypeptide, peptide histidine–isoleucine, gastrin-releasing peptide and
neuromedin C) (Albers et al 1991) had no apparent effect on the amount or
precision of circadian running-wheel activity (Fig. 1, upper right).

One peptide behaved exactly as predicted for a SCN locomotor inhibitory
factor. Transforming growth factor α (TGFα), previously noted to be localized
in the SCN (Ma et al 1992), produced a complete blockade of running-wheel
activity during the $\sim$3-week infusion. Upon cessation of the infusion, the
running-wheel activity rhythm quickly reappeared with the expected phase and
period ($n=6$) (Fig. 1, lower left). The only known receptor for TGFα is the
epidermal growth factor receptor (EGFR), which is also activated by EGF (Lee
et al 1995). To determine whether TGFα was acting through the EGFR, we next
tested EGF, which is not detectably expressed in adult hypothalamus (unpublished
data). EGF produced virtually the same reversible blockade of running-wheel
activity as TGFα ($n=4$) (Fig. 1, lower right), implicating the EGFR as the
relevant receptor for TGFα.

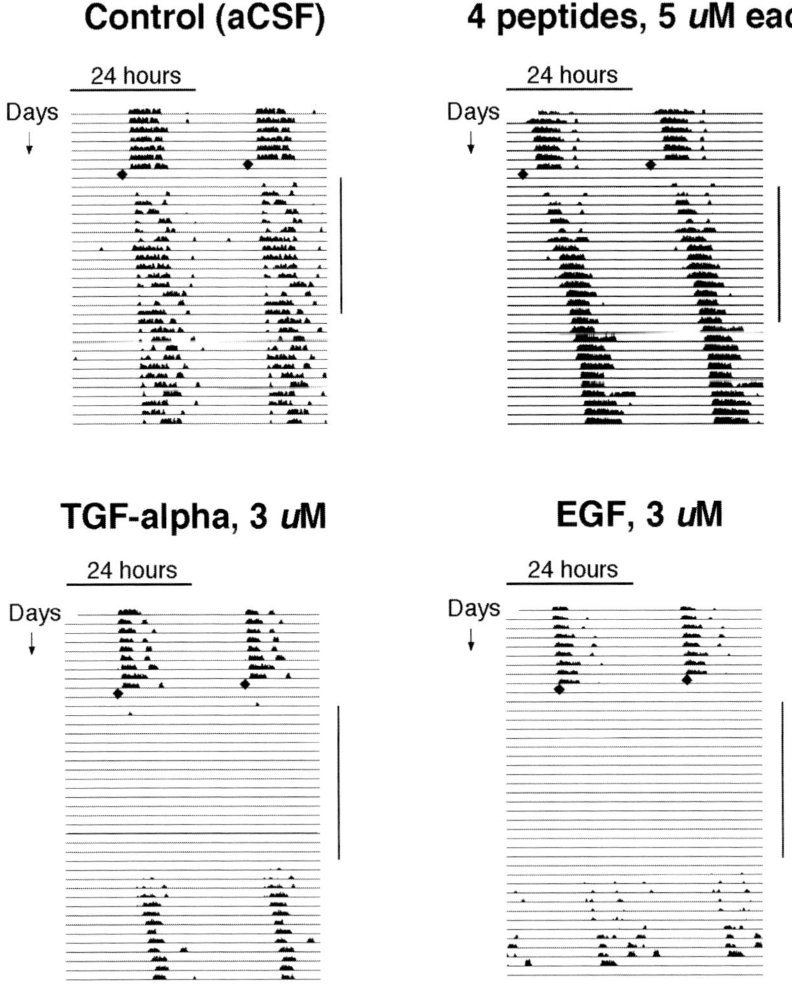

FIG. 1.   Reversible inhibition of locomotor activity by TGFα. Double-plotted running-wheel activity records of hamsters in constant dim light are examples from a behavioural screen for secreted SCN locomotor factors. Factors were infused into the 3rd ventricle (0.5 μl/h) for 2–3 weeks via an implanted cannula with an osmotic minipump, and locomotor activity was monitored. Two days are represented horizontally, and lines on the vertical axis represent successive days. Tick marks, running-wheel revolutions (>0 per 10 min bin); height of each mark, number of revolutions; closed diamond, time of cannulation; and bar at the right of the record, period of infusion. Control infusions of artificial CSF (aCSF) caused at most a modest reduction in activity, and the pool of four SCN neuropeptides produced no substantial effect. TGFα produced a reversible inhibition of locomotor activity without affecting the phase or period of the circadian clock ($n = 6$). EGF ($n = 4$) had an identical effect.

TGF*α* and EGFR in the hypothalamus

We next examined the expression of *TGFα* mRNA in the SCN at different circadian times (CT), as described (Morris et al 1998). Strong expression was detected in the SCN and piriform cortex, with somewhat weaker expression in the caudate-putamen and the supraoptic nuclei (Fig. 2A). In the SCN, *TGFα* mRNA showed a circadian rhythm (Fig. 2B) comparable in amplitude to that of *Cry1* and *Cry2* measured by the same method (Okamura et al 1999). The phase of the *TGFα* rhythm agrees with that expected for a locomotor inhibitory factor—its peak (CT6) corresponds to the time of locomotor quiescence and its trough (CT18) to a time of locomotor activity. As expected, TGFα protein was detected (Kramer et al 2001) in SCN cells (Fig. 2C).

We next performed immunohistochemistry to test whether the EGFR is localized in adult brain as predicted for receptors of SCN locomotor factors. Scattered immunoreactive cells were detected in multiple areas, as reported (Gomez-Pinilla et al 1988), but in addition we found a dense concentration in the hypothalamus flanking the 3rd ventricle, corresponding to the SPZ, the major target field of the SCN (Fig. 2D). The majority of labelled cells had neuronal morphology when viewed at high magnification (Fig. 2E). The EGFR is thus localized as predicted by SCN transplant and lesion studies for a receptor regulating circadian locomotor activity. TGFα and its receptor, the EGFR, therefore satisfy pharmacological, temporal and anatomical predictions for playing a role in circadian inhibition of locomotor activity by the SCN.

Effect of TGF*α* on the sleep–wake cycle

To evaluate the physiological effects of TGFα in greater detail, we monitored the electroencephalogram (EEG), the electromyogram (EMG), bodily movement and body temperature of hamsters kept in constant darkness during control or TGFα 3rd ventricle infusions (Kramer et al 2001). During control infusions, episodes of waking, non-REM sleep and REM sleep were normal for hamsters (Naylor et al 1998), and the circadian component of sleep–wake regulation was evident (Fig. 3, top left). As expected, there was a circadian rhythm of bodily movement (Fig. 3, middle left), an assay which measures positional changes and exploratory behaviour, unlike running-wheel activity, which reports only strong locomotor drive. Also evident was a circadian rhythm of body temperature (Fig. 3, bottom left), which in rodents is mainly a consequence of physical activity rather than an independent rhythm (DeCoursey et al 1998).

During TGFα infusions, episodes of waking, non-REM sleep, and REM sleep were normal in appearance, amount and duration (unpublished data). Thus the blockade of running-wheel behaviour by TGFα was not due to hypersomnolence

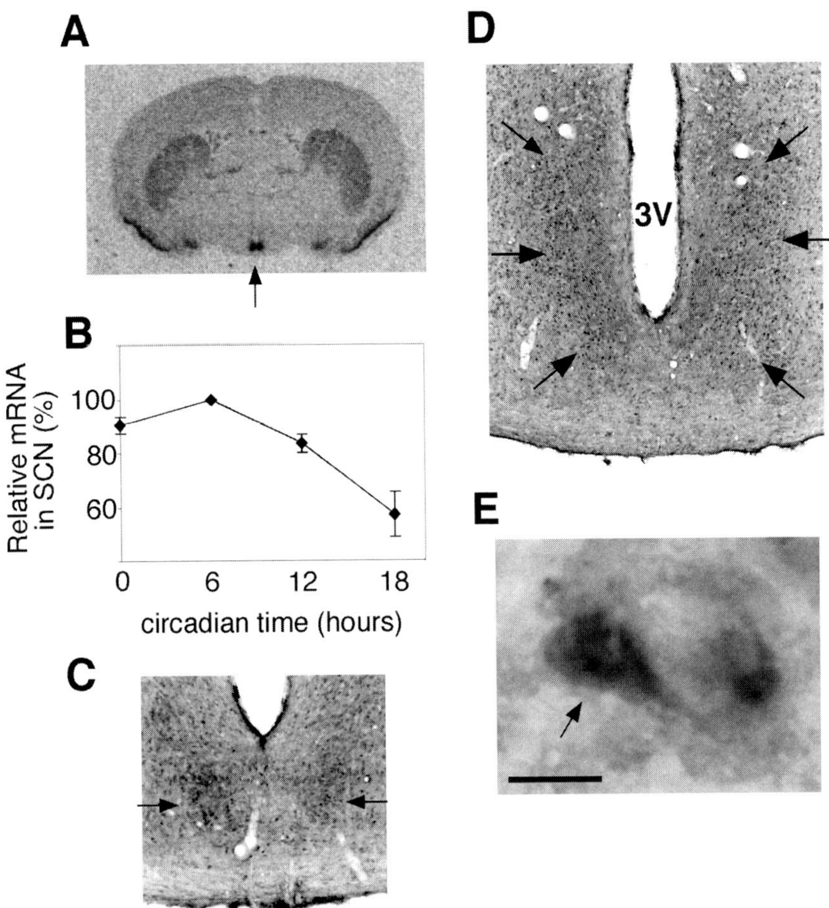

FIG. 2.   TGFα and the EGFR in the hypothalamus. (A) *In situ* hybridization showing *TGFα* mRNA in a coronal section of hamster brain obtained at CT6. Arrow, SCN. (B) Densitometric analysis of *TGFα* mRNA levels in the SCN at different circadian times, plotted relative to the maximum signal ($n = 5$ animals for each timepoint; error bars, SEM). A rhythm with the same phase and amplitude was found in animals under LD cycles (not shown). (C) Immunolabelled hamster brain section showing TGFα protein in SCN neurons. Arrows, bilateral SCN. (D) Low-magnification view of immunolabelled coronal hamster brain section showing the EGFR in neurons (dark cell bodies) in the SPZ (roughly delimited by arrows). Far fewer immunolabelled cells were seen in the lateral hypothalamus, and virtually none were seen in the thalamus (not shown). 3V, 3rd ventricle. (E) Typical neuronal morphology of EGFR-expressing cells in the SPZ. Note large triangular cell body and prominent proximal dendrite. Scale bar, 10 μm.

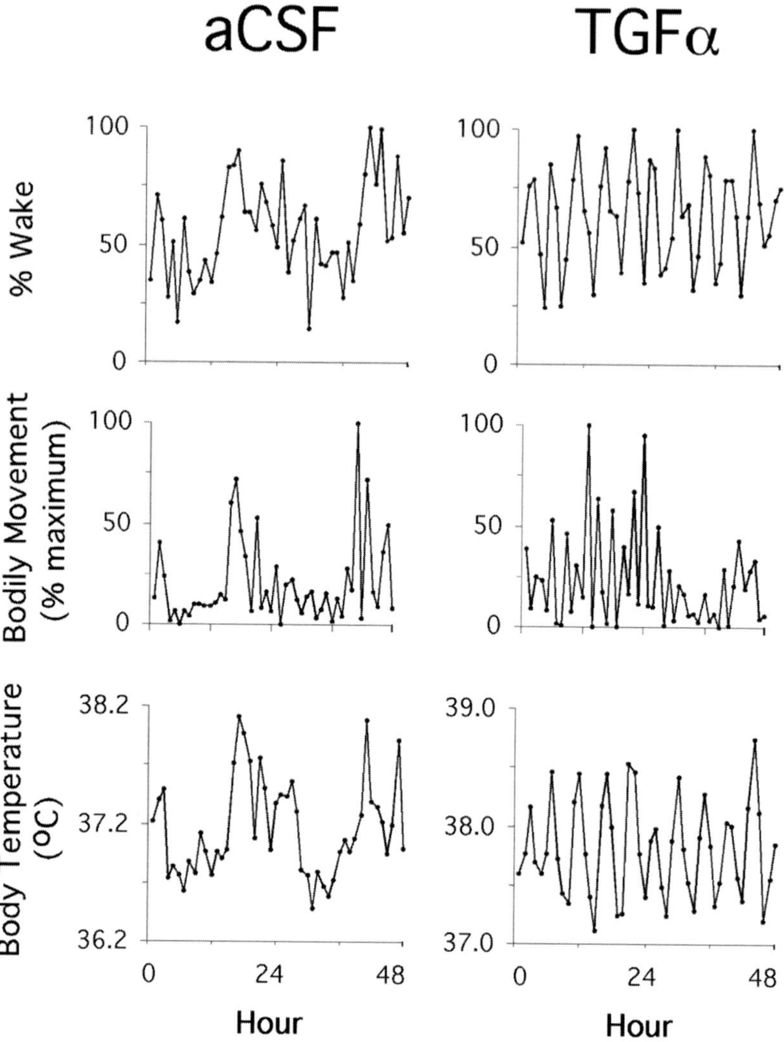

FIG. 3.   Disruption of the circadian rhythm of sleep–wake behaviour by infusion of TGFα into the 3rd ventricle. The effects were highly reproducible (controls, $n=3$; TGFα, $n=5$). Data are displayed in 1 h bins.

or other gross disturbances of cortical physiology but rather to a more discrete action. However, TGFα infusion did produce a marked alteration in the timing of the sleep–wake cycle; the normal circadian regulation was lost, and in its place appeared a highly regular and reproducible ultradian rhythm of 5 to 6 cycles per day (Fig. 3, top right). Like controls, animals infused with TGFα were physically active

(Fig. 3, middle right), indicating that the blockade of running-wheel activity was not due to a general blockade of motor function. Nonetheless, the circadian rhythm of bodily movement was disrupted or diminished (Fig. 3, middle right), suggesting that TGFα plays a role in the circadian regulation of activities less vigorous than wheel-running without acting as an inhibitor. Body temperature showed an ultradian rhythm that precisely followed the sleep–wake rhythm (Fig. 3, bottom right) with a lag of ~ 30 minutes, providing an independent measure of the ultradian physiological oscillation produced by TGFα.

The ultradian sleep–wake and temperature rhythm produced by 3rd ventricle infusion of TGFα closely resembles the effect of a focal excitotoxic lesion of SPZ neurons (Lu et al 2001). This ultradian rhythm is normally suppressed by circadian control and is disinhibited when SPZ neurons fail to relay SCN circadian information to sleep–wake circuits. Our results indicate that chronic TGFα administration uncouples SPZ neurons from sleep-regulatory circuits and that SPZ neurons expressing the EGFR transmit circadian information from the SCN to sleep–wake centres, in addition to likely regulating circadian locomotor activity.

Genetic analysis of the role of the EGFR in locomotor activity

If the EGFR mediates circadian inhibition of locomotor activity in a non-redundant manner, then mice with a loss-of-function mutation in the EGFR should exhibit excessive activity during the light period (day) in light–dark cycles and during subjective day in constant darkness (times when mice are normally quiescent). We monitored the running-wheel behaviour of mice with *waved-2*, a point mutation in the EGFR that causes an 80–95% decrease in ligand-stimulated receptor tyrosine kinase activity (Luetteke et al 1994). Unlike EGFR null mutants, which die in the early postnatal period or earlier (Threadgill et al 1995), *waved-2* mice develop into viable and essentially normal adults.

The running-wheel activity of *waved-2* mutant mice showed entrainment to a 12:12 h light–dark cycle (LD) and had an appropriate circadian period in constant darkness (DD) (Fig. 4A), indicating that the fundamental properties of the SCN circadian clock were normal. However, in light–dark cycles *waved-2* mice were abnormally active during the day compared to wild-type mice of identical genetic background, and this abnormal activity substantially degraded the precision of activity onsets at night (Fig. 4A). As expected, wild-type mice had very little daytime running-wheel activity, only 1.4 ± 0.57 (SEM)% of the total, whereas mutants had $11.5\pm4.75\%$. Heterozygotes were similar to wild-types ($3.9\pm0.72\%$), and overall there was a significant effect of genotype (ANOVA, $P<0.02$) (Kramer et al 2001). These results demonstrate that the EGFR mediates inhibition of locomotor activity, as predicted by the results of TGFα infusion, at least under light–dark cycles.

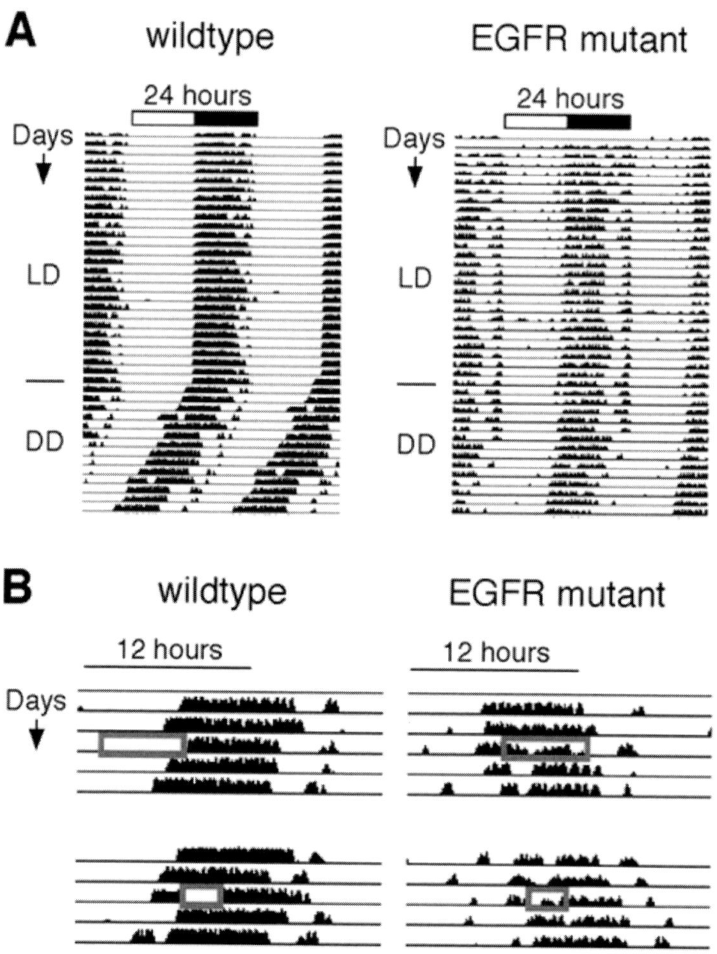

FIG. 4.   Abnormal locomotor behaviour of EGFR mutant mice (*waved-2*, a partial loss-of-function mutation in the EGFR). (A) Double-plotted running-wheel records of littermate wild-type and homozygous *waved-2* mutant mice housed under the same conditions. LD, 12:12 h light–dark cycle, represented by bar at top. Horizontal line at the left of each record marks the transition from LD into constant darkness, DD. Note excessive daytime activity in mutant. (B) Defect in the masking response to light in EGFR mutant mice. Single-plotted running-wheel records magnified to show acute responses to 3 and 6 h light pulses (represented by boxes). Records are from the same two mice as in (A), exposed together to the same light-pulses. In this example, the 6 h light pulse fell at different circadian phases for the two mice because of a small difference in their circadian periods. The phase of light-pulse administration was not a factor in the masking responses of wild-type or mutant mice.

When the mice were in constant darkness, we could detect no statistically significant difference among the genotypes in the amount of running-wheel activity during subjective day or in the distribution of activity during subjective night (although abnormalities observed in the mutants in LD cycles often appeared to persist). Thus it is possible that the EGFR does not play a role in the circadian inhibition of locomotor activity, but is somehow restricted to acting only under LD cycles. Alternatively, this partial loss-of-function mutation might not produce a strong locomotor phenotype in constant darkness because of redundancy in the circadian control of locomotor inhibition. The latter seems more likely given the broad evidence for the involvement of TGFα and the EGFR in the circadian inhibition of locomotor activity and sleep (Figs 1–3).

Why is the locomotor phenotype of *waved-2* mutants different under LD and constant darkness? To address this question, we monitored masking of running-wheel behaviour in response to 3 and 6 h light pulses during subjective night. As expected, wild-type mice showed essentially complete inhibition of running-wheel activity during the light pulses, whereas *waved-2* mutants showed little inhibition (Fig. 4B) (inhibition, 95%$\pm$1.1, 90%$\pm$1.9, and 53%$\pm$20.9 [SEM] for wild-type, heterozygous and homozygous mice, respectively; $P < 0.01$, ANOVA) (Kramer et al 2001). These results demonstrate that EGFR activity is required for normal masking responses and consequently for the proper organization of daily locomotor activity in a 24 h LD cycle, in addition to any role in the circadian regulation of locomotor activity.

Because *waved–2* mutants entrain to light–dark cycles (Fig. 4A) and show appropriate phase-shifts to light-pulses (unpublished data), the retinal photoreceptors thought to underlie both circadian phase-shifting and masking (Mrosovsky et al 1999), and the transmission of luminance information by the RHT to the hypothalamus, must be intact. Thus the defect is very likely manifested within the hypothalamus or in downstream circuits. Taken together, our results implicate EGFRs on hypothalamic SPZ neurons in the inhibitory regulation of locomotor activity, likely in response both to light and to the circadian secretion of TGFα from the SCN.

TGFα and EGF in the retina

Because EGFR signalling is required for masking (Fig. 4B), and masking does not require an intact SCN (Redlin et al 1999), the ligands for the EGFR that mediate masking must come from a source outside the SCN. If the documented projection from retinal ganglion cells to the SPZ mediates masking, then our results predict that at least a subset of retinal ganglion cells might be expected to express one or more ligands for the EGFR. TGFα and EGF are found in adult human retina, with TGFα immunoreactivity observed in Müller glia and ganglion cells and weak EGF

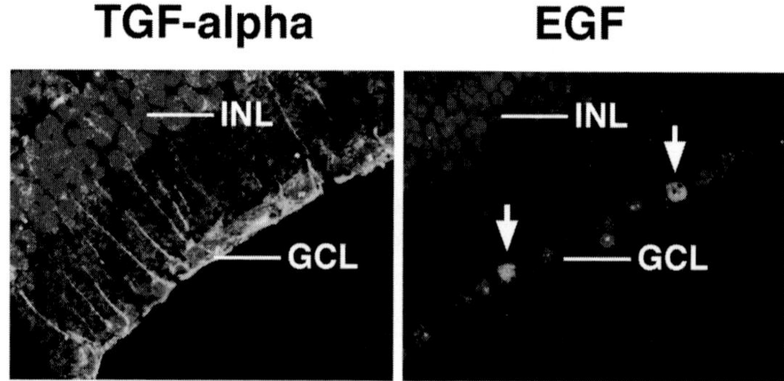

FIG. 5. TGFα and EGF in the retina. Confocal immunofluorescence images with the green channel (seen here as light grey) representing TGFα or EGF immunofluorescence, as indicated, and the red channel (seen here as dark grey) representing fluorescence from ethidium bromide, which labels all cell nuclei. INL, inner nuclear layer; GCL, ganglion cell layer. TGFα is expressed in Müller glia (parallel processes running between the INL and the GCL) and throughout the GCL. EGF is expressed in rare INL cells (not seen in this section) and in a small, widely distributed subset of GCL cells (arrows).

immunoreactivity reported throughout the retina (Patel et al 1994). To confirm and refine this view, we performed immunohistochemistry (Kramer et al 2001) for TGFα and EGF on sections from adult mouse retinas. As expected, TGFα was expressed in Müller glia and throughout the ganglion cell layer (Fig. 5, left). In contrast, EGF expression was confined to a few cells in the inner nuclear layer (not shown) and to a small, widely distributed subset of cells in the ganglion cell layer (Fig. 5, right). In number and distribution, this subset closely resembles the small subset of retinal ganglion cells that give rise to the RHT (Moore et al 1995). These results are consistent with direct regulation of the EGFR on SPZ neurons by retinal TGFα or, perhaps more likely, EGF.

Hypothalamic EGFR signalling
and the daily regulation of behavioural activity

In the nervous system, TGFα and EGFR have been implicated in diverse developmental processes, such as astrocyte differentiation and neuronal survival, but far less is known about their actions in the adult nervous system, examples of which are neural stem cell proliferation and cellular responses to brain injury (Xian & Zhou 1999). Our results strongly suggest that TGFα is a secreted SCN factor involved in the circadian regulation of locomotor activity and sleep and that EGFR signalling in SPZ neurons mediates this regulation. Genetic analysis demonstrated that EGFR activity is required for the acute inhibition of locomotor activity in response to light, in addition to a likely role in the circadian inhibition of locomotor activity.

Our results suggest that the independent regulation of behavioural activity by light and by the SCN converge upon EGFR signalling in SPZ neurons. According to this view, luminance information from photoreceptors in the inner retina is transmitted by the RHT to the SCN, where it mediates clock resetting, and to the SPZ, where it mediates masking. EGF (or TGFα) from retinal ganglion cells mediates masking by activating the EGFR on SPZ neurons, inhibiting locomotor activity. TGFα, secreted in a circadian fashion from the SCN, activates the EGFR on the same SPZ neurons, contributing to the circadian inhibition of locomotor activity. Thus regulation of behaviour by light and by the SCN can be considered as two different inputs to a single hypothalamic circuit that has evolved to regulate behaviour precisely in relation to the natural 24 h light–dark cycle.

References

Albers HE, Liou S-Y, Stopa EG, Zoeller RT 1991 Interaction of colocalized neuropeptides: functional significance in the circadian timing system. J Neurosci 11:846–851

Davis FC, Menaker M 1980 Hamsters through time's window: temporal structure of hamster locomotor rhythmicity. Am J Physiol 239:R149–R155

Decoursey PJ, Pius S, Sandlin C, Wethey D, Schull J 1998 Relationship of circadian temperature and activity rhythms in two rodent species. Physiol Behav 65:457–463

Earnest DJ, Liang FQ, Ratcliff M, Cassone VM 1999 Immortal time: circadian clock properties of rat suprachiasmatic cell lines. Science 283:693–695

Freedman MS, Lucas RJ, Soni B et al 1999 Regulation of mammalian circadian behavior by non-rod, non-cone, ocular photoreceptors. Science 284:502–504

Gomez-Pinilla F, Knauer DJ, Nieto-Sampedro 1988 Epidermal growth factor receptor immunoreactivity in rat brain. Development and cellular localization. Brain Res 438: 385–390

Johnson RF, Morin LP, Moore RY 1988 Retinohypothalamic projections in the hamster and rat demonstrated using cholera toxin. Brain Res 462:301–312

Kita H, Omamura Y 1982 An anterograde HRP study of retinal projections to the hypothalamus in the rat. Brain Res Bull 8:249–253

Klein DC, Moore RY, Reppert SM 1991 Suprachiasmatic nucleus: the mind's clock. Oxford University Press, New York

Klein RD, Gu Q, Goddard A, Rosenthal A 1996 Selection for genes encoding secreted proteins and receptors. Proc Natl Acad Sci USA 93:7108–7113

Kramer A, Yang F-C, Snodgrass P et al 2001 Regulation of daily locomotor activity and sleep by hypothalamic EGF receptor signaling. Science 294:2511–2515

Lee DC, Fenton SE, Berkowitz EA, Hissong MA 1995 Transforming growth factor alpha: expression, regulation, and biological activities. Pharmacol Rev 47:51–85

LeSauter J, Silver R 1998 Output signals of the SCN. Chronobiol Int 15:535–550

Lu J, Zhang YH, Chou TC et al 2001 Contrasting effects of ibotenate lesions of the paraventricular nucleus and subparaventricular zone on sleep–wake cycle and temperature regulation. J Neurosci 21:4864–4874

Luetteke NC, Phillips HK, Qiu TH et al 1994 The mouse waved-2 phenotype results from a point mutation in the EGF receptor tyrosine kinase. Genes Dev 8:399–413

Ma YJ, Junier MP, Costa ME, Ojeda SR 1992 Transforming growth factor-alpha gene expression in the hypothalamus is developmentally regulated and linked to sexual maturation. Neuron 9:657–670

Miller JD, Morin LP, Schwartz WJ, Moore RY 1996 New insights into the mammalian circadian clock. Sleep 19:641–667

Moore RY, Speh JC, Card JP 1995 The retinohypothalamic tract originates from a distinct subset of retinal ganglion cells. J Comp Neurol 352:351–366

Morris ME, Viswanathan N, Kuhlman S, Davis FC, Weitz CJ 1998 A screen for genes induced in the suprachiasmatic nucleus by light. Science 279:1544–1547

Mrosovsky N 1999 Masking: history, definitions, and measurement. Chronobiol Int 16: 415–429

Mrosovsky N, Foster RG, Salmon PA Thresholds for masking responses to light in three strains of retinally degenerate mice.1999 J Comp Physiol [A] 184:423–428

Naylor E, Buxton OM, Bergmann BM, Easton A, Zee PC, Turek FW 1998 Effects of aging on sleep in the golden hamster. Sleep 21:687–693

Okamura H et al 1999 Photic induction of mPer1 and mPer2 in cry-deficient mice lacking a biological clock. Science 286:2531–2534

Patel B, Hiscott P, Charteris D, Mather J, McLeod D, Boulton M 1994 Retinal and preretinal localisation of epidermal growth factor, transforming growth factor alpha, and their receptor in proliferative diabetic retinopathy. Br J Ophthalmol 78:714–718

Pittendrigh CS 1993 Temporal organization: reflections of a Darwinian clock-watcher. Annu Rev Physiol 55:17–54

Pittendrigh CS, Daan S 1976a A functional analysis of circadian pacemakers in nocturnal rodents. I. The stability and lability of spontaneous frequency. J Comp Physiol A 106:223–252

Pittendrigh CS, Daan S 1976b A functional analysis of circadian pacemakers in nocturnal rodents. IV. Entrainment: pacemaker as clock. J Comp Physiol [A] 106:291–331

Ralph MR, Foster RG, Davis FC, Menaker M 1990 Transplanted suprachiasmatic nucleus determines circadian period. Science 247:975–978

Redlin U, Mrosovsky N 1999 Masking by light in hamsters with SCN lesions. J Comp Physiol [A] 184:439–448

Silver R, LeSauter J, Tresco PA, Lehman MN 1996 A diffusible coupling signal from the transplanted suprachiasmatic nucleus controlling circadian locomotor rhythms. Nature 382:810–813

Threadgill DW, Dlugosz AA, Hansen LA et al 1995 Targeted disruption of mouse EGF receptor: effect of genetic background on mutant phenotype. Science 269:230–234

van der Horst GT, Muijtjens M, Kobayashi K et al 1999 Mammalian Cry1 and Cry2 are essential for maintenance of circadian rhythms. Nature 398:627–630

Vitaterna MH, King DP, Chang AM et al 1994 Mutagenesis and mapping of a mouse gene, Clock, essential for circadian behavior. Science 264:719–725

Vogelbaum MA, Menaker M 1992 Temporal chimeras produced by hypothalamic transplants. J Neurosci 12:3619–3627

Watts AG, Swanson LW 1987 Efferent projections of the suprachiasmatic nucleus: II. Studies using retrograde transport of fluorescent dyes and simultaneous peptide immuno-histochemistry in the rat. J Comp Neurol 258:230–252

Xian CJ, Zhou X-F 1999 Roles of transforming growth factor-alpha and related molecules in the nervous system. Mol Neurobiol 20:157–183

DISCUSSION

Loros: It looked like when you released your EGF receptor mutants into free-run they had a long period.

Weitz: That is just a typical inter-individual difference. When we look at large numbers of animals, there is no difference in period as a function of genotype.

Loros: Does the mutant have a larger standard deviation in its period?

Weitz: I don't think so. There is a fair amount of individual variability in period length. These mice had been in DD for a long time. Under those conditions there is a lot of variability. The total amount of running wheel activity, and the period and phasing of the animals is not detectably different as a function of genotype. In LD they look very different because 10% of the activity of mutant animals is during the light part of the cycle as compared with about 1% in the wild-types.

Hastings: We don't see the world through a hamster's eye, so we don't understand why they run in a wheel. This has something to do with motivation, presumably. What about motivational aspects to your pharmacological treatment? These animals are clearly healthy at the end of that treatment, so what about these other rhythmic behavioural parameters that are normally linked into this motivated behaviour of wheel running?

Weitz: I don't know about those. When we first saw the records, until you see them come back you don't know whether they are alive. We didn't have video cameras; we were just doing a simple screen. Subsequently, we worked with Tom Scammell to examine various physiological parameters during TGFα infusion. We looked at low-level activity with a minimeter system, i.e. translational movements with respect to a grid on the cage, and we looked at continuous body temperature and sleep EEGs (electroencephalograms). What we found was that there was no difference in the amounts that the animals were moving, but their sleep–wake behaviour was disturbed in an interesting way. No one knows whether the sleep control from the SCN and locomotor control share components or circuitry, or whether they are utterly divergent outputs. The suggestion was that they might share some aspect of EGFR signalling, because with the TGFα infusion the number of bouts of REM, non-REM and waking by EEG criteria were quite similar, but their circadian organization was utterly disturbed.

Hastings: Do the orxin/hypocretin neurons in the dorsomedial hypothalamus have EGFR expression?

Weitz: We don't know.

Hastings: Would you like to comment on prokineticin? This is a parallel story (Cheng et al 2002). Did this show up in your screens?

Weitz: Once we saw that paper we looked for this in our collection of secreted SCN factors, but didn't find it. It is a very interesting peptide, whatever it is, and it is probably very important. From their experiments it isn't clear whether it is a locomotor factor that acts downstream or whether it acts on the SCN. They did an acute injection. Their data showed an acute suppression of locomotor activity and then the abnormal appearance of locomotor activity during the daytime phase of the next day, which they called a 'rebound effect'. Another way to consider this 'rebound' is as a transient phase shift of the system, implying an action on the pacemaker itself.

Hastings: Then it very quickly transiently went back the other way. I would probably go with the receptor desensitization interpretation.

Weitz: It could well be. And certainly I don't expect there to be just one or two of these factors. From our screen we have a whole other class of factors not related to TGFα. They have virtually identical effects in our assays, with a complete inhibition of behaviour for the duration of the infusion, and return of the behaviour with the right phase and period at the end. One of the receptor components looks like it is around the 3rd ventricle.

Menaker: I have a suggestion for an experiment. If this is a specific SCN output factor which regulates locomotion, and if there are extra-SCN oscillators that control food anticipatory activity (FAA), it ought to be possible to entrain these animals to restricted food during the time when your compound is blocking locomotor activity. They should still anticipate a food restriction with locomotor activity.

Weitz: I can think of two possibilities. One is that the food entrainable oscillator actually converges onto the same circuit. You wouldn't see it during this period and then it would re-emerge at the end. One reason I like this idea is that when one looks at these FAA experiments, although by and large the SCN-driven band of activity is intact, one often sees subtle changes. For instance, nocturnal activity often drops when the FAA starts. This is fairly common. It has been suggested that there is a competition for output between the two oscillators.

Menaker: They can run out of phase.

Weitz: Yes, the output is not a clock. Just like your temporal chimeras, you get two different clocks driving a common substrate. The other possibility would be that they actually damp down different circuits, in which case your suggestion would have a positive result. I think it is really worth doing.

Van Gelder: Have you looked in any of the clock mutants at where the TGFα levels are pegged when the clock is non-functional genetically?

Weitz: No, and the amplitude is not very high to begin with. We have really been trying to look at protein levels to see what is going on with the protein. We don't know. The complication there is that TGFα and its relatives are synthesized as transmembrane precursor proteins that are cleaved by a metalloprotease to release the active peptide. The antibodies only see unreleased TGFα. We have tried sampling through a ventricle and from the foramen magnum, and we can measure some TGFα but we don't see any rhythms yet.

Van Gelder: The *Bmal* mutant mice are hypoactive, but it is unclear whether this is secondary to skeletal problems. You can imagine that the signal would be pegged differently depending on where the clock is frozen in, say, a cryptochrome double mutant versus a *Bmal* mutant.

Weitz: I don't think there is any reason to think that this is the only thing of primary importance. We are more interested in uncovering the variety of components that have activity in the system.

Dunlap: The masking data suggest that some of these inhibitory factors might be acutely light induced.

Weitz: But where? Since masking requires neither a clock nor a functional SCN, the source of the ligands is going to be somewhere else. Since there is a direct projection from the retina to the subparaventricular zone, the simplest hypothesis is that the source is the retina, and there is an acute release of factors from the retina that converge onto the SCN targets. We have been looking at this and can detect immunoreactivity for a couple of the EGFR ligands in the retina, but these antibodies aren't ideal. We don't know about light induction in the SCN, but this isn't the first thing we would look for on the basis of the lesion experiments.

Kay: Perhaps a follow on to that is that with John Hogenesch and Ben Rusak we did look at light induction in the SCN. A couple of things that came out of this were oxytocin and galanin. They heavily labelled the ependymal layer in response to light, including those ciliated cells that have been studied in rabbits. We could also see them coming on really well in the SPZ, but it was nothing to do with the core SCN.

Weitz: There is an intermediate possibility, which is that the SCN is not required for masking, but perhaps contributes to masking anyway. This is not an unreasonable suggestion. Fred Davis has data showing that while SCN-lesioned animals are perfectly normal in masking with the acute bright-light exposure, under dim-light LD cycles they show some funny behavioural breakthroughs. There might be a SCN component.

Schibler: Did you try to infuse TGFα antibodies?

Weitz: We did this but we didn't see any effect. We tried immunoneutralizing antibodies to a number of factors on the grounds that if these factors diffuse through the third ventricle on the way to their targets, you might be able to tie them up. We have not seen any significant effects on locomotor behaviour.

Foster: On the first day we were discussing about the differential contributions of the novel receptors, rods and cones as an input system. If there are these positive and negative factors, what you might imagine is that the phasing of the positive and negative factors might be under differential rod/cone novel control. This might explain the different phasing effects we saw in the rodless–coneless mouse. It will be interesting to model this and see whether you could overlay masking on top of this.

Weitz: From the heterozygote knock-in animals that are labelled with β galactosidase and show where the melanopsin axons go, there are some hints that a small number of fibres actually by-pass the SCN and go towards the SPZ. This hasn't been worked out yet, but there is an idea that both the SPZ projection and the SCN projection may be melanopsinergic. There could be differential projections.

Hastings: You made the point that the relative phasings of the putative positive and negative factors give you less effective or better definition to transitions from activity to rest and vice versa. In the EGFR mutant animal, it seemed that activity offset was really well defined, especially for a hamster. Was that a common finding?

Weitz: No.

Hastings: So the breakthrough activity that you saw in that individual was common to the genotype, but the definition of offset wasn't.

Weitz: That is correct. We noticed this and were very curious about it.

Menaker: I don't want to make your life more difficult, but it is possible to tether hamsters and infuse rhythmically. You don't have to use the minipump. It would be very convincing if you could drive the activity with a rhythmic infusion, rather than simply block it with continuous infusion.

Weitz: That is more of a biology experiment. We are doing a screen of a lot of factors, and this would be unpractical in a screen. If the goal is to attempt to reconstitute circadian control of locomotor activity in a lesioned animal, I would want to wait until we had more information about which factors are involved.

Menaker: Along those lines, what do you make of the phase difference between electrical activity in the bed nucleus and electrical activity in the rest of the brain? It is a strong result which has been ignored. It is a consequence of Shin Yamazaki's persistence for more than two years to get electrical activity recordings from awake hamsters in different parts of the brain. As other people have found, he finds that electrical activity outside of the SCN is antiphase to the SCN except in the bed nucleus. In the bed nucleus the electrical activity is not only in phase with the SCN, but also it is almost 1:1 correlated with it. It looks like there might be a monosynaptic projection. It looks like a direct connection.

Reference

Cheng MY, Bullock CM, Li C et al 2002 Prokineticin 2 transmits the behavioural circadian rhythm of the suprachiasmatic nucleus. Nature 417:405–410

CK1 and GSK3 in the *Drosophila* and mammalian circadian clock

Emily Harms, Michael W. Young[1] and Lino Saez

Laboratory of Genetics, The Rockefeller University, 1230 York Avenue, Box 288, New York, NY 10021, USA

Abstract. Two kinases, DOUBLETIME and SHAGGY, have been shown to play a role in the circadian clock. DOUBLETIME, the *Drosophila* orthologue of casein kinase 1, can phosphorylate PERIOD in the cytoplasm and in the nucleus. This phosphorylation destabilizes PERIOD in both locations and sets patterns of both cytoplasmic accumulation and nuclear turnover. Cytoplasmic phosphorylation postpones accumulation of PERIOD and affects timing of nuclear accumulation of PERIOD/ TIMELESS complexes. SHAGGY, the *Drosophila* orthologue of glycogen synthase kinase 3, phosphorylates TIMELESS and promotes nuclear translocation of PERIOD/ TIMELESS complexes. Thus, the opposing effects of these two kinases in the cytoplasm are crucial for establishing the ~ 24 h period of circadian rhythmicity in *Drosophila*. Casein Kinase 1 has been shown to be a component of the circadian clock in mammals. Recent studies are also pointing to a role for glycogen synthase kinase 3 in the mammalian clock.

2003 Molecular clocks and light signalling. Wiley, Chichester (Novartis Foundation Symposium 253) p 267–279

Many behaviours, including feeding and locomotor activity, are produced with a circadian (or ~ 24 h) rhythm generated by an internal, self-sustained molecular clock. The circadian clock is composed of a network of autoregulatory genetic interactions. The clocks arise at the level of single cells, acting autonomously. At the core of the molecular oscillator in *Drosophila* are two interlocked and sequentially acting feedback loops (reviewed in Panda et al 2002, Williams & Sehgal 2001, Young & Kay 2001). In the first, a *period* (*per*)/*timeless* (*tim*) loop, two transcription factors, CLOCK (CLK) and CYCLE (CYC), activate the transcription of the genes *per* and *tim*. PERIOD (PER) and TIMELESS (TIM) proteins in turn repress their own expression through physical association with, and inhibition of, CLK and CYC. This repression occurs in the nucleus. However, PER and TIM must heterodimerize to allow their stable nuclear

---

[1]This paper was presented at the symposium by Michael W. Young to whom all correspondence should be addressed.

accumulation. This fosters delays and oscillations in their activity. In the second loop, *clock* transcription is directly repressed by accumulation of VRILLE (VRI), a protein whose accumulation is controlled by CLK/CYC-dependent *vri* transcription (Blau & Young 1999, Cyran et al 2003, Glossop et al 2003). Since *per*, *tim* and *vri* are transcribed together in response to CLK/CYC, and only PER and TIM accumulation are subject to extensive post-transcriptional delays, Clk RNA and protein levels are initially regulated within each circadian cycle by VRI, while CLK activity is subsequently controlled by association with PER/ TIM (Young & Kay 2001, Cyran et al 2003, Glossop et al 2003).

PER and TIM are phosphorylated by two different serine/threonine kinases (Young & Kay 2001). PER is phosphorylated by DOUBLETIME (DBT) in the cytoplasm and nucleus in the absence of TIM. At the beginning of each cycle of *per* and *tim* transcription, phosphorylation promotes the degradation of newly formed, cytoplasmic PER. This is thought to delay formation of PER/TIM complexes, and retard nuclear accumulation. TIM is phosphorylated by SHAGGY (SGG). Increasing SGG activity accelerates nuclear accumulation of PER/TIM complexes, perhaps due to its modification of TIM. TIM is degraded in response to light, allowing photo-entrainment of the clock. This response is influenced by CRYPTOCHROME (CRY), SGG and an unidentified tyrosine kinase. Light may induce the binding of TIM and CRY, promoting the phosphorylation of TIM by a tyrosine kinase, followed by ubiquitination and degradation via the proteosome (Panda et al 2002, Williams & Sehgal 2001, Young & Kay 2001). SGG appears to prime TIM for this response, as hyperphosphorylated forms of TIM produced by action of SGG are preferentially lost upon exposure to daylight (Martinek et al 2001). CLK is also post-translationally modified by phosphorylation (Lee et al 1998), but the kinase(s) responsible for that modification, and the role of such phosphorylation, have not been established (see below).

Kinases as clock components

A genetic screen for novel clock mutations in *Drosophila* led to the discovery of the first kinase that plays a role in circadian rhythmicity (Price et al 1998). *Doubletime* (*dbt*) is a member of the casein kinase 1 (CK1) family, and is 86% identical to human casein kinase 1ε (CK1ε) within the kinase domain (Kloss et al 1998). Originally, two mutations were isolated that produce short (18 h; *dbt^S*) and long (27 h; *dbt^L*) behavioural rhythms. Since then, many period-altering mutations and loss-of-function alleles have been described. The majority of these mutations reside in the kinase domain of *dbt*, altering its ability to bind and/or phosphorylate its substrate (Fig. 1). All of the DBT point mutations analysed alter the phosphorylation state and accumulation of PER, indicating that DBT affects

dco[3]   dbt[h]   dbt[S]

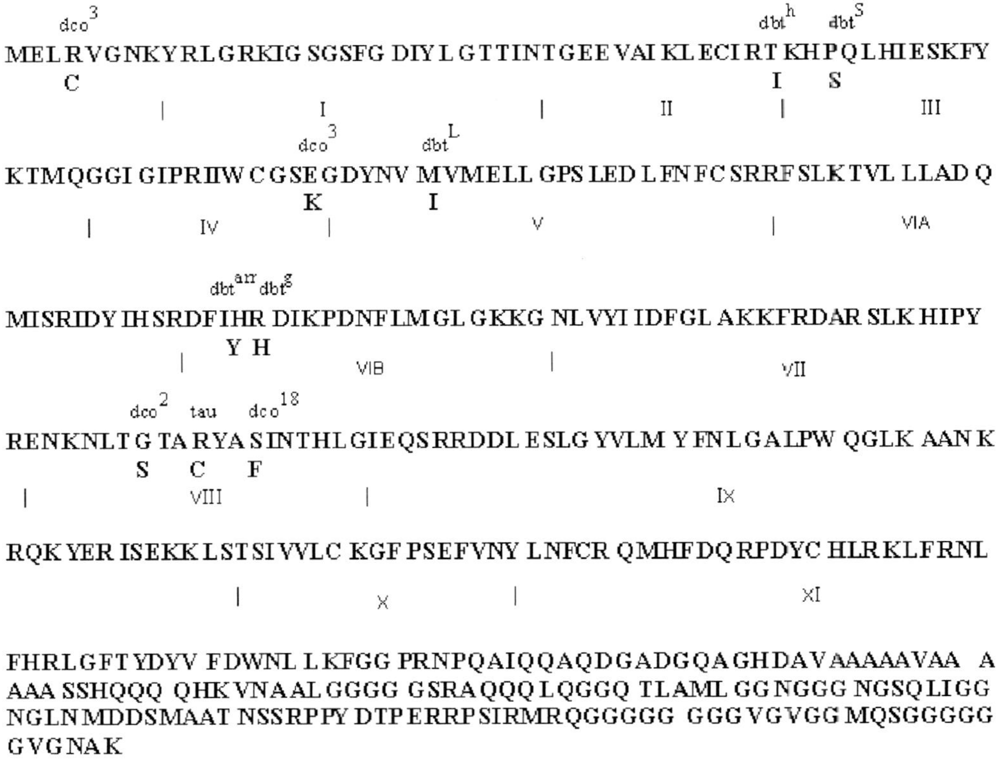

MEL RVGNKY RL GRKIG SGSFG DIY L G TTINTGEE VAI KL ECI RT KHPQ LHI ESKFY
C   I   S
          | I | II | III

                    dco[3]   dbt[L]
KTMQGGI GIPRIIW CG SE GDYNV MVMELL GPS LED L FN FC SRRF SLK TVL LLAD Q
                              K   I
          | IV | V | VIA

          dbt[ar] dbt[g]
MISRIDY IHSRDFIHR DIKPDNFLMGL GKKG NL VYI IDFGL AKKFRDAR SLK HIPY
                    Y H
          | VIB | VII

dco[2]   tau   dco[18]
RENKNLT G TA RYA SINTHL GIEQSRRDDL ESLG YVLM Y FNLGALPW QGLK AAN K
          S   C   F
| VIII | IX

RQK YER ISEKK LSTSIVVLC KGF PSEFVNY LNFCR QMHFDQ RPDYC HLRKLFRNL
          | X | XI

FHRLGFT YDYV FDWNL L KFGG PRNPQAIQQAQDGADGQA GHDAV AAAAAVAA A
AAA SSHQQQ QHKVNAAL GGGG GSRAQQQ LQGGQ TLAML GG NGGG NGSQ LIGG
NGLNMDDSMAAT NSSRPPY DTP ERRPSIRMRQGGGGG GGG VGVGG MQSGGGGG
GVGNAK

FIG. 1.   *Doubletime* mutations. The sequence of DBT and its mutations are shown. The DBT kinase domain (residues 1–292) was deduced by comparison to CK1 (Gross & Anderson, 1998). The kinase sub-domains are given in roman numerals. *dbt*[ar] and *dbt*[g] mutants map to the catalytic domain. *dco*[2] and *dco*[18] are embryonic lethal mutations. *dco*[3] is a pupal lethal associated with overgrowth of imaginal discs (Zilian et al 1999). The *tau* mutation, which arose in the Syrian hamster, affects the residue equivalent to 178 of DBT as shown.

PER protein stability through phosphorylation (Price et al 1998, Suri et al 2000, Rothenfluh et al 2000).

DBT has been shown to bind to *Drosophila* PER both *in vitro* and *in vivo* (Kloss et al 1998, 2001), and this interaction persists throughout the circadian cycle, resulting in rhythmic changes in subcellular localization of DBT (Kloss et al 2001, Fig. 2). With respect to the clock, it appears that DBT has somewhat different roles in the cytoplasm and in the nucleus, although using a single molecular mechanism. In the cytoplasm, it promotes delays in the circadian cycle by destabilizing PER in the absence of its partner TIM, slowing down the increase and nuclear accumulation of PER/TIM complexes. In the nucleus, DBT function seems to define the end of the molecular cycle by degrading PER. Nuclear TIM delays DBT's phosphorylation and degradation of nuclear PER, lengthening the circadian cycle. Thus, PER–TIM interactions in both the cytoplasm and nucleus influence the period length of the cycle.

The serine-threonine kinase SGG is the *Drosophila* orthologue of glycogen synthase kinase 3 (GSK3) (Bourouis et al 1990, Siegfried et al 1990). Overexpression of SGG in *Drosophila* pacemaker cells shortens the locomotor activity rhythms of the fly. Conversely, a reduction in SGG expression lengthens *Drosophila* locomotor activity rhythms. These period-shortening and period-lengthening phenotypes are associated with hyper- and hypo-phosphorylation of TIM, respectively, indicating that TIM may be a substrate of SGG phosphorylation. In support of this hypothesis, GSK3 has been shown to phosphorylate TIM *in vitro* (Martinek et al 2001). Rates of PER/TIM nuclear accumulation are also altered in association with these period length changes. The phosphorylation of TIM by SGG appears to promote the nuclear translocation of PER/TIM complexes (Martinek et al 2001). Like DBT, SGG also has an effect on nuclear TIM. TIM appears to be phosphorylated by SGG throughout the subjective night (Martinek et al 2001). This phosphorylation could also be responsible for the dissociation of TIM from the PER/TIM/DBT complex, facilitating the phosphorylation of PER by DBT and ending the circadian cycle. Loss-of-function *sgg* mutants are not viable as adults, so the effect of a complete loss of *sgg* function on the circadian clock is not yet known.

Although a substantial body of evidence indicates that the functions of DBT and SGG depend on PER and TIM phosphorylation respectively, these may not be the only substrates in the *Drosophila* clock. For example, in mammals it was recently found that casein kinase 1 phosphorylates CRY and BMAL1 (the orthologue of CYC) in addition to PER (see below) (Eide et al 2002).

Kinases in the mammalian clock

The first mutation observed to alter circadian behaviour in mammals arose spontaneously in the Syrian hamster. It was given the name *tau*. This mutation

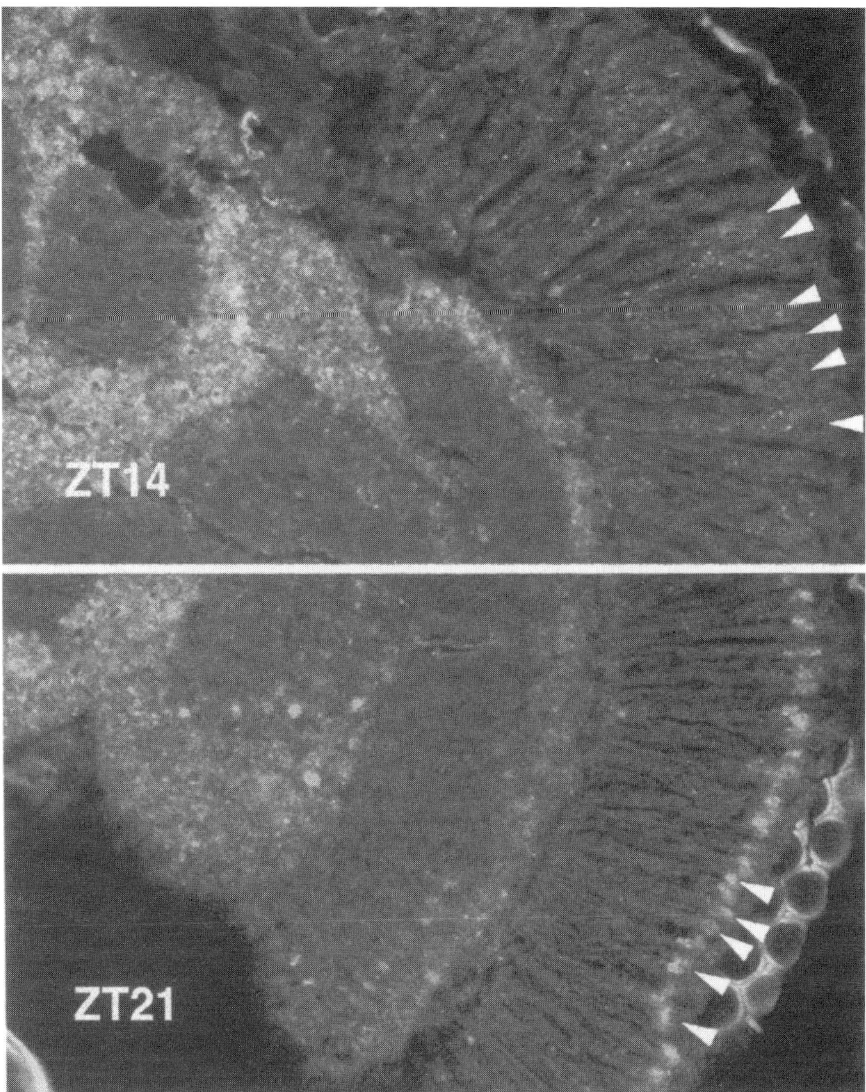

FIG. 2. Doubletime subcellular localization rhythms. The pattern of accumulation of DOUBLETIME protein in fly heads was determined by immunocytochemistry. DBT is most abundant in the cytoplasm of the photoreceptor cells at the beginning of the night (ZT14 is shown, two hours after lights off in a 12 h:12 h light–dark cycle), but shifts to a predominantly nuclear location just before dawn (ZT21 depicted, three hours before lights on). Arrowheads indicate the photoreceptor cell nuclei.

shortens the period length of circadian rhythms to $\sim 20\,$h in homozygotes (Ralph & Menaker 1988). Genetic linkage mapping indicated that a DBT orthologue, casein kinase 1ε, was a candidate for *tau*. Subsequent cloning revealed a single base-pair change within *CK1ε* in *tau* mutants (Lowrey et al 2000). This produces an amino acid substitution at a conserved residue in the kinase (see Fig. 1). While it was found that both the wild-type and *tau* mutant CK1ε enzymes could bind to mammalian PER1 and PER2 *in vitro*, the mutant CK1ε phosphorylated these PER proteins much less efficiently than the wild-type kinase (Lowrey et al 2000).

Further genetic evidence for an involvement of CK1ε in the mammalian clock came with the mapping of a mutation associated with a human sleep disorder, familial advanced-sleep-phase syndrome (FASPS). Individuals with FASPS have advanced sleep rhythms, in which they fall asleep and awaken approximately 4 h earlier than unaffected individuals (Jones et al 1999). Through extensive linkage analysis, FASPS was associated with the human PER2 gene in one large kindred (Toh et al 2001). These FASPS individuals have a single amino acid substitution within the CK1ε binding region of human PER2. *In vitro*, this mutation leads to hypophosphorylation of human PER2 by CK1ε (Toh et al 2001). These results underscore the importance of CK1ε in the mammalian circadian clock and also indicate that hPER2 is a physiological substrate for CK1ε *in vivo*.

In addition to CK1ε, a second casein kinase 1 orthologue, CK1δ, has been implicated in the mammalian circadian clock. CK1δ and CK1ε both bind and phosphorylate mammalian PER proteins *in vitro* (Keesler et al 2000, Vielhaber et al 2000, Camacho et al 2001), and are physically associated with PER and CRY *in vivo* (Lee et al 2001). In *tau* mutants, PER proteins continue to be phosphorylated in spite of the lowered function measured for CK1ε *in vitro* (Lee et al 2001). It has been suggested that the residual phosphorylation might be supplied by PER-associated CK1δ in the mutants, and that CK1ε and CK1δ have overlapping functions in the mammalian circadian system (Lee et al 2001).

The mammalian counterpart of SGG is GSK3. Unlike CK1, a role for GSK3 in the mammalian circadian clock has not yet been demonstrated. However lithium, which is known to be a potent inhibitor of GSK3 *in vitro* and *in vivo*, has long been known to alter the period of mammalian circadian rhythms (Abe et al 2000). There are two forms of GSK3 found in mammals: GSK3α and GSK3β. As in *Drosophila*, the *gsk3β* knockout mouse is embryonic lethal (Hoeflich et al 2000), and therefore cannot be readily analysed for defects in circadian rhythmicity. Nevertheless, *gsk3β* $^{-/-}$ mouse embryonic fibroblasts (MEFs) have been generated from *gsk3β* knockout embryos (Hoeflich et al 2000), and can be used to test for abnormal circadian rhythmicity in a cultured cell assay. It is possible that GSK3α may be able to functionally compensate for the lack of GSK3β. Therefore, MEFs from embryos deficient for both *gsk3α* and *gsk3β* may be required to detect a role for

GSK3 in mammalian clocks. As of this writing, *gsk3α* knockout mice have not been reported.

TIM, the substrate of SGG in *Drosophila*, has not been shown to have a circadian function in mammals: null mutations of the mammalian orthologue of TIM produce early embryonic lethality that precludes behavioural tests for rhythmicity. Therefore, the mammalian TIM protein may or may not be a relevant substrate for GSK3 phosphorylation in the mammalian clock. It is also possible that GSK3 plays a role in the mammalian clock, but has other protein targets. For instance, the mammalian cryptochromes (CRY1 and CRY2) have taken over some of the roles played by TIM in the *Drosophila* clock. These include binding and stabilizing the PER proteins, and regulating their nuclear accumulation (Kume at al 1999, Lee et al 2001). Preliminary studies have indicated that the mammalian CRY1 protein is in fact an *in vitro* substrate of GSK3 (E. Harms, unpublished observation).

A role for GSK3 in the mammalian circadian clock, if established, could be relevant for the understanding and treatment of important psychiatric diseases. Bipolar, or manic–depressive, disorder is a severe, recurrent mood disorder affecting 1–2% of the human population. Disturbed sleep is a common symptom of patients with depressive illnesses. More importantly, sleep deprivation and phase-advance treatments that should affect circadian rhythmicity, rapidly and dependably elevate mood in manic depressives (Bunney & Bunney 2000). Currently, the most effective long-term treatment for bipolar disorder is administration of lithium (Muller-Oerlinghausen et al 2002), which, as indicated above, has been shown to directly inhibit GSK3β activity (Klein & Melton 1996). GSK3 activity is also reduced by valproate, perhaps the second-most widely applied drug for stabilizing mood (Mitchell & Malhi 2002). Since GSK3 is a common target of both mood stabilizers, it may contribute to their beneficial action. These findings provide a further impetus for exploring the possible role of this enzyme in mammalian behavioural rhythms.

Additional phosphorylations in the circadian clock

In the *Drosophila* circadian clock, three proteins are rhythmically phosphorylated throughout the circadian cycle: PER, TIM and dCLK. The electrophoretic mobility of these three proteins all undergo changes during the circadian day by Western analysis. In all cases, phosphatase treatment reduced or eliminated the slower migrating bands, suggesting that the mobility shifts are due to phosphorylation (Edery et al 1994, Zeng et al 1996, Lee et al 1998). While there is evidence that DBT phosphorylates PER and SGG phosphorylates TIM, it is possible that other kinases phosphorylate these proteins as well. TIM, for instance, is phosphorylated by a tyrosine kinase before it is ubiquitinated and degraded by the proteasome (Naidoo et al 1999). Since SGG is a serine-threonine kinase, there is likely another kinase responsible for this tyrosine phosphorylation of TIM. PER also may be phosphorylated by a kinase other than DBT. Both *dbt^P* and *dbt^ar* are arrhythmic mutants in which PER over-accumulates, suggesting decreased DBT kinase activity (Price et al 1998, Rothenfluh et al 2000). However, while PER is hypo-phosphorylated in *dbt^P* mutants, it is still phosphorylated to an intermediate level in *dbt^ar* mutants. Whether this residual phosphorylation in *dbt^ar* mutants is the product of DBT activity or the activity of another kinase is not yet known.

Cycling protein phosphorylation plays a role in the mammalian circadian clock as well. PER1, PER2 and BMAL all show temporal changes in electrophoretic mobility that are eliminated by phosphatase treatment (Lee et al 2001). Although these PER phosphorylations are likely to reflect CK1 activity they may not be the only clock-related substrates of this enzyme family. CRY1 and CRY2, for instance, can be phosphorylated by CK1ε *in vitro* when present in a CRY/PER/CK1ε complex (Eide et al 2002). Two isoforms of mammalian CLOCK (orthologue of *Drosophila* CLK) also appear to be phosphorylated, resulting in mobility shifts by Western analysis (Lee et al 2001). The kinase(s) responsible for CLOCK phosphorylation is (are) unknown.

Processive phosphorylation may be the rule for some of these substrates. As PER proteins accumulate overnight and during the early morning, decreasing mobilities and multiple species emerge following electrophoresis. CK1ε preferentially phosphorylates substrates with the sequence phospho-S-X-X-S, where a serine is phosphorylated at an upstream position, priming the phosphorylation of a serine three amino acids downstream. The mutation in human PER2 found in individuals affected with FASPS is a serine to glycine mutation (Toh et al 2001). Several putative CK1ε phosphorylation sites are found downstream of this serine and are spaced at three amino acid intervals. Since CK1ε-dependent phosphorylation of downstream serines is suppressed by the mutation *in vitro*, it has been suggested that human PER2 is multiply and sequentially phosphorylated according to this pattern (Toh et al 2001). Similarly spaced motifs are found in mammalian PER1 and PER3, indicating that processive phosphorylation may be a common feature of all PER proteins. GSK3 can also sequentially modify its substrates. Unlike CK1ε, GSK3 processively phosphorylates in a C-terminal to N-terminal direction, and modifies serines or threonines spaced at four amino acid intervals rather than three. Phosphorylation site mutants that might affect a processive modification of TIM are now under investigation in *Drosophila* (E. Harms, unpublished results). It is also possible that a separate kinase may provide a priming phosphorylation event, which in turn could allow CK1ε or GSK3 to initiate a cascade of protein phosphorylation.

Lastly, there are phosphorylation events in both *Drosophila* and mammals for which the function and associated kinases have not been identified: as mentioned above, the kinase(s) that rhythmically phosphorylates *Drosophila* and mammalian CLK is not yet known, and it will be important to determine whether phosphorylation of CLK influences its transcriptional activity or its ability to bind to DNA. Although CK1ε modifies mammalian CRY1, CRY2 and BMAL1 *in vitro* (Eide et al 2002), and CK1δ will similarly phosphorylate some mammalian PER proteins (Camacho et al 2001), none of these activities have been securely established *in vivo*.

Circadian regulation of kinase activity

Unlike other clock proteins, neither DBT/CK1 nor SGG/GSK3 is rhythmically expressed. Their substrates, however, are rhythmically phosphorylated throughout the circadian day. How, then, are the activities of these kinases regulated?

While the abundance of DBT does not oscillate, the subcellular localization of DBT does change throughout the circadian day (Fig. 2). In *Drosophila*, the subcellular distribution of DBT in the lateral neurons of the brain and in photoreceptor cells of the eye largely follows the changing localization of PER (Kloss et al 2001). In mammals, the pattern of CK1 accumulation is also under circadian control: the kinase appears to associate initially with mPER and mCRY in the cytoplasm, but it is also found in nuclear complexes and may regulate their movement to the nucleus (Lee et al 2001).

Unlike DBT/CK1, there is currently no evidence that the subcellular localization of SGG/GSK3 is rhythmically controlled. It is known, however, that SGG/GSK3 activity can be regulated by phosphorylation. Two different phosphorylation events have been shown to influence the activity of GSK3. Phosphorylation of Tyr216 in the mammalian enzyme GSK3β leads to an increase in GSK3 activity, presumably by facilitating substrate binding (Dajani et al 2001). Conversely, phosphorylation at position Ser 9 of GSK3β (Ser 21 of GSK3α) negatively regulates the activity of the kinase (Stambolic & Woodgett 1994, Sutherland et al 1993). Preliminary work indicates that some of these phosphorylations are controlled by the circadian clock (S. Kaladchibachi, A. Manoukian, E. Harms and L. Saez, unpublished data).

Acknowledgements

This work was supported by NIH GM 54339 (M.W.Y.) and by funds from The Rockefeller University's *Women & Science* Fellowship Program (E.H.).

References

Abe M, Herzog ED, Block GD 2000 Lithium lengthens the circadian period of individual suprachiasmatic nucleus neurons. Neuroreport 11:3261–3264

Blau J, Young MW 1999 Cycling vrille expression is required for a functional *Drosophila* clock. Cell 99:661–671

Bourouis M, Moore P, Ruel L, Grau Y, Heitzler P, Simpson P 1990 An early embryonic product of the gene shaggy encodes a serine/threonine protein kinase related to the CDC28/cdc2+ subfamily. EMBO J 9:2877–2884

Bunney WE, Bunney BG 2000 Molecular clock genes in man and lower animals: possible implications for circadian abnormalities in depression. Neuropsychopharmacology 22: 335–345

Camacho F, Cilio M, Guo Y et al 2001 Human casein kinase Idelta phosphorylation of human circadian clock proteins period 1 and 2. FEBS Lett 489:159–165

Cyran SA, Buchsbaum AM, Reddy KL et al 2003 vrille, Pdp1, and dClock form a second feedback loop in the *Drosophila* circadian clock. Cell 112:329–341

Dajani R, Fraser E, Roe SM et al 2001 Crystal structure of glycogen synthase kinase 3β: structural basis for phosphate-primed substrate specificity and autoinhibition. Cell 105:721–132

Edery I, Zwiebel LJ, Dembinska ME, Rosbash M 1994 Temporal phosphorylation of the *Drosophila* period protein. Proc Natl Acad Sci USA 91:2260–2264

Eide EJ, Vielhaber EL, Hinz WA, Virshup DM 2002 The circadian regulatory proteins BMAL1 and cryptochromes are substrates of casein kinase Iepsilon. J Biol Chem 277:17248–17254

Glossop NR, Houl JH, Zheng H, Ng FS, Dudek SM, Hardin PE 2003 VRILLE feeds back to control circadian transcription of Clock in the *Drosophila* circadian oscillator. Neuron 37:249–261

Gross SD, Anderson RA 1998 Casein kinase I: spatial organization and positioning of a multifunctional protein kinase family cell signal. Cell Signal 10:699–711

Hoeflich KP, Luo J, Rubie EA, Tsao MS, Jin O, Woodgett JR 2000 Requirement for glycogen synthase kinase-3beta in cell survival and NF-κB activation. Nature 406:86–90

Jones CR, Campbell SS, Zone SE et al 1999 Familial advanced sleep-phase syndrome: a short-period circadian rhythm variant in humans. Nat Med 5:1062–1065

Keesler GA, Camacho F, Guo Y, Virshup D, Mondadori C, Yao Z 2000 Phosphorylation and destabilization of human period I clock protein by human casein kinase Iε. Neuroreport 11:951–955

Klein PS, Melton DA 1996 A molecular mechanism for the effect of lithium on development. Proc Natl Acad Sci USA 93:8455–8459

Kloss B, Price JL, Saez L et al 1998 The *Drosophila* clock gene double-time encodes a protein closely related to human casein kinase Iε. Cell 94:97–107

Kloss B, Rothenfluh A, Young MW, Saez L 2001 Phosphorylation of period is influenced by cycling physical associations of double-time, period, and timeless in the Drosophila clock. Neuron 30:699–706

Kume K, Zylka MJ, Sriram S et al 1999 mCRY1 and mCRY2 are essential components of the negative limb of the circadian clock feedback loop. Cell 98:193–205

Lee C, Bae K, Edery I 1998 The *Drosophila* CLOCK protein undergoes daily rhythms in abundance, phosphorylation, and interactions with the PER–TIM complex. Neuron 21:857–867

Lee C, Etchegaray JP, Cagampang FR, Loudon AS, Reppert SM 2001 Posttranslational mechanisms regulate the mammalian circadian clock. Cell 107:855–867

Lowrey PL, Shimomura K, Antoch MP et al 2000 Positional syntenic cloning and functional characterization of the mammalian circadian mutation tau. Science 288:483–492

Martinek S, Inonog S, Manoukian AS, Young MW 2001 A role for the segment polarity gene shaggy/GSK-3 in the *Drosophila* circadian clock. Cell 105:769–779

Mitchell PB, Malhi GS 2002 The expanding pharmacopoeia for bipolar disorder. Annu Rev Med 53:173–188

Muller-Oerlinghausen B, Berghofer A, Bauer M 2002 Bipolar disorder. Lancet 359:241–247

Naidoo N, Song W, Hunter-Ensor M, Sehgal A 1999 A role for the proteasome in the light
    response of the timeless clock protein. Science 285:1737–1741
Panda S, Hogenesch JB, Kay SA 2002 Circadian rhythms from flies to human. Nature 417:
    329–335
Price JL, Blau J, Rothenfluh A, Abodeely M, Kloss B, Young MW 1998 double-time is a novel
    Drosophila clock gene that regulates PERIOD protein accumulation. Cell 94:83–95
Ralph MR, Menaker M 1988 A mutation of the circadian system in golden hamsters. Science
    241:1225–1227
Rothenfluh A, Abodeely M, Young MW 2000 Short-period mutations of per affect a double-
    time-dependent step in the *Drosophila* circadian clock. Curr Biol 10:1399–1402
Siegfried E, Perkins LA, Capaci TM, Perrimon N 1990 Putative protein kinase product of the
    Drosophila segment-polarity gene zeste-white3. Nature 345:825–829
Stambolic V, Woodgett JR 1994 Mitogen inactivation of glycogen synthase kinase-3β in intact
    cells via serine 9 phosphorylation. Biochem J 303:701–704
Suri V, Hall JC, Rosbash M 2000 Two novel doubletime mutants alter circadian properties and
    eliminate the delay between RNA and protein in *Drosophila*. J Neurosci 20:7547–7555
Sutherland C, Leighton IA, Cohen P 1993 Inactivation of glycogen synthase kinase-3β by
    phosphorylation: new kinase connections in insulin and growth-factor signalling. Biochem
    J 296:15–19
Toh KL, Jones CR, He Y et al 2001 An hPer2 phosphorylation site mutation in familial advanced
    sleep phase syndrome. Science 291:1040–1043
Vielhaber E, Eide E, Rivers A, Gao ZH, Virshup DM 2000 Nuclear entry of the circadian
    regulator mPER1 is controlled by mammalian casein kinase I epsilon. Mol Cell Biol
    20:4888–4899
Williams JA, Sehgal A 2001 Molecular components of the circadian system in *Drosophila*. Annu
    Rev Physiol 63:729–755
Young MW, Kay SA 2001 Time zones: a comparative genetics of circadian clocks. Nat Rev
    Genet 2:702–715
Zeng H, Qian Z, Myers MP, Rosbash M 1996 A light-entrainment mechanism for the
    Drosophila circadian clock. Nature 380:129–135
Zilian O, Frei E, Burke R et al 1999 *double-time* is identical to *discs overgrown*, which is required for
    cell survival, proliferation and growth arrest in *Drosophila* imaginal discs. Development
    126:5409–5420

DISCUSSION

Sehgal: In the Rat1 cells have you tried to manipulate GSK3 using any method
other than lithium?

Young: No, but we certainly want to use valproate on whole animals and on the
cultured cell tests for rhythmicity.

Sehgal: What about over-expressing GSK3?

Young: If we didn't have the GSK3β knockout MEFs we would do that, but our
plan is to look at the mutant MEFs first. We are currently in the process of looking
at these. We have also got GSK3α knockout MEFs on the way. We are going to try
to make double knockouts from heterozygous parents. GSK3β single knockouts
die at about embryonic day 14 or 15. If the doubles also survive that long then we
will be able to do the experiment in the best possible fashion. If not, we may have to

do some mixed RNAi tests with one of the mutants. It could be intriguing if we found that both of these had independent contributions to clock function.

Sehgal: Have you looked for armadillo repeats in genes that cycle in the liver?

Young: No.

Dunlap: In FRET experiments, from the number, size and location of the particles can you make any estimate as to what they might be or where they are within the cell?

Young: I don't know. There have been reports of nuclear speckling. These are very curious. It is as if there is an assembly site relevant to cytoplasmic accumulation and then something very different takes place in the nuclei. These are cultured cells. There aren't any rhythms in these cells. We have been working with them over the years, and there have been some indications that we may be able to generate rhythms in these cells with the expression of the right components, but I still think that you have to take the specifics with a grain of salt until we move these tests back into transgenic flies. Unlike the MEFs we don't have an operational clock. The things that we are seeing grossly resemble what we see *in vivo*. This gives us something to look for *in vivo* when the tools become available.

Stanewsky: There is a recent paper showing that PER can enter the nucleus before TIM (Shafer et al 2002). Do you ever see PER going into the nucleus in the S2 cells without TIM?

Young: We haven't seen that, but in mammals an interesting story has emerged about nuclear–cytoplasmic shuttling. I wonder whether something similar to that might be going on in flies. We don't have anything that rules that out. However, when you transfect S2 cells and you just put PER or TIM in alone, you just don't see any evidence for nuclear accumulation. Both are needed for either to accumulate in the nucleus.

Rosbash: They are not mutually exclusive. This doesn't exclude the fact that this is dependent on prior PER–TIM interactions. The lack of temporal coordination of PER and TIM's nuclear entry could be dependent on a prior cytoplasmic association. This would be very reasonable.

Stanewsky: Do they always go into the nucleus together in S2 cells?

Young: Yes, the kinetics appear to be the same.

Ishida: You showed *Cry1* mRNA production in the Rat1 cell after lithium treatment. What is the mechanism for this *Cry1* induction?

Young: The model we have in mind is one in which GSK3 is a clock component, not necessarily providing exactly the same function as in flies. Mammalian TIM is still a mystery: there is no compelling evidence for it being a needed target in the mammalian system, but it is another target for GSK3 that is part of the clock. All of these things are now showing a response because the oscillator as a whole has now changed kinetics.

Sassone-Corsi: The results with the inhibitor were fascinating. However, at that concentration you are very likely targeting also additional kinases such as PKC. My question is, do you know whether at that concentration the phosphorylation of GSK3 is blocked?

Young: We don't know that. The concentration chosen was deliberately high enough to do just what you are saying. Now we see some effect at this concentration the important thing is to back this down.

Rosbash: You mentioned briefly inositol monophosphatase. My impression from the literature, and in particular the recent *Nature* paper (Williams et al 2002), is that somewhat more than half of the people interested in brain neurochemistry and lithium would vote for the inositol pathway rather than GSK3. In any case, it is up for grabs.

Young: Yes. There is a peculiar spin in that particular paper because the same kind of argument was made for GSK3 several years ago when it was found that inositol monophosphatase had nothing to do with the effects on lithium on yet another lithium and GSK3-dependent developmental pathway in *Xenopus*—the Wnt pathway. No one is looking at mood here: they are looking at a cell differentiation result. This is one of the dangers in trying to figure out the mechanism for the phenotype you are interested in by doing biochemistry on unrelated cells that are giving a growth or developmental phenotype in response to that drug: none of these are necessarily related to the mechanism underlying the behavioural phenotype.

Rosbash: It definitely hits both enzymes.

Sassone-Corsi: PIK is also upstream, as is PKC.

Young: At this point we will be satisfied just to understand whether or not there is a role for this kinase in circadian rhythmicity, and then we'll take the other steps as they come.

Takahashi: From lots of cell culture work and pharmacology, we've seen that 20 mM changes can be really non-specific. You'd have to at least put in 20 mM NaCl. This is like a 7% increase in ionic strength. In our experience even changes of 1% ionic strength in cell culture can perturb pineal cells in culture. Lithium should really be working in the submillimolar range.

Young: We started with 20 mM because historically this is the concentration that has been used for many years in developmental assays aimed at totally blocking GSK3 function. Also our first assay was on mPER1 protein and it looked like there was a lengthening of period, which was not nearly the same thing as just killing everything. When we begin to see that the same concentration was giving blockade or altered expression levels in other components of the clock then our concern rose. We'll back the concentration down.

Rosbash: I have a student in my lab who is interested in mood disorders. Over the course of last summer he did experiments with lithium in flies. He did dose–

response curves of lithium with increased sodium as a control. We got no period lengthening at concentrations that eventually destroyed rhythmicity and with effects on locomotor activity.

Sehgal: We have looked at the effect of lithium on flies. We found that the flies became arrhythmic at high enough concentrations. However, there was no consistent or significant lengthening.

Menaker: There is a similar effect in rodents. You can get effects on mice and rats with lithium in the drinking water, but you cannot get effects in hamsters. The reason for this is that because hamsters are desert animals they have an incredibly powerful kidney, and they simply don't allow the lithium concentration in the blood to rise to a level where it will do anything.

Rosbash: In flies we do get effects, just not period lengthening. I'm sceptical of these results.

Young: This is what we have to do until we have the β knockouts to do the right experiments with.

References

Shafer OT, Rosbash M, Truman JW 2002 Sequential nuclear accumulation of the clock proteins Period and Timeless in the pacemaker neurons of *Drosophila melanogaster*. J Neurosci 22:5946–5954

Williams RS, Cheng L, Mudge AW, Harwood AJ 2002 A common mechanism of action for three mood-stabilizing drugs. Nature 417:292–295

Final general discussion

Menaker: In this final discussion, I am going to open the floor for questions that have arisen at any point in the meeting.

Hastings: I'd like to focus on the relationship between the molecular loop and membrane function in excitable cells. The electrical silencing in *Drosophila* VPDH neurons not only knocks out behaviour, but also stops the molecular loop (Nitabach et al 2002). Hitoshi Okamura's work with TTX effects on gene expression also touches on this. In our v\VPAC knockout mice, you could argue that GABAergic suppression is responsible for the phenotype. This is another model where electrical functions are kicked out. Is this a general emerging theme? It may be a common feature for clocks: it is not just that the molecular loop tells the excitable membrane what to do, but that the molecular loop requires the excitable membrane to sustain it

Menaker: It is clearly an important issue for us as a field to investigate.

Young: Do excitable membranes exist in the liver? How does this relate to peripheral clocks?

Menaker: I was thinking of the pineal, as well.

Rosbash: One way of rationalizing those two is to think about self-sustainment as being aided by membrane components.

Young: In the pineal is there any indication that there are circadian rhythms of membrane potential or any other electrical activity?

Menaker: Joe Takahashi, do we have any reason to believe there are electrical correlates of pineal cell rhythmicity?

Takahashi: I am not aware of any.

Weitz: Don't pinealocytes project processes deep into the brain in some species?

Menaker: No.

Young: Joe Takahashi, in the cultured pineal cell experiments that you did, were the rhythms self-sustaining?

Takahashi: No, it is more like a week. What happens is that the individual cells are uncoupled and they dissociate from one another.

Menaker: Because of the pineal transplant experiments, you can demonstrate that it is a persistent oscillator which carries phase with it.

Rosbash: But not as a single cell, necessarily. The fly experiments that were published recently (Nitabach et al 2002) show membrane function as opposed to some electrical activity. It seems to me that there is the potential for interpreting the data to argue that the free running dark–dark (DD) rhythms are more fragile or more subject to complications that arise when membrane potential is disrupted in some way or other. The light–dark (LD) kick makes it less susceptible to these kinds of problems.

Weitz: One caveat in all of these experiments is that it is very difficult to silence cells electrically without doing something to their resting potential or entire ionic gradients. In the *Drosophila* case you mentioned this is a really interesting paper but they used these various channels that are leaky to silence the cells. They do silence them electrically, but what else do they do? Because of the variety of species and manipulations, though, we have to take seriously the idea that electrical activity somehow feeds back into the molecular machinery to sustain it. It could also be that in all these cases the cells used are pushed into a regime where the oscillatory solution is not robust. This is the curmudgeonly generic accusation against almost every good clock experiment!

Van Gelder: One of the genes that showed up on just about everyone's clock-related gene list was *Slob* (slowpoke binding protein), which presumably has something to do with K^+ conductance. Steve Kay, you showed some early data on *Slob* a while back: is there anything new to report?

Kay: We've looked more at *slowpoke* than *Slob*. BigK (the mammalian version of slowpoke) also cycles in the SCN. This is a bit more complicated because of differential splicing, but what Fernanda Ceriani in my lab showed is that SLOWPOKE cycles at the protein level, and it stains neurons that are adjacent to the LNs. *Slowpoke* mutants are pretty much arrhythmic even in LD. There is a little bit of a LD effect which looks like a startle effect, but they have the same amount of total activity in a given 24 h period. If you look at other channel mutants, you can see activity cycling but overall activity is greatly reduced.

Rosbash: What happens to the clock RNAs and proteins?

Kay: We are looking at this now in the lateral neurons (LNs). As you might suspect, we can still see some cycling protein activity there. *Slowpoke* does look to me like it is providing some necessary function for gating or rhythmically guiding activity.

Rosbash: It could be all output, right? In that context this is the key question.

Kay: I was pretty shocked by how much it cycles at the protein level. Channels are usually pretty stable proteins. But yes, our current working hypothesis is it is a primary output pathway for locomotion behaviour.

Menaker: A lot of the models at the molecular level depend on, or at least include, events that involve transfer between the cytoplasm and the nucleus. There is one system in which it has been reported that rhythmicity persists in the absence of the nucleus. These are the old experiments on *Acetabularia* (e.g. Karakashian &

Schweiger 1976, Schweiger et al 1986). I am not sure how reliable they were, but many papers came from this work showing that you can lose the nucleus and rhythmicity persists. If this is true, what does it say about our models, or what is special about *Acetabularia*?

Loros: Of course, *Synecococcus* doesn't have a nucleus but it displays perfectly functional rhythms. There may be something there. Who has replicated the *Acetabularia* work? Another perfectly reasonable explanation of the *Acetabularia* phenomenon is that parts of the clock requiring daily transcription are encoded for in the chloroplast. It could be a plastid clock.

Menaker: But on the other hand, something has to be happening there that is parallel to the nuclear cytoplasmic exchange.

Weitz: It has unusual organelles, so there is the possibility that these organelles with their own genomes could be involved.

Dunlap: When the nucleus is taken out the rhythms get pretty ragged and the period lengths are much less confined to the 'circadian range'. There is a lot of drifting of period, so precision can be lost. One possibility is that *Acetabularia* without a nucleus could be analogous to *Drosophila* in the experiments Amita Sehgal did in which both *per* and *tim* were expressed at a constant rate. With no more rhythmic transcription, the cells are running on constant RNA and re-making protein more-or-less rhythmically.

Stanewsky: In insects there are examples in moth and *Musca* where PER and TIM don't go into the nucleus.

Loros: Do you really believe that there is no PER that goes into the nucleus? What I saw was that a substantial proportion of PER doesn't go into the nucleus, but it doesn't mean that there is no PER. We need on the order of 15 molecules per cell of FRQ in the nucleus to have a perfectly functional circadian rhythm. This is well below the resolution of the kinds of experiments that were published.

Young: You would also like to ask whether you want to throw out your faith in the likelihood that these are transcription factors. This would almost certainly require them to have a function in the nucleus.

Rosbash: There have always been iconoclasts, and once in a while they turn out to be right.

Weitz: Part of the discussion concerns what we need to make an oscillator; this is different from what we need to make a biological clock. Above and beyond generating oscillations, a clock has to control multiple outputs in most organisms. Also, it has to be entrained with a phase-response curve relevant to the organism's survival. It may turn out that the transcriptional component of circadian oscillations is not *per se* an absolute requirement for the presence of the oscillation, but it might nonetheless be required for proper regulation of phase angle or distributing outputs.

References

Karakashian MW, Schweiger HG 1976 Circadian properties of the rhythmic system in individual nucleated and enucleated cells of *Acetabularia mediterranea*. Exp Cell Res 97: 366–377

Nitabach MN, Blau J, Holmes TC 2002 Electrical silencing of *Drosophila* pacemaker neurons stops the free-running circadian clock. Cell 109:485–495

Schweiger HG, Berger S, Kretschmer H et al 1986 Evidence for a circaseptan and a circasemiseptan growth response to light/dark cycle shifts in nucleated and enucleated *Acetabularia* cells, respectively. Proc Natl Acad Sci USA 83:8619–8623

Closing remarks

Michael Menaker

Department of Biology, University of Virginia, Charlottesville, VA 22903, USA

This has been a great meeting, and it reflects the current status of this field, which couldn't be more exciting. I don't think we have solved anything here, but we have sharpened some of the questions. This seems to me particularly apparent in cases of photic input, but applies to other questions as well. The field at the moment is on the verge of realizing what has been its potential since its inception. Now that the questions are getting sharper and new techniques are available, there is a great deal of work for all of us to do on many different levels. My personal view is that the future of a great deal of biology is going to be in integrating what we are rapidly learning about molecular mechanisms with the physiology and behaviour that has been so long unexplained in mechanistic terms. Although this is a general trend, the field of circadian rhythmicity is poised to be a model for that integration. This is because we have information at so many different levels of organization, the questions are sharp and specific, and the system is so clearly defined relative to so many of the other big questions in biology. We are going to find molecular genetic explanations of circadian physiology and behaviour, and the pay-off at the biomedical level is going to be great because the circadian system is so pervasive and such an important part of overall biological organization. At the organismal level we are going to learn a great deal about how organisms adapt to their environments as this field moves forward. I am already looking forward to the next Novartis Foundation meeting, five or six years from now, at which we will be able to chart our progress in those directions. I thank you all for your contributions and participation: it has been a great three days.

Index of contributors

Non-participating co-authors are indicated by asterisks. Entries in bold indicate papers; other entries refer to discussion contributions.

A

*Albus, H. **56**
*Allada, R. **223**
*Appleford, J. M. **3**

B

*Bellingham, J. **3**
*Bonnefont, X. **56**
*Brown, S. **89**

C

Cahill, G. M. 27, 50
*Cardone, L. **126**
Cermakian, N. 71, **126**, 138, 150
*Chahad-Ehlers, S. **203**
*Colot, H. **184**

D

*Damiola, F. **89**
*Davidson, A. J. **110**
*Davis, F. C. **250**
*Denault, D. **184**
*Doi, M. **126**
*Dryer, S. E. **140**
Dunlap, J. C. 27, 28, 50, 83, 84, 87, 123, 155, 157, 181, **184**, 199, 200, 201, 202, 265, 278, 283

F

Foster, R. G. **3**, 23, 24, 25, 26, 27, 28, 29, 30, 44, 45, 47, 48, 49, 50, 52, 53, 103, 104, 105, 108, 198, 265
*Froehlich, A. C. **184**

G

*Garabette, M. **203**
*Glossop, N. R. J. **140**
Golden, S. 87
Green, C. B. 26, 136, 138, 159

H

*Hankins, M. **3**
Hardin, P. E. 46, 100, **140**, 150, 151, 152, 153, 154, 155, 159, 183, 234
*Harms, E. **267**
Hastings, M. H. 46, 69, 70, 71, 84, 101, 103, 104, 123, 124, 180, **203**, 218, 219, 220, 221, 222, 249, 263, 264, 266, 281
*Hogenesch, J. B. **73**, **171**
*Houl, J. H. **140**

I

Ishida, N. 46, 71, 84, 160, **238**, 249, 278

J

*Jenkins, A. **3**

K

Kay, S. A. 25, 26, 43, 50, 51, 52, **73**, 83, 84, 85, 86, 87, 104, 107, 150, 156, 157, **171**, 181, 200, 265, 282
*King, V. M. **203**
*Kramer, A. **250**
*Krishnan, B. **140**
Kyriacou, C. P. 150, 151, 158, 233, 237

L

*Le-Minh, N. **89**
Lee, C. C. 42, 107, 218
*Lee, K. **184**
*Li, X. **250**
Loros, J. J. 23, 24, 45, 68, 72, 84, 85, 86, 87,
    101, 104, 124, 183, **184**, 199, 200, 263,
    283
*Lucas, R. J. **3**

M

*Maywood, E. S. **203**
*McDonald, M. **223**
*Meijer, J. H. **56**
Menaker, M. **1**, 24, 25, 30, 44, 45, 47, 48, 49,
    50, 52, 53, 54, 66, 67, 69, 70, 71, 84, 86,
    100, 101, 102, 103, 104, 105, 106, 107,
    108, **110**, 121, 122, 123, 124, 125. 137,
    138, 152, 154, 156, 157, 159, 180, 181,
    182, 199, 218, 221, 222, 237, 264, 266,
    280, 281, 282, 283, **285**
*Mezaki, M. **238**
*Miyazaki, K. **238**
*Morse, D. **126**
*Muñoz, M. **3**

N

*Ng, F. S. **140**
*Nowrousian, M. **184**

O

*O'Brien, J. **203**
Okamura, H. 100, 103, 136, **161**

P

*Panda, S. **73**, **171**
*Pando, M. P. **126**
*Peng, Y. **223**
*Pregueiro, A. **184**
*Preitner, N. **89**

R

*Reddy, A. B. **203**
*Ripperger, J. **89**
Rosbash, M. 24, 26, 43, 44, 45, 47, 49,
    51, 52, 67, 68, 71, 72, 83, 87, 101, 102,
    103, 104, 105, 106, 107, 108, 121, 122,
    124, 137, 138, 152, 153, 154, 155,
    156, 157, 158, 159, 180, 182, 183,
    200, 218, 219, 220, 221, **223**, 233,
    234, 235, 236, 237, 278, 279, 280,
    281, 282, 283

S

*Saez, L. **267**
*Sancar, A. **31**
Sassone-Corsi, P. 28, 43, 55, 69, 71, 84, 85,
    125, **126**, 136, 137, 138, 150, 156, 200,
    235, 279
*Scammell, T. E. **250**
Schibler, U. 24, 28, 43, 50, 54, 55, 70, **89**, 99,
    100, 101, 102, 103, 122, 125, 137, 138,
    152, 153, 156, 157, 159, 182, 200, 201,
    218, 220, 221, 265
Sehgal, A. 42, 70, 99, 136, 137, 150, 151,
    152, 153, 154, 155, 158, 159, 183, 199,
    233, 277, 278, 280
*Snodgrass, P. **250**
Stanewsky, R. 27, 67, 106, 107, 123, 154,
    155, 158, 199, 234, 278, 283

T

Takahashi, J. S. 47, 48, 49, 72, 83, 85, 86,
    104, 139, 151, 154, 157, 158, 159, **171**,
    181, 182, 183, 220, 279, 281
*Thompson, S. **3**

V

van der Horst, G. T. J. **56**, 67, 68, 69, 70, 71,
    102, 103, 136
van Gelder, R. N. 29, **31**, 42, 43, 44, 45,
    46, 47, 48, 49, 50, 51, 52, 53, 54, 67,
    68, 70, 85, 101, 102, 103, 104, 105,
    106, 107, 108, 124, 125, 138, 151,
    153, 154, 157, 158, 159, 182, 183,
    200, 201, 202, 219, 220, 221, 235,
    236, 237, 264, 265, 282

W

Weitz, C. J. 54, 67, 69, 70, 71, 72, 83, 86,
    101, 105, 107, 108, 122, 124, 137, 153,
    156, 157, 159, 181, 199, 218, 219, 221,

232, 236, **250**, 263, 264, 265, 266, 281, 282, 283

Y

*Yamazaki, S. **110**
*Yang, F.-C. **250**
Young, M. W. 25, 29, 43, 45, 46, 53, 69, 88, 100, 103, 107, 153, 154, 155, 157, 158,

181, 201, 218, 220, 222, 233, 234, 236, 249, **267**, 277, 278, 279, 280, 281, 283

*Yujnovsky, I. **126**

Z

*Zhao, J. **223**

*Zheng, H. **140**

Subject index

A

absorption spectrum   24–25
Acetabularia   282–283
action spectrum   8, 9, 24–25, 40, 48
actographs   53–54
albino hamster   49
algae   27
angiotensin II   167
antennae
   electroantennagram (EAG)   146, 150–151
   olfaction rhythm control   148
aorta, temporal profiling   181
Arabidopsis
   cryptochromes   32, 37
   flowering timing   83–84
   fluence rate response curves   74
aromatic L-amino acid decarboxylase   165
arousal   208
artificial cerebrospinal fluid (aCSF)   252
Atlantic salmon   5

B

bacterial alkaline phosphatase (BAP)   244
bed nucleus   266
bees, *zeitgedachtnis*   123–124
behavioural responses
   non-rod, non-cone photoreception   9
   suprachiasmatic nucleus electrical activity
      66–67
   see also locomotor activity
BigK   282
'biological clock'   2
bipolar disorder   273
blind cave fish   101
blind mole   101
blindness, daytime sleepiness   53–54
bloodflow, cyclic control   181
BMAL1
   CLOCK/BMAL1 heterodimers   162, 165
   CRY   91
   PER   91
   WC-1 similarity   87

Bmal1
   loop   142
   PER2   63
   REV-ERBα   63, 90, 91
   transcription activators   92–93
body temperature *see* temperature
brain, photosensitivity   4, 44

C

C-Box   190, 192
c-*fos*   36, 80
Ca/Cam dependent kinase (CAMK-1)   187
cAMP responsive elements (CREs)   63, 176
candidate genes   17, 19–21
carp   5
   Cyprinus carpio   5
casein kinase 1 (CK1)
   CK-1a and CK-1b   187
   nuclear complexes   275
   phosphorylation role   187, 270
casein kinase 1δ (CK1δ)   272
   PER phosphorylation   275
   suprachiasmatic nucleus   220
casein kinase 1ε (CK1ε)   164, 239, 272, 274
casein kinase II (CK2)   87, 187, 225
cave fish   101
CCA1   75–76
Ccr4   174
Cds1   196
cell cycle   196
chemotherapy, circadian-linked efficacy   196
chimeras   158, 182
CHK orthologues   196
Chk2   196
Chk2   196
cholesterol biosynthesis   177
chromatophores   14
circadian clock   57–59
   resetting   31
circadian rhythms
   Drosophila   223–232, 267–268
   mammals   57–59, 161–170

circadian rhythms (*cont.*)
  Neurospora 185–187
  opsins 5
  orphan nuclear receptors 91–95
  rods and cones 9, 12
CK2 230
CK2^{Tik} 230
CLOCK (CLK) 224
  CLK-CYC, *Clk* repression 141, 142
  CLK-GR fusion gene 226
  CLK/MOP3 complex 172, 174–176
  CLOCK/BMAL1 heterodimers 162, 165
  CRY 91
  PER 91
  phosphorylation 268, 274, 275
  suprachiasmatic nucleus 204
Clock (Clk)
  CLK-CYC repression 141, 142
  Drosophila 140, 225, 228–230, 235
  expression pattern 233
  mutant mice 61, 63
  REV-ERBα 63, 90, 91
  transcription activators 92–93
  VRI repression 142, 146, 148
Clock-Box (C-Box) 190, 192
cones 9, 12, 80
conservation
  opsins 5, 8
  PERs 241
CONSTANS (CO) 83–84
constant RNA 158
CORCOS 173
CREB-mediated transcription 81
CREM transcription factor 132–133
cry^{baby} 76, 78
Cry double knockouts
  masking 34, 61, 67, 103
  phase inversions 104
  pupillary responses 35
  suprachiasmatic nucleus transplants 71
CRY1 162
  action spectrum 25
  circadian clock 57–58
  GSK3 substrate 273
  phosphorylation 274
  plants 74–75
  suprachiasmatic nucleus 204
CRY2 162
  circadian clock 57–58
  phosphorylation 274
  plants 75

suprachiasmatic nucleus 204
cryptochromeopsin 105
cryptochromes (CRY) 268
  action spectrum/absorption spectrum
    matching 25
  bleaching time-course 50
  BMAL1 expression 91
  circadian clock 57–59
  CLOCK expression 91
  Drosophila 32, 36, 37, 76, 78, 143, 145
  PER and 91, 164–165
  peripheral tissues 145
  photic signalling to suprachiasmatic
    nucleus 33–34
  photoentrainment 80
  photopigment role 20, 31–42, 49, 52–53,
    58, 76, 78
  PIAS interaction 36, 46
  plants 74
  pupillary responses 35, 47–48
  retinal 39, 80
  suprachiasmatic nucleus 59, 80, 204, 206
  transcription rhythm 100–101
  ubiquitination 164–165
  zebrafish 127
CYCLE (CYC) 224
  CLK-CYC, *Clk* repression 141, 142
cycloheximide 226, 242

D

Danio rerio see zebrafish
Dartnall curves 48, 49
D-element Binding Protein (DBP) 90, 162
Dbp/dbp 90, 162
Dbp/Hlf/Tef triple knockout 99–100
Dbt/dbt 230, 268
deep sea animals 101
delayed sleep phase syndrome 239, 246–247
delta-aminolevulinate synthase 178
depressive illness 273
dexamethasone 129, 226
diurnal species 208, 210
Dmnk 196
Doubletime (DBT) 225, 268, 270, 275
Doubletime (dbt/Dbt) 230, 268
Drosophila
  central and peripheral oscillators 127,
    140–150, 155
  circadian rhythm 223–232, 267–268
  Clk 140, 225, 228–230, 235

cryptochrome   32, 36, 37, 76, 78, 143, 145
DBT   268, 270, 275
lifespan   159, 160
light input   76–78
locomotor activity   76, 141, 145
olfaction rhythm   141, 145–146, 148
par domain protein 1   99
SGG   270

E

E4BP4   142, 150, 162, 164, 165
electroantennagram (EAG)   146, 150–151
electrophoretic mobility shift assays (EMSA)   188
ELF3   75, 85
ELISA   242
endothelin   167
epidermal growth factor (EGF)   252, 259–260
epidermal growth factor receptor (EGFR)   173, 252, 254, 257–259, 260–261
ER24   183
ethyl methane-sulfonate mutagenesis (EMS)   228
evolution   86–87
excitable cells   281
extraretinal photoreceptors   1–2

F

familial advanced sleep phase syndrome (FASPS)   239, 246–247, 272, 274
FHA domain   194, 196
fibroblasts
    endothelin   167
    PER phosphorylation and degradation   238–248, 249
    serum shock   129, 167, 241–242
    see also mouse embryo fibroblasts
fish   4, 5–8, 17
flavin-based photopigment   31–32
flavin mononucleotide (FAD)   190
FLO (see FRQ-less oscillator)   184, 194
FLOWERING LOCUS T (FT) cycling   83, 84
fluence rate response curves (FRC)   74
follicle-stimulating hormone (FSH)   133
food, peripheral clock entrainment   96, 129–131, 168

food-anticipatory activity (FAA)   116–119, 121–125, 264
food-entrained oscillator (FEO)   116–119, 121–125
forkhead associated (FHA) domain   194, 196
forskolin   129
Fos induction   108
frequency (frq)   185, 188
FRQ
    complex size   201
    long period mutants   183
    PEST sequences   187
    phosphorylation   187
    temperature regulation   188
    WC-1/WC-2 interaction   185–186
    WCC/C-Box binding   192
FRQ-less oscillator (FLO)   184, 194

G

GABA (gamma amino butyric acid)   210, 219
GAL4   234
GAL4/UAS system   146
galanin   265
gastrointestinal hormones   96
gastrointestinal tract, food-entrained oscillator (FEO)   116–119, 122
GET effect   196
glass   78
glial cells   234
glucocorticoids   64, 96, 101, 129, 130
    CLK-GR fusion gene   226
glucose   96, 131
glutamate   176
glutamatergic gene   206, 207
glutaminergic transmission   80
glycogen synthase kinase 3 (GSK3)   239, 270, 272–273, 274, 275
    GSK3α   272
    GSK3β   272, 273, 275
gonadotrophin synthesis   132
Gonyaulax   153
ground squirrel   208

H

habenular region, opsin   5
hamster   45, 47
    albino   49
    tau   270, 272
heavy water   125

HERG channel   87
hierarchical structure   168
histamine   176
HLF   162
HMG-CoA reductase   177, 178
Hofbauer–Buchner (H–B) eyelets   78
'hour-glass' clock   64, 68
humoral signals   129
hypothalamus, ventromedial, food-entrained
        oscillator (FEO)   123

I

inner nuclear layer, CRY expression   80
inner retina, photoreception   4, 5–8, 9, 12–13,
        31–42
inositol monophosphatase   279
iridophores   4
Israelian blind mole   101

J

Jak/Stat pathway   46
Jun   80

L

lamprey P opsin   8
lateral neurons
    ablation   227
    DBT   275
    locomotor activity   141
    oscillations   157–158
    peripheral oscillators and   155–156
LHY   75–76
light response elements (LREs)   188, 190, 200
light signalling, history   1–2
lithium   272, 273, 279–280
liver
    food-entrainment   116–119, 122, 168
    gene cycling   174
    principal role   177
    rate-limiting steps   178
    REV-ERBα and REV-ERBβ   92
    suprachiasmatic nucleus transplant   69–71
    transcriptional cycling   214
    xenobiotic metabolism   177
locomotor activity
    Drosophila   76, 141, 145
    epidermal growth factor receptor (EGFR)
        173, 257–259
    non-rod, non-cone photoreception   9

    small ventral lateral neurons   141
    suprachiasmatic nucleus   251, 252
    TGFα   173
LOV domain   190
luteinizing hormone   133

M

manic–depressive disorder   273
masking   35, 251
    Cry double knock-outs   34, 61, 67, 103
    suprachiasmatic nucleus lesions   220–222
    two forms   68
math5⁻/⁻ mice   102, 103, 124
melanophores, *Xenopus laevis*   4, 14, 17
melanopsin
    co-expression with rod-opsin   17
    functional properties   19, 20
    inner retina   13
    knock-out mouse   23–24
    photosensor/photoisomerase role   17,
        25–27, 29–30, 80
    retinal ganglion cells   14, 15, 19, 32, 57, 64,
        80
    somatic cells   28
melatonin   49
metallothionein 1 activator   177–178
methamphetamine-induced rhythmicity
        116
methyltransferases   177
microarray concordance   235
mitogen activated protein kinase (MAPK)
        63, 81
    suprachiasmatic nucleus   206
    Z3 cells   128
MOP3   172, 174–176
motivated behaviour   263
motor cortex   70
mouse embryo fibroblasts (MEFs)   131,
        136–137, 138, 272
mRNA cycling   141, 142, 148, 174, 226
multiple measures corrected minus β (MMCβ)
        173–174
multiunit electrode activity (MUA)   59

N

neuronal signalling   176–177, 178
neuropeptide release   176–177
neuropeptide Y (NPY)-positive afferents
        208

Neurospora 184–198
  circadian system 185–187
  FRQ-less oscillator (FLO) 194
  FRQ mutants 183
  light regulation/photoreceptors 188–192, 198–200
  light response elements (LREs) 188, 190, 200
  temperature regulation 187–188
nicotinamide adenine dinucleotide (NAD) 96
NIH 3T3, suprachiasmatic nucleus co-culture 129
nocturnal species 208, 210
Nocturnin 174
norpA 76
nuclear complexes, CK1 275
nuclear export 165
nuclear speckles 84–85
nucleoli size and morphology 178
nucleus
  CK1 275
  PER and TIM 278

O

olfaction rhythms 141, 145–146, 148
Opn4 80
opsins
  algae 27
  circadian rhythm 5
  conservation 5, 8
  habenular region 5
  photosensors/photoisomerases 16–17
  pineal organ 5
  rodents 9, 12–13
  tmt-opsin 28
  vertebrate ancient (VA) 5–8, 17
  vitamin A-based chromophore binding 3, 4–5
orphan nuclear receptors 90, 91–95
outer nuclear layer, CRY expression 80
oxytocin 265

P

PAC1 64
PACAP
  retinal ganglion cells 15, 44, 64, 80
  VPAC2 receptor 219
par domain protein 1 99

PAR (Proline Amino acid Rich) proteins 162, 164, 165
PAS A 241
PAS B 241
PAS domain evolution 87
pCREB 207
PDF 227, 229, 232–233
Pdf 154
PDP1 226
peptidergic signalling 176, 210, 212
PER
  BMAL1 expression 91
  CLOCK expression 91
  conservation 241
  CRY 91, 164–165
  degradation 238–248
  Drosophila 225, 267–268
  history 224
  nucleus 278
  phosphorylation 164, 225, 238–248, 249, 268, 270, 274, 275
  proteasome-dependent degradation 164
  sleep phase syndromes 239, 241
  suprachiasmatic nucleus 204, 206, 208
  ubiquitination 164
per
  diurnal versus nocturnal species 208, 210
  Drosophila 140–141, 225, 230
  pineal 168
Per/Cry loop 142
Per-luc 111, 113, 116
PER1 162
  phosphorylation 164, 274
  suprachiasmatic nucleus 204, 206
Per1 81, 111–113, 162
  entrainment 206–207
  peripheral clocks 131
  suprachiasmatic nucleus 165–167
  testis 132
PER2 63, 162
  familial advanced sleep phase syndrome 239, 272, 274
  phosphorylation 164, 274
  suprachiasmatic nucleus 204, 206
Per2 81, 162
  entrainment 206–207
  suprachiasmatic nucleus 165–167
PER3 162, 241
  phosphorylation 274
PERIOD *see* PER

peripheral clocks 167–168, 180–181
    Drosophila 127, 140–150, 155
        entrainment 96–98, 129–131, 168
        lateral neuron clock and 155–156
        suprachiasmatic nucleus 64, 69–71,
            113–115, 126–136, 214
peripheral tissues
    CRY 145
    temperature compensation 181
peropsin 17
PEST sequences, FRQ 187
Petromyzon marinus 8
phase response curves (PRCs) 152, 153
phosphorylation
    Chk2 196
    CLK 268, 274, 275
    CRY1/CRY2 274
    degradation and 249
    FRQ 187
    PER 164, 225, 238–248, 249, 268, 270,
        274, 275
    TIM 225, 239, 268, 270, 273–274
    Tyr216 275
photic noise 75
photolyases 40, 43, 57
phototropins 40
PHYA 74–75
PHYB 75
phytochromes 74
PIAS3 43, 45
PIF3 76
pineal system
    electrical correlates of rhythmicity 281
    melatonin 49
    multiple photopigments 4
    opsin 5
    Per 168
    photoreception 4
pituitary adenylate cyclase-activating peptide
    (PACAP)
    retinal ganglion cells 15, 44, 64, 80
    VPAC2 receptor 219
plants
    HMG-CoA 178
    light input 74–76
Plecoglossus altivelis 8
PLR 9
PRD-4 196
prd-4 194, 196
prokineticin 263–264
prokineticin 2 173

proteasome-dependent degradation, PER
        protein 164
protein degradation 238–248, 249
protein folds 87–88
protein inhibitor of activated STAT (PIAS)
        36, 46
pseudo-receiver domain 88
psychiatric disease 273
pupillary response
    continuous light 106
    cryptochrome-dependence 35, 47–48
    math5$^{-/-}$ mice 103
    non-rod, non-cone photoreception 9
    outer retina 44

R

RAD53p 196
rate-limiting steps 178
rebound effect 264
redundancy 105–106
retina
    epidermal growth factor (EGF) 259–260
    inhibitory factors 265
    inner, photoreception 4, 5–8, 9, 12–13,
        31–42
    pupillary response 44
    suprachiasmatic nucleus tuning 115
    TGFα 259–260
retinal G protein-coupled receptor (RGR) 17
retinal ganglion cells
    cryptochromes 39, 80
    dendritic arbors 37
    melanopsin 14, 15, 19, 32, 57, 64, 80
    PACAP 15, 44, 64, 80
    photoresponsive cells 32
retinoic acid 129
REV-ERB 90
REV-ERBα 143
    Bmal1 and *Clock* 63, 90, 91
    cooperative binding 100
    functions 93
    liver 92
Rev-Erbα 142, 162, 176
REV-ERBβ 90, 92
reverse transcriptase-polymerase chain
        reaction (RT-PCR) 229, 235–236, 241,
        242
RNA cycling 151, 157, 158
roach 8, 17
    Rutulis rutulis 8

rods 9, 12, 80
ROR (Retinoic-acid receptor related Orphan
    Receptor) 90, 93, 100, 176
RORα 90, 91, 92
RORβ 90, 92
ROREs (ROR elements) 90, 91
RORγ 90, 92
RT-PCR 229, 235–236, 241, 242

S

salmon (*Salmo salar*) 5
secretogranin III 177
secretory granule neuroendocrine protein 1
    176–177
Ser513, FRQ phosphorylation 187
serotonergic-positive afferents 208
serum shock 129, 167, 241–242
Shaggy (SGG) 225, 268, 270, 275
shaggy 230, 239
sleep syndromes 239, 241, 246–247, 272, 274
sleep–wake cycle, TGFα 254, 256–257
Slob 282
Slowpoke 177
small ventral lateral neurons 141
smelt fish 8
somatic cell melanopsin 28
somatostatin 176
spermatogenesis 132
Spermophilus 208
STAT3 45
subparaventricular zone (SPZ) 251, 266
suprachiasmatic nucleus (SCN) 32, 78,
    110–121, 161, 171–173, 203–217
    aromatic L-amino acid decarboxylase 165
    casein kinase 1δ 220
    clock proteins 204–206
    cryptochrome 59, 80, 204, 206
    cycling genes 174–176
    diurnal versus nocturnal species 208, 210
    electrical rhythms 57, 59–61, 66–67
    food entrainment 102–103
    light entrainment 206–208
    locomotor activity 251, 252
    masking 220–222
    metallothionein 1 activator 177–178
    mRNA cycling 174
    neuronal signalling 176–177
    neuropeptide release 176–177
    NIH 3T3 co-culture 129
    non-photic cues 208

peptidergic signalling 176, 210, 212
PER 204, 206, 208
Per1/Per2 165–167
peripheral clocks 64, 69–71, 113–115,
    126–136, 214
phosphorylated MAP kinase 206
photic signalling to 33–34
photoresponse in vitamin A-depleted
    mutants 53
retinal role 115
rod and cone loss 12
somatostatin 176
TGFα 254
transplants 66–67, 69–71, 136, 251
vasopressin 176
Synecococcus 283

T

tau 270, 272
TEF 162
teleost fish 4, 5–8, 17
temperature
    peripheral clock entrainment 96–98
    peripheral tissue compensation 181
    regulation, *Neurospora* 187–188
testis 132–133
TIM 162
    circadian function in mammals 273
    Drosophila 225, 267–268
    nucleus 278
    phosphorylation 225, 239, 268, 270,
        273–274
tim 140–141, 225, 230
Timeless 32, 36 *see also* TIM
tmt-opsin 28
TOC1 76, 85
transcriptional profiling 171–180
transducin 5
transforming growth factor α (TGFα) 173,
    252, 254–257, 259–260
transparency 159
twilight 45
Tyr216 phosphorylation 275

U

ubiquitin-proteasome pathway 244, 246
ubiquitination 164–165
ultraradian rhythm 257

V

VA (vertebrate ancient) opsin   8
  VAL   8
  VAM   8
valproate   273
vascular smooth muscle cells   167
vasopressin (AVP)   176
vertebrate ancient (VA) opsin   5–8, 17
VIP   210, 219
vitamin A-based chromophore   3, 4–5
vitamin A binding proteins   30
VIVID   186–187, 199
voles   101
VPAC2   64, 210, 212, 219
Vpac2 knock-out mice   218
VRILLE (VRI)   142, 146, 148, 150, 154–155,
      226, 268
VVD   186–187, 199
vvd   187

W

waved-2   257, 259
WC-1 (White Collar 1)
  Bmal similarity   87
  C-Box binding   192
  FRQ interaction   185–186
  LOV domain   190
  LRE-bound complexes   188, 190
  photoreceptor role   190, 198–199, 201–202
wc-1 (white collar 1)   185

WC-2 (White Collar 2)
  C-Box binding   192
  FRQ interaction   185–186
  LRE-bound complexes   188, 190
  period length mutants   183
wc-2 (white collar 2)   185
white collar complex (WCC)   185
  C-Box binding   192

X

xenobiotic metabolism   177
Xenopus laevis
  melanophores   4, 14, 17
  melanopsin   15, 17, 27

Y

yeast two-hybrid screen   36, 50–51

Z

Z3   127–129
zebrafish
  melanopsin   15, 27
  peripheral clocks   127
  vertebrate ancient opsin   5
  Z3   127–129
zeitgeber, feeding times   96
zeitgedachtnis, bees   123–124
Zif268   80